Milk Proteins: Chemistry and Biology

Milk Proteins: Chemistry and Biology

Editor: Kinley Crews

STATES
ACADEMIC PRESS
www.statesacademicpress.com

States Academic Press,
109 South 5th Street,
Brooklyn, NY 11249, USA

Visit us on the World Wide Web at:
www.statesacademicpress.com

ISBN: 978-1-63989-356-0 (Hardback)

Cataloging-in-Publication Data

Milk proteins : chemistry and biology / edited by Kinley Crews.
p. cm.
Includes bibliographical references and index.
ISBN 978-1-63989-356-0
1. Milk proteins. 2. Milk--Composition. 3. Milk proteins--Analysis.
4. Biology. I. Crews, Kinley.
QP552.M54 M55 2022
612.015 75--dc23

Table of Contents

Preface VII

Chapter 1 **Milk Proteins: Processing of Bioactive Fractions and Effects on Gut Health** 1
Anindya Mukhopadhya and Torres Sweeney

Chapter 2 **Donkey Milk Proteins: Digestibility and Nutritional Significance** 18
Donata Marletta, Flavio Tidona and Salvatore Bordonaro

Chapter 3 **Bioactive Lactoferrin-Derived Peptides** 29
Adham M. Abdou and Hend A. Elbarbary

Chapter 4 **Allergenic Milk Proteins: Friend or Foe Nutritional Proteins?** 52
Guillermo Docena, Paola Smaldini, Renata Curciarello and Angela María Candreva

Chapter 5 **Bioactive Peptides from Milk** 73
R. Anusha and O.S. Bindhu

Chapter 6 **Usefulness of Faecal Markers in Cow's Milk Protein Immunomediated Reactions** 112
Maria Elisabetta Baldassarre, Raffaella Panza and Nicola Laforgia

Chapter 7 **Casein Proteins: Structural and Functional Aspects** 142
Mohd Younus Bhat, Tanveer Ali Dar and Laishram Rajendrakumar Singh

Chapter 8 **The Protein Component of Sow Colostrum and Milk** 158
P.K. Theil and W.L. Hurley

Chapter 9 **Human Milk Lactoferrin and Antibodies: Catalytic Activities, Complexes, and Other Features** 174
Sergey E. Sedykh, Valentina N. Buneva and Georgy A. Nevinsky

Chapter 10 **An Important Milk Enzyme: Lactoperoxidase** 204
Zeynep Koksal, Ilhami Gulcin and Hasan Ozdemir

Chapter 11 **Measurement of Casein Micelle Size in Raw Dairy Cattle Milk by Dynamic Light Scattering** 219
Peter Hristov, Ivan Mitkov, Daniela Sirakova, Ivan Mehandgiiski and Georgi Radoslavov

Permissions

List of Contributors

Index

Preface

This book was inspired by the evolution of our times; to answer the curiosity of inquisitive minds. Many developments have occurred across the globe in the recent past which has transformed the progress in the field.

Proteins are chains of amino acids and molecules which are connected by peptide bonds. Proteins that are found in milk and associated dairy products are called milk proteins. Milk contains all the nine essential amino acids required by the human body. It is synthesized in the mammary gland of animals. The total milk protein content and amino acid composition varies with the animal. Milk proteins are broadly defined into two categories based on their chemical composition and physical properties. The casein group contains phosphorous while the serum proteins do not contain phosphorous. The casein proteins are not sensitive to heat but high temperature can cause denaturation in serum proteins. This book includes some of the vital pieces of work being conducted across the world, on various topics related to milk proteins. It aims to shed light on some of the unexplored aspects of the chemistry and biology of milk proteins and the recent researches in this field. Researchers and students in this field will be assisted by this book.

This book was developed from a mere concept to drafts to chapters and finally compiled together as a complete text to benefit the readers across all nations. To ensure the quality of the content we instilled two significant steps in our procedure. The first was to appoint an editorial team that would verify the data and statistics provided in the book and also select the most appropriate and valuable contributions from the plentiful contributions we received from authors worldwide. The next step was to appoint an expert of the topic as the Editor-in-Chief, who would head the project and finally make the necessary amendments and modifications to make the text reader-friendly. I was then commissioned to examine all the material to present the topics in the most comprehensible and productive format.

I would like to take this opportunity to thank all the contributing authors who were supportive enough to contribute their time and knowledge to this project. I also wish to convey my regards to my family who have been extremely supportive during the entire project.

Editor

1

Milk Proteins: Processing of Bioactive Fractions and Effects on Gut Health

Anindya Mukhopadhya and Torres Sweeney

Abstract

Milk is nature's most complete food. While milk clearly provides basic nutritional requirements, bioactive components within milk also impart a wide range of additional health benefits to both the neonate and the adult. However, human milk is compositionally different from cow's milk, and certain protein components of cow's milk can act as allergens to susceptible humans. One way of extracting the benefits of cow's milk proteins, while eliminating the risk of allergenicity in humans, is to hydrolyse the milk proteins. Hydrolysis of milk proteins generates smaller peptide sequences from their parent protein that can be biologically active when released. At an industrial scale, hydrolysis of milk proteins can be achieved through either enzymatic hydrolysis or fermentation. An alternative process of generating similar sized peptides is by *in silico* synthesis. These compounds can subsequently be developed as fortifying food agents.

A number of milk-derived bioactives have been characterised with a variety of health benefits in the gastrointestinal tract. These biological activities include supporting the establishment of a healthy commensal microbiome, suppressing the colonization of pathogenic bacteria and supporting barrier function. Hydrolysates of casein and whey also impart anti-inflammatory and immunomodulatory activity. This chapter gives an overview on the future potential of food grade milk hydrolysates to support homeostasis in the gastrointestinal tract.

Keywords: milk hydrolysates, anti-inflammatory, gut health, gut microbiota, gut homeostasis

1. Overall composition of milk

The overall composition of milk depends on a range of factors including genetics (species and breed), physiological state (age and stage of lactation) and environment (food and climate) [1–5]. While water is the main constituent of milk, comprising ~87% of the total volume, the remainder is composed of carbohydrates, fats and proteins in varying volumes across different species [6–8]. Among the numerous nutritional benefits of milk, milk proteins have gathered enormous attention for being a 'complete' protein as they provide all nine essential amino acids (leucine, isoleucine, valine, phenylalanine, tryptophan, histidine, threonine, methionine, lysine) required by humans [9]. The proteins in milk are categorised into major proteins that include casein and whey fractions [1] and minor proteins that include lactoferrin, lactoperoxidases, lipases, lactase [6, 10] and miscellaneous proteins (cytokines, immunoglobulins, etc.) [11].

2. Milk protein hydrolysates

The process of breaking down milk proteins to shorter peptide sequences is termed 'hydrolysis'. This process happens naturally in the gastrointestinal tract and can be simulated in the laboratory or on an industrial scale. During the normal transit through the gastrointestinal tract, milk proteins are exposed to proteinases such as pepsin, trypsin and chymotrypsin which break them down into smaller peptides. These peptides are further digested by brush border peptidases present at the surface of intestinal epithelial cells where they produce amino acids; however, some oligopeptides still remain intact [12]. In laboratory or at an industrial scale, milk hydrolysates are released either by treatment of milk proteins with food grade enzymes or through fermentation with bacteria, which is described in detail in the following sections.

The shorter peptide sequences often possess bioactive properties beyond their nutritional contribution along with eradicating any protein-specific allergenicity [13, 14]. Processing and enriching for food grade bioactive peptides is a goal for the functional food industry. A functional food can be described as:

'a food that can beneficially affect one or more target functions in the body beyond adequate nutritional effects in a way that is relevant to either an improved state of health and well-being and/or a reduction of risk of disease' [15].

Once the hydrolysates are released, they can potentially have bioactive properties which can exert their effects in receptive cells, including those present in the gastrointestinal tract [16]. The bioactivities of the resulting hydrolysates are variable depending on a range of factors, including the enzyme used, the processing conditions and the final size of the peptide sequence following hydrolysis [17]. The degree of hydrolysis (DH) is defined as the percentage of cleaved peptide bonds, i.e. the number of hydrolysed bonds per total number of peptide bonds in the protein [18]. This affects the size and amino acid composition of the peptides, which subsequently determines the biological activity of the peptide. Hence, DH is an important consideration from the perspective of functional food research [19].

2.1. Enzymatic hydrolysis

The enzymatic hydrolysis process is conducted under mild conditions (pH 6–8, temperature 40–60°C) to minimise side reactions and to retain the amino acid composition similar to the starting material [17]. Enzymatic hydrolysis improves the solubility and heat stability of peptides, which is of benefit to the food industry. However, consumption of certain enzymes leads to allergic or toxic responses; hence, consumer safety is an important factor and requires the regulation of enzymes used for hydrolysis [20]. Enzymes that obtain 'generally recognised as safe' (GRAS) status and special approval of 'food grade' quality are legally considered as safe [20]. The food grade enzymes generally used to hydrolyse milk proteins into hydrolysates include pepsin, trypsin and chymotrypsin [21, 22]. In addition, food grade proteolytic enzymes, derived from microorganisms, can also be used to generate hydrolysates [23]. Proteolytic enzymes are of two types, depending upon their hydrolysing mechanism: endopeptidases which hydrolyse peptide bonds within protein molecules and exoproteases which hydrolyse N or C terminal peptide bonds. Post enzymatic hydrolysis, the hydrolysates usually need an additional treatment. The most common procedures include ultrafiltration, heat treatment and/or activated carbon treatment to control molecular size and elimination of bitterness in the hydrolysates [17].

2.2. Hydrolysis through microbial fermentation

Fermentation of milk proteins with proteolytic starter culture is another method of bulk production of hydrolysates. Safety measures should be considered with regard to toxicity and pathogenicity associated with the microorganisms used for fermentation. Food grade microorganisms with no related toxigenic and pathogenic response in humans are widely used. During microbial fermentation, milk proteins are subjected to 'splitting' as they are broken down by the proteolytic system of microorganisms [24]. Bacterial cultures of *Lactobacilli* spp., *Lactococci* spp. and *Streptococci* spp. are commonly used to generate hydrolysates from milk [25]. The proteolytic system of lactic acid bacteria (LAB) contains cell envelope-associated proteinases, endopeptidases, aminopeptidases, tripeptidases and dipeptidases for the production of hydrolysates [26]. LAB requires free amino acids and peptides for their growth, which they obtain from milk proteins by degradation [27]. The peptides or hydrolysates not utilised by bacteria can promote various bioactivities. LAB proteinases hydrolyse more than 40% of the peptide bonds in α-S1 and β-caseins, producing oligopeptides ranging from 4 to 40 amino acid residues [28]. Fermentation parameters such as enzyme/substrate ratio, composition of medium, heat treatment, temperature, pH and carbon/nitrogen ratio influence the release of hydrolysates from milk proteins. An alternative strategy used by food industry is starter LAB culture along with food grade enzyme to hydrolyse milk proteins. This strategy not only increases the peptide content of the hydrolysate but also diversifies the bioactivity of the hydrolysate [27].

2.3. Peptide synthesis

The *in silico* synthesis is an alternative process for generating peptide sequences modelled on various milk protein hydrolysates. Depending upon the length and quality of the desired peptide, either recombinant DNA technology or chemical synthesis methodologies can be used to synthesise peptides [29]. The application of recombinant DNA technology is preferable if the objective is to generate large peptides consisting up to several hundred amino acids. However, this is a long and expensive process. Chemical synthesis is currently used for laboratory scale peptide synthesis, especially for peptides used in therapy [30]. However, chemical synthesis uses toxic reagents that may contribute to environmental pollution and generates unwanted peptide by-products. Instead, the solid-phase synthesis approach, which is a variant of chemical synthesis, can generate peptides composed of 10 to over 100 residues in small scale, using lower amounts of chemicals [31]. Modification of peptide function by substitution of a particular amino acid in the sequence is easily done in solid-phase synthesis [30].

3. Models used for bioactivity evaluation and challenges

After the generation of milk hydrolysates, their bioactivity profile needs to be determined. In laboratory or at an industrial scale, the primary screening for the bioactivity is performed on *in vitro* platforms, using various cell culture models. Due to the difficulty in growing primary intestinal epithelial cells, cell lines derived from human intestinal tissues have been used extensively such as Caco-2, HT-29 and T-84 cells, as reviewed by Shimizu et al. [32]. Caco-2, HT-29 and T-84 cell cultures functionally resemble colonic enterocytes; however, these cultures only have a single cell type. This limits the understanding of the mechanistic effects of hydrolysates in a heterogeneous cell environment such as that of *in vivo*. This limitation is overcome by the use of intestinal tissue explants. Following the idea that pigs are good models for human research, our group used porcine intestinal tissue explants to test the bioactivity of seaweed extracts in an *ex vivo* system [33]. Further studies were carried out to establish the *ex vivo* model by our group, and Bahar et al. concluded from his study that explants from porcine gastrointestinal tract can be used to test the bioactivity of test compounds up to 3 h postmortem [34]. Hence, the cellular heterogeneity of the tissue explants helps in better understanding of the effects of hydrolysates on cell growth, differentiation and functionality [35, 36]. Our group has applied these *in vitro* and *ex vivo* models for screening and testing the anti-inflammatory activities of milk hydrolysates [37, 38].

Although cell- and tissue-based model systems are an alternative to animal experiments, they do not reflect the *in vivo* conditions of cells and tissues in their natural state in the organism. The experimental culturing environment lacks the effect of endocrine and nervous systems that are involved in the homeostatic regulation *in vivo*, and hence animal and human trials are necessary for the full evaluation of bioactives. The difficulty to characterise the 'pharmacokinetic' and 'pharmacodynamic' properties of hydrolysates creates a challenge for researchers [39]. *In vitro* studies do not consider the degradation of the peptides by gastric, pancreatic and small intestinal brush border enzymes. Most therapeutic peptides act systemically; however, only nano-molar or pico-molar quantities may be transported to the circulatory system due to enzymatic degradation in the gastrointestinal tract [39, 40]. Hence, findings between *in vitro, ex vivo* and *in vivo* experiments can be inconsistent. Overcoming these inconsistencies requires standardized methodologies for the analysis and the application of robust clinical trials in order to evaluate both the efficacy and metabolic fate of a particular hydrolysate [40].

4. Functionality of bioactive hydrolysates

To date, the bioactivities of a wide variety of hydrolysates have been characterised using *in vitro, ex vivo* and *in vivo* experimental platforms (**Table 1**). Hydrolysates are easily ingested in functional food offerings; however, their bioavailability might be affected due to post-ingestion break down. As seen in an *in vivo* trial by our group, milk hydrolysates with potent anti-inflammatory activity in *in vitro* and *ex vivo* platforms [37], lost this activity in the gastrointestinal tract of weaning piglets, most likely due to breakdown in the stomach (Mukhopadhya et al. *In Press*). These milk hydrolysates, combined with a viscous β-glucan, travelled to the distal section of gut with intact bioactivity (Mukhopadhya et al. *In Press*). The following sections review the potential for a spectrum of different milk hydrolysates to support the activities of the gastrointestinal tract microbiota, the gastrointestinal tract immune system and the gastrointestinal tract mucosal barrier to support homeostasis in the gastrointestinal tract.

Protein hydrolysates	Approximate content in cow milk (g/l)	Hydrolysis process	Related bioactivity in gastrointestinal tract	References
Total casein	~80%	EH	Bifidogenic	[51, 52]
		EH	Antimicrobial	[62]
		EH	↑Mucin	[72, 77]
		EH	↑IgG, ↑IgA	[88]
		EH, Fermentation, PS	Immunomodulation	[37, 38, 92, 93, 99]
α-S1 casein	9.1	EH	Antimicrobial	[62]
		EH, Fermentation	↑IgG, ↑IgA	[86, 87]
α-S2 casein	2.4	EH	Antimicrobial	[62]
		EH, Fermentation	↑IgG, ↑IgA	[86, 87]
β-Casein	8.5	EH	↑Mucin	[74, 75]
		EH, Fermentation	↑IgG, ↑IgA	[86, 87]

Protein hydrolysates	Approximate content in cow milk (g/l)	Hydrolysis process	Related bioactivity in gastrointestinal tract	References
k-Casein	3.0	EH	Antimicrobial	[64]
		EH	Immunomodulation	[94, 100, 101]
Total whey	**~18%**	Fermentation	Bifidogenic	[53]
		EH	Antimicrobial	[60]
		EH	↑Mucin	[78, 79]
		EH	↑IgG, ↑IgA	[89]
		EH	Immunomodulation	[89]
α-Lactalbumin	1.1	EH	Antimicrobial	[60, 65]
		EH	↑Mucin	[73, 80]
β-Lactoglobulin	2.8	EH, Fermentation	Bifidogenic	[54]
		EH	Antimicrobial	[60]
		EH	↑Mucin	[76]
Lactoferrin, lactoperoxidase, lysozyme, proteose-peptone, glycomacropeptide	~3%	EH	Bifidogenic	[55, 56, 58]
		EH, PS	Antimicrobial	[67, 68, 69]
		EH, PS	↑IgG, ↑IgA, ↑IgM	[90]

EH, enzyme hydrolysis; PS, peptide synthesis

Table 1. Milk protein hydrolysates, content in cow's milk, hydrolysis process used and their related bioactivity.

4.1. Prebioitc activity of milk hydrolysates

The World Health Organisation now recommends breastfeeding for up to 6 months, as breast milk has a major positive impact on the health and growth of the infant [41]. One of the most important benefits of breastfeeding the newborn is the colonisation of the gut by 'healthy' microbiota. 'Healthy' gut microbiota confers nutritive, metabolic and protective functions that affect intestinal physiology, immunity and whole-body metabolism. The establishment of a 'healthy' microflora in the gut during early life is crucial for the healthy development of a balanced immune regulatory network in the gut, a feature which affects the overall health of the individual [42]. Beneficial gut microorganisms aid gut health by releasing growth substrates from milk [43], improving vaccine responses [44] and decreasing gut permeability [45, 46]. After birth, the gut is colonised with bacteria from four main phyla namely *Bacteroides, Proteobacteria, Firmicutes* and *Actinobacteria,* and these phyla in turn influence the development of the gut-associated lymphoid tissue (GALT) [47, 48].

Milk hydrolysates show bifidogenic activity, i.e. they support the growth of Gram-positive anaerobic bacteria namely *Bifidobateria* spp., in the gut [49, 50]. The prebiotic characteristics of milk hydrolysates are outlined in **Table 1**. For example, a proteolytic casein hydrolysate [51, 52] and hydrolysates of whey proteins fermented with *Lactobacillus casei* strains [53] effectively stimulate *Bifidobacteria* spp. growth. Interestingly, proteolytic digestion of β-lactoglobulin, a major whey protein of the bovine milk that is absent in human milk, generated peptides supporting both *Bifidobacterium* and *Lactobacillus* spp. growth [54]. *In vitro* peptic digestion of human and bovine lactoferrin [55, 56] and a synthetic peptide modelled on lactoferrin [57] have confirmed bifidogenic activity. Not only hydrolysates of major proteins but hydrolysates of the minor milk proteins, proteose-peptone, also supported the proliferation of *Bifidobacterium animalis* [58].

4.2. Antimicrobial activity of milk hydrolysates

Antimicrobial milk peptides prevent attachment and invasion of pathogens by either directly interacting with the pathogen and killing them or changing the host environment, leading to the inhibition of growth of microorganisms [59, 60]. The direct interaction of antimicrobial milk hydrolysates with microorganisms is specific, as they show affinity towards polarised bacterial membranes rather than dipolar membranes of eukaryotic cells [50]. There is growing evidence that the antimicrobial property of milk hydrolysates is related to the formation of α-helical structure of the peptides. The modifications of peptide's secondary and tertiary structures by phosphorylation of specific amino acid or chemical modification of C or N terminal dramatically affects the antimicrobial activity [50, 61]. Another mode of action for antimicrobial peptides is by aggregating in the cytoplasmic membrane, disrupting the membrane permeability of bacteria, and causing cell death [13, 30]. On the contrary, the indirect antimicrobial activity of milk hydrolysates is achieved by decreasing the host intestinal pH and thus limiting the growth of pathogenic microorganisms. This mechanism is also known as 'colonisation resistance' [57].

The antimicrobial effects of milk hydrolysates have been listed in **Table 1**. Casein is a major source of antimicrobial peptides, and hydrolysates of α-S1 casein exert protective effects against *Staphylococcus aureus, Streptococcus pyogenes* and *Listeria monocytogenes* [62]. Peptic hydrolysis of α-S2 casein generated two hydrolysates with antimicrobial activity against Gram-positive and Gram-negative bacteria, specifically *Saccharomyces thermophilus* and *E. coli*, respectively [62]. α-S2 casein is not present in human milk, hence hydrolysates from bovine α-S2 casein have interesting potential as a novel human gastrointestinal tract microbiota modulating agent [63]. The minor subunit of casein, κ-casein, inhibits the adhesion of *Helicobacter pylori*, an early-life pathogen, to human gastric mucosa [64]. The digestion of major whey proteins, α-lactalbumin and β-lactoglobulin, by trypsin and chymotrypsin releases peptides with antimicrobial activity against specific Gram-positive bacteria [60]. However, hydrolysates of α-lactalbumin, generated by protease treatment, have antimicrobial properties with regard to *E. coli, Klebsiella pneumoniae, Staphylococcus aureus, Staphylococcus epidermidis, Staphylococci* and *Candida albicans* [65]. Milk products fermented with *Lactobacillus paracasei* CBA L74 have anti-inflammatory activities in response to *Salmonella typhimurium* infection in mice [66]. Lactoferrin has gathered most attention as a source of antimicrobial peptides [67].

Pepsin digestion of bovine lactoferrin generated hydrolysates that are amphiphilic and displayed antimicrobial activity against a broad range of Gram-positive and Gram-negative bacteria including *Listeria, E. coli, Salmonella* and *Campylobacter* but not against several strains of *Bifidobacterium* [68]. Lactoferrin hydrolysates also act against the most important fungal pathogen in humans, *Candida albicans* [69].

4.3. Milk hydrolysates preserve gastrointestinal mucosal integrity

The intestinal epithelial cell layer of the gastrointestinal tract lies at the border between the gut-associated lymphoid tissue (GALT), which is the most abundant accumulation of lymphocytes in the body, and the intestinal lumen which contains a high number of dietary antigens and a varied commensal microbiota [70]. The intestinal epithelial cell layer is covered by a mucus gel, which functions as a protective layer for the gastrointestinal system. This barrier function includes the prevention of entry of pathogenic microorganisms, toxins and allergens. The mucus gel is composed of glycoproteins called mucins, with up to 20 mucin genes identified. Mucin genes are expressed by specific cells (goblet cells and enterocytes) and categorised as gel-forming secretory mucins (*Muc2, Muc5AC, Muc5B, Muc6,* etc.) or membrane-bound mucins (*Muc1, Muc3, Muc4, Muc13,* etc.) [71]. Interestingly, mice deficient in the *Muc2* gene develop spontaneous colitis and consequently colon cancer.

Milk hydrolysates can influence the expression and secretion of mucins, as outlined in **Table 1**. The modulation of mucin production by milk hydrolysates may assist in the development of dietary strategies to enhance and protect the mucus layer. In rats, jejunal *ex vivo* studies, hydrolysates of casein and α-lactalbumin increased mucin secretion, whereas the native casein did not have any effect [72, 73]. An enzymatic hydrolysate of β-casein enhanced mucin secretion by up-regulating *Muc2* and *Muc3* genes in rat intestinal cells (DHE) and *Muc5AC* gene in human intestinal cells (HT29-MTX) [74, 75]. In a similar experimental model, the enzymatic hydrolysates of β-lactoglobulin were even more effective in increasing mucus production compared to a β-casein hydrolysate [76]. Rats receiving a casein hydrolysate supplemented diet had up-regulated the expression of *Muc3* and *Muc4* genes in the small intestine and colon [77]. This β-casein fragment, which up-regulated *Muc5AC* in HT29-MTX cells, also up-regulated *Muc2, Muc4, defensin 5* and *lysozyme* expression in rat ileum [75]. In a dextran sulphate sodium (DSS)-induced colitis model in rats, diets supplemented with either cheese whey protein [78] or whey protein isolate [79] or α-lactalbumin [80] displayed gut protective effects by increased mucus production. The protection of the gut from DSS challenge by cheese whey protein was possibly mediated by its high threonine and cysteine content. In fact, specific amino acids such as threonine, cysteine, proline and serine increase the number of Muc2-containing goblet cells, up-regulate mucin synthesis and eventually help to restore *Enterobacteriaceae, Enterococcus* and *Lactobacillus* populations in the gastrointestinal tract, thus enforcing the gut defence mechanism and mucosal healing in DSS challenged rats [81].

4.4. Milk hydrolysates can modulate the gastrointestinal immune system

The intestinal mucosa exists in a non-pathological state of continuous 'physiological inflammation'. This low level of inflammation is required to prime the GALT for potential pathogenic bacteria [82]. The mucosal immune system features immune cells including neutrophils, monocyte/macrophages, dendritic cells, mast cells, B and T cells. The crosstalk between intestinal epithelial cells, gut microbiota and local immune cells is essential to maintain intestinal homeostasis, whereas, dysregulation leads to chronic intestinal inflammation [83]. Much of the experimental data come from model organisms such as mice and rats; however, a number of studies have been carried out in humans. Several examples of anti-inflammatory activity exhibited by a variety of milk hydrolysates across a range of experimental models are listed elsewhere [59, 60, 82, 84].

Several milk protein hydrolysates enhance immune cell function by increasing secretion of immunoglobulins, as outlined in **Table 1**. Immunoglobulins are glycoprotein molecules that specifically recognise antigens from bacteria or viruses and aid in their destruction through a highly complex and specific immune response [85]. Hydrolysates of αs1-casein, αs2-casein and β-casein stimulated the immune system through the enhancement of immunoglobulin G (IgG) and IgA concentrations [86, 87]. Casein hydrolysates conferred protective effects against pathogenic microorganisms in mice challenged with bacterial endotoxin, LPS, by increasing intestinal and faecal IgA and anti-LPS IgA levels [88]. Similar modulation of immune response was recorded in mice against *E. coli* infection, when receiving trypsin/chymotrypsin-digested whey protein fractions [89]. Not only strengthening the immune response but post-weaning complications were also eradicated in piglets supplemented with bovine lactoferricin and lactoferrampin fusion peptide by increasing serum levels of IgA, IgG and IgM concentrations and improved diarrhoeal scores [90].

Immune cells, such as monocytes and macrophages, play an important role in inflammatory responses and tissue repair and remodelling by either interacting directly with microorganisms during infections and/or secretion of cytokines that mediate biological effects [91]. Milk hydrolysates can modulate the gastrointestinal immune system by modulating proliferation and maturation of localised immune cells; the immunomodulatory activities of milk hydrolysates are outlined in **Table 1**. Casein peptides induced innate host immune responses in humans, by stimulating the proliferation of lymphocytes and macrophages, [92] and in mice, by activating monocytes and macrophages [93]. On the contrary, rennin-digested κ-casein fragments inhibited the proliferation of mouse spleen lymphocyte and rabbit Peyer's patch cells [94]. The mechanisms of this κ-casein fragment include acting either as an anti-IL-1 antibody or suppressing IL-2 receptor expression on CD4+ T-cells [95]. Functionally, the phagocytic activity of inflamed murine macrophages was increased by *Lactobacillus helveticus*-fermented skim milk through increasing *TNF-α* production [96].

Particular milk hydrolysates modulate the MAP kinase and NF-κβ pathways that consequently control the secretion of several cytokines that can induce inflammatory responses and strengthen the host defence mechanisms [97]. Mice supplemented with *Lactobacillus helveticus* R389-fermented milk peptides had increased circulatory intestinal calceneurin enzyme, an activator of the gastrointestinal immune system and cytokine *IL-6* [98]. Interestingly, the immune response against LPS was inhibited by casein-derived peptides in Balb/c mice diet by increasing circulatory anti-inflammatory cytokines IL-10 and IL-14 and suppressing proinflammatory cytokines TNFα and IFNγ [99]. Similar suppression of a pro-inflammatory response by down-regulation of IL-8 in inflamed Caco-2 cells was recorded by casein hydrolysate and its size fractions [37, 38]. Further validation of the anti-inflammatory activity was performed in porcine colonic explants and the casein hydrolysate, and its size fractions down-regulated *IL-1*α, *IL-1*β and *IL-8* expression [37, 38]. Bovine κ- casein hydrolysate inhibited circulatory IFNγ secretion and suppressed *IL-10* and *FoxP3* expression in concanavalin A (ConA)-stimulated rat splenocytes [100]. This bovine κ-casein, in a human macrophage cell line U937, was associated with the suppression of circulatory pro-inflammatory cytokines IL-1β, TNFα and IL-8 production [101]. However, whey protein hydrolysate up-regulated the immune response in *E. coli* infected mice by increasing circulatory serum transforming growth factor (TGF)-β secretion [89].

5. Future developments

The potential health benefits of milk hydrolysates are a subject of growing commercial interest from a health-promoting functional-food perspective. Several commercial products are currently available in the market, and this trend is likely to continue. There are three major areas where developments can be made. The generation of milk hydrolysates is the first area of development. The generation and processing of food grade milk hydrolysates should be carefully designed to yield hydrolysates with diverse bioactivities. Novel technologies can be developed, focussing on the process of enrichment of the hydrolysates with active peptides from milk proteins. The second area of development is the research technologies used to evaluate the bioactivity of milk hydrolysates. The investigation of biochemical properties using newly developed modern analytical technologies is required to understand the cross reactivity between milk hydrolysates and the carrier food matrix. Third, robust platforms should be developed to study the molecular mechanisms by which the bioactives exert their activities. This area is the most challenging research area as the outcome from these studies forms the basis of tailored dietary formulations.

Author details

Anindya Mukhopadhya* and Torres Sweeney

*Address all correspondence to: anindya.mukhopadhya@ucd.ie

School of Veterinary Medicine, University College Dublin, Dublin, Ireland

References

[1] Fox, P.F., Milk proteins as food ingredients. *Int. J. Dairy Technol.* 2001, *54*, 41–55.

[2] Donovan, S.M., Role of human milk components in gastrointestinal development: Current knowledge and future NEEDS. *J. Pediatr.* 2006, *149*, S49–S61.

[3] Pouliot, Y., Gauthier, S.F., Milk growth factors as health products: Some technological aspects. *Int. Dairy J.* 2006, *16*, 1415–1420.

[4] Clare, D.A., Swaisgood, H.E., Bioactive milk peptides: A prospectus. *J. Dairy Sci.* 2000, *83*, 1187–95.

[5] Sabahelkhier, M.K., Faten, M.M., Omer, F.I., Comparative determination of biochemical constituents between animals (goat, sheep, cow and camel) milk with human milk. *Res. J. Recent Sci.* 2012, *1*, 69–71.

[6] Fox, P.F., Uniacke-Lowe, T., McSweeney, P.L.H., O'Mahony, J.A., Dairy chemistry and biochemistry. Springer International Publishing, Cham; 2015.

[7] Fox, P.F., McSweeney, P.L.H., Dairy chemistry and biochemistry. Springer Science & Business Media; 1998.

[8] Wong, N.P., Jenness, R., Keeney, M., Marth, E.H. (Eds.), Fundamentals of dairy chemistry. Springer US, Boston, MA; 1995.

[9] McGregor, R.A., Poppitt, S.D., Milk protein for improved metabolic health: A review of the evidence. *Nutr. Metab.* 2013, *10*, 46.

[10] Haug, A., Høstmark, A.T., Harstad, O.M., Bovine milk in human nutrition—review. *Lipids Health Dis.* 2007, *6*, 25.

[11] Ebringer, L., Ferencík, M., Krajcovic, J., Beneficial health effects of milk and fermented dairy products—review. *Folia Microbiol. (Praha).* 2008, *53*, 378–394.

[12] Segura-Campos, M., Chel-Guerrero, L., Betancur-Ancona, D., Hernandez-Escalante, V.M., Bioavailability of bioactive peptides. *Food Rev. Int.* 2011, *27*, 213–226.

[13] Hartmann, R., Meisel, H., Food-derived peptides with biological activity: From research to food applications. *Curr. Opin. Biotechnol.* 2007, *18*, 163–169.

[14] Kitts, D., Weiler, K., Bioactive proteins and peptides from food sources. Applications of bioprocesses used in isolation and recovery. *Curr. Pharm. Des.* 2003, *9*, 1309–1323.

[15] ILSI Europe, Scientific concepts of functional foods in Europe: Concensus document. *Br. J. Nutr.* 2002, *81*, 1S–27S.

[16] Ricci-Cabello, I., Herrera, M.O., Artacho, R., Possible role of milk-derived bioactive peptides in the treatment and prevention of metabolic syndrome. *Nutr. Rev.* 2012, *70*, 241–255.

[17] Clemente, A., Enzymatic protein hydrolysates in human nutrition. *Trends Food Sci. Technol.* 2000, *11*, 254–262.

[18] Nielsen, P.M., Petersen, D., Dambmann, C., Improved method for determining food protein degree of hydrolysis. *J. Food Sci.* 2001, *66*, 642–646.

[19] Jamdar, S.N., Rajalakshmi, V., Pednekar, M.D., Juan, F., et al., Influence of degree of hydrolysis on functional properties, antioxidant activity and ACE inhibitory activity of peanut protein hydrolysate. *Food Chem.* 2010, *121*, 178–184.

[20] Holzapfel, W.H., Geisen, R., Schillinger, U., Biological preservation of foods with reference to protective cultures, bacteriocins and food-grade enzymes. *Int. J. Food Microbiol.* 1995, *24*, 343–362.

[21] Yamamoto, N., Akino, A., Takano, T., Antihypertensive effect of the peptides derived from casein by an extracellular proteinase from *Lactobacillus helveticus* CP790. *J. Dairy Sci.* 1994, *77*, 917–922.

[22] Korhonen, H., Milk-derived bioactive peptides: From science to applications. *J. Funct. Foods.* 2009, *1*, 177–187.

[23] Choi, J., Sabikhi, L., Hassan, A., Anand, S., Bioactive peptides in dairy products. *Int. J. Dairy Technol.* 2012, *65*, 1–12.

[24] Hayes, M., Stanton, C., Fitzgerald, G.F., Ross, R.P., Putting microbes to work: Dairy fermentation, cell factories and bioactive peptides. Part II: Bioactive peptide functions. *Biotechnol. J.* 2007, *2*, 435–449.

[25] Fitzgerald, R.J., Murray, B.A., Bioactive peptides and lactic fermentations. *Int. J. Dairy Technol.* 2006, *59*, 118–125.

[26] Agyei, D., Danquah, M.K., Rethinking food-derived bioactive peptides for antimicrobial and immunomodulatory activities. *Trends Food Sci. Technol.* 2012, *23*, 62–69.

[27] Hafeez, Z., Cakir-Kiefer, C., Roux, E., Perrin, C., et al., Strategies of producing bioactive peptides from milk proteins to functionalize fermented milk products. *Food Res. Int.* 2014, *63*, 71–80.

[28] Phelan, M., Kerins, D., The potential role of milk-derived peptides in cardiovascular disease. *Food Funct.* 2011, *2*, 153–167.

[29] Korhonen, H.J.T., Pihlanto, A., Bioactive peptides: Novel applications for milk proteins. 2004.

[30] Zambrowicz, A., Timmer, M., Polanowski, A., Lubec, G., et al., Manufacturing of peptides exhibiting biological activity. *Amino Acids.* 2013, *44*, 315–320.

[31] Schlimme, E., Meisel, H., Bioactive peptides derived from milk proteins. Structural, physiological and analytical aspects. *Food/Nahrung.* 1995, *39*, 1–20.

[32] Shimizu, M., Interaction between food substances and the intestinal epithelium. *Biosci. Biotechnol. Biochem.* 2010, *74*, 232–241.

[33] Leonard, S.G., Sweeney, T., Bahar, B., O'Doherty, J. V, Effect of maternal seaweed extract supplementation on suckling piglet growth, humoral immunity, selected microflora, and immune response after an ex vivo lipopolysaccharide challenge. *J. Anim. Sci.* 2012, *90*, 505–514.

[34] Bahar, B., O'Doherty, J.V, Sweeney, T., Assessment of RNA integrity in the postmortem pig colonic tissue ex vivo. *J. Anim. Sci.* 2012, *90*(Suppl 4), 22–24.

[35] Groothuis, G.M.M., de Graaf, I.A.M., Precision-cut intestinal slices as in vitro tool for studies on drug metabolism. *Curr. Drug Metab.* 2013, *14*, 112–119.

[36] Randall, K.J., Turton, J., Foster, J.R., Explant culture of gastrointestinal tissue: A review of methods and applications. *Cell Biol. Toxicol.* 2011, *27*, 267–284.

[37] Mukhopadhya, A., N. Noronha, B. Bahar, M. T. Ryan, B. A. Murray, P. M. Kelly, I. B. O'Loughlin, J. V. O'Doherty, and T. Sweeney. "The anti-inflammatory potential of a moderately hydrolysed casein and its 5 kDa fraction in in vitro and ex vivo models of the gastrointestinal tract." *Food & function* 6, no. 2 2015: 612–621.

[38] Mukhopadhya, A., Noronha, N., Bahar, B., Ryan, M.T., et al., Anti-inflammatory effects of a casein hydrolysate and its peptide-enriched fractions on TNFα-challenged Caco-2 cells and LPS-challenged porcine colonic explants. *Food Sci. Nutr.* 2014, *2*, 712–723.

[39] Picariello, G., Mamone, G., Nitride, C., Addeo, F., et al., Protein digestomics: Integrated platforms to study food-protein digestion and derived functional and active peptides. *TrAC Trends Anal. Chem.* 2013, *52*, 120–134.

[40] Li-Chan, E.C., Bioactive peptides and protein hydrolysates: Research trends and challenges for application as nutraceuticals and functional food ingredients. *Curr. Opin. Food Sci.* 2015, *1*, 28–37.

[41] Kramer M, Kakuma R.The Optimal Duration of Exclusive Breastfeeding: A Systematic Review. Geneva, Switzerland: World Health Organization; 2002

[42] Canny GO, McCormick BA. Bacteria in the intestine, helpful residents or enemies from within?. *Infection and immunity*. 2008 Aug 1;76(8):3360–73.

[43] Subramanian, S., Blanton, L.V., Frese, S.A., Charbonneau, M., et al., Cultivating healthy growth and nutrition through the gut microbiota. *Cell.* 2015, *161*, 36–48.

[44] Huda, M.N., Lewis, Z., Kalanetra, K.M., Rashid, M., et al., Stool microbiota and vaccine responses of infants. *Pediatrics*. 2014, *134*, e362–e372.

[45] Ewaschuk JB, Diaz H, Meddings L, Diederichs B, Dmytrash A, Backer J, Looijer-van Langen M, Madsen KL. Secreted bioactive factors from Bifidobacterium infantis enhance epithelial cell barrier function. *American Journal of Physiology–Gastrointestinal and Liver Physiology*. 2008 Nov 1;295(5):G1025-34.

[46] Weng, M., Ganguli, K., Zhu, W., Shi, H.N., et al., Conditioned medium from *Bifidobacteria infantis* protects against *Cronobacter sakazakii*-induced intestinal inflammation in newborn mice. *Am. J. Physiol. Gastrointest. Liver Physiol.* 2014, *306*, G779–G787.

[47] Koropatkin, N.M., Cameron, E.A., Martens, E.C., How glycan metabolism shapes the human gut microbiota. *Nat. Rev. Microbiol.* 2012, *10*, 323–335.

[48] Palmer, C., Bik, E.M., DiGiulio, D.B., Relman, D.A., et al., Development of the human infant intestinal microbiota. *PLoS Biol.* 2007, *5*, e177.

[49] Pacheco, A.R., Barile, D., Underwood, M.A., Mills, D.A., The impact of the milk glycobiome on the neonate gut microbiota. *Annu. Rev. Anim. Biosci.* 2015, *3*, 419–445.

[50] Benkerroum, N., Antimicrobial peptides generated from milk proteins: A survey and prospects for application in the food industry. A review. *Int. J. Dairy Technol.* 2010, *63*, 320–338.

[51] Kehagias, C., Jao, Y.C., Mikolajcik, E.M., Hansen, P.M.T., Growth response of *Bifidobacterium bifidum* to a hydrolytic product isolated from bovine casein. *J. Food Sci.* 1977, *42*, 146–150.

[52] Raynaud M, Bizzini B. Purification et propriétés du facteur bifidus 2. France Cent Nat Coord Etudes Rech Nutr Aliment Journees Sci. 1971.

[53] Moon, G.-S., Bifidobacterial growth stimulation by *lactobacillus casei* via whey fermentation. *J. Food Sci. Nutr.* 2009, *14*, 265–268.

[54] Poltronieri P, Cappello MS, D'Urso OF Bioactive peptides with health benefit and their differential content in whey of different origin. Whey: Types, Composition and Health Implications 2012

[55] Liepke, C., Adermann, K., Raida, M., Mägert, H.-J., et al., Human milk provides peptides highly stimulating the growth of bifidobacteria. *Eur. J. Biochem.* 2002, *269*, 712–718.

[56] Kim, W.-S., Rahman, M.M., Kumura, H., Shimazaki, K., Comparison of growth promoting effects on *Bifidobacterium* spp. by bovine lactoferrin hydrolysates. *Biosci. Microflora* 2005, *24*, 119–123.

[57] Lonnerdal, B., Nutritional and physiologic significance of human milk proteins. *Am J Clin Nutr.* 2003, *77*, 1537S–1543.

[58] Etienne, L., Girardet, J.M., Linden, G., Growth promotion of *Bifidobacterium animalis* by bovine milk proteose-peptone. *Lait.* 1994, *74*, 313–323.

[59] Gauthier, S.F., Pouliot, Y., Saint-Sauveur, D., Immunomodulatory peptides obtained by the enzymatic hydrolysis of whey proteins. *Int. Dairy J.* 2006, *16*, 1315–1323.

[60] Hernández-Ledesma, B., García-Nebot, M.J., Fernández-Tomé, S., Amigo, L., et al., Dairy protein hydrolysates: Peptides for health benefits. *Int. Dairy J.* 2014, *38*, 82–100.

[61] Dashper, S.G., Liu, S.W., Reynolds, E.C., Antimicrobial peptides and their potential as oral therapeutic agents. *Int. J. Pept. Res. Ther.* 2007, *13*, 505–516.

[62] Nagpal, R., Behare, P., Rana, R., Kumar, A., et al., Bioactive peptides derived from milk proteins and their health beneficial potentials: An update. *Food Funct.* 2011, *2*, 18–27.

[63] Lahov, E., Regelson, W., Antibacterial and immunostimulating casein-derived substances from milk: Casecidin, isracidin peptides. *Food Chem. Toxicol.* 1996, *34*, 131–145.

[64] Strömqvist, M., Falk, P., Hansson, S.B.L., Lönnerdal, B., et al., Human Milk K-casein and inhibition of *helicobacter pylori* adhesion to human gastric mucosa. *J. Pediatr. Gastroenterol. Nutr.* 1995, *21*(3), 288–296.

[65] Pellegrini, A., Isolation and identification of three bactericidal domains in the bovine Î ±-lactalbumin molecule. *Biochim. Biophys. Acta: Gen. Subj.* 1999, *1426*, 439–448.

[66] Zagato, E., Mileti, E., Massimiliano, L., Fasano, F., et al., Lactobacillus paracasei CBA L74 metabolic products and fermented milk for infant formula have anti-inflammatory activity on dendritic cells in vitro and protective effects against colitis and an enteric pathogen in vivo. *PLoS One.* 2014, *9*, e87615.

[67] Zivkovic, A.M., Lewis, Z.T., German, J.B., Mills, D.A., Establishment of a milk-oriented microbiota (MOM) in early life: How babies meet their MOMs. *Funct. Food Rev.* 2013, *5*, 3–12.

[68] Quintieri, L., Pistillo, B.R., Caputo, L., Favia, P., et al., Bovine lactoferrin and lactoferricin on plasma-deposited coating against spoilage *Pseudomonas* spp. *Innov. Food Sci. Emerg. Technol.* 2013, *20*, 215–222.

[69] Shah, N.P., Effects of milk-derived bioactives: An overview. *Br. J. Nutr.* 2007, *84*, 3–10.

[70] Dahan, S., Roth-Walter, F., Arnaboldi, P., Agarwal, S., et al., Epithelia: Lymphocyte interactions in the gut. *Immunol. Rev.* 2007, *215*, 243–253.

[71] Kim, Y.S., Ho, S.B., Intestinal goblet cells and mucins in health and disease: Recent insights and progress. *Curr. Gastroenterol. Rep.* 2010, *12*, 319–330.

[72] Claustre, J., Toumi, F., Trompette, A., Jourdan, G., et al., Effects of peptides derived from dietary proteins on mucus secretion in rat jejunum. *Am. J. Physiol. Gastrointest. Liver Physiol.* 2002, *283*, G521–G528.

[73] Trompette, A., Claustre, J., Caillon, F., Jourdan, G., et al., Milk bioactive peptides and {beta}-casomorphins induce mucus release in rat jejunum. *J. Nutr.* 2003, *133*, 3499–3503.

[74] Zoghbi, S., Trompette, A., Claustre, J., El Homsi, M., et al., beta-Casomorphin-7 regulates the secretion and expression of gastrointestinal mucins through a mu-opioid pathway. *Am. J. Physiol. Gastrointest. Liver Physiol.* 2006, *290*, G1105–G1113.

[75] Plaisancié, P., Claustre, J., Estienne, M., Henry, G., et al., A novel bioactive peptide from yoghurts modulates expression of the gel-forming MUC2 mucin as well as popula-

tion of goblet cells and Paneth cells along the small intestine. *J. Nutr. Biochem.* 2013, *24*, 213–221.

[76] Martínez-Maqueda, D., Miralles, B., Cruz-Huerta, E., Recio, I., Casein hydrolysate and derived peptides stimulate mucin secretion and gene expression in human intestinal cells. *Int. Dairy J.* 2013, *32*, 13–19.

[77] Han, K.-S., Deglaire, A., Sengupta, R., Moughan, P.J., Hydrolyzed casein influences intestinal mucin gene expression in the rat. *J. Agric. Food Chem.* 2008, *56*, 5572–5576.

[78] Sprong, R.C., Schonewille, A.J., van der Meer, R., Dietary cheese whey protein protects rats against mild dextran sulfate sodium-induced colitis: Role of mucin and microbiota. *J. Dairy Sci.* 2010, *93*, 1364–1371.

[79] Castro, G.A., Carvalho, J.E., Tinti, S.V, Possenti, A., et al., Anti-ulcerogenic effect of a whey protein isolate and collagen hydrolysates against ethanol ulcerative lesions on oral administration to rats. *J. Med. Food.* 2010, *13*, 83–90.

[80] Mezzaroba, L.F.H., Carvalho, J.E., Ponezi, A.N., Antônio, M.A., et al., Antiulcerative properties of bovine α-lactalbumin. *Int. Dairy J.* 2006, *16*, 1005–1012.

[81] Faure, M., Mettraux, C., Moennoz, D., Godin, J.-P., et al., Specific amino acids increase mucin synthesis and microbiota in dextran sulfate sodium-treated rats. *J. Nutr.* 2006, *136*, 1558–1564.

[82] Martínez-Augustin, O., Rivero-Gutiérrez, B., Mascaraque, C., Sánchez de Medina, F., Food derived bioactive peptides and intestinal barrier function. *Int. J. Mol. Sci.* 2014, *15*, 22857–22873.

[83] Ozdemir, O., Any benefit of probiotics for autoimmune gastrointestinal diseases? *J. Pediatr. Sci.* 2012, *4*.

[84] Haug, A., Høstmark, A.T., Harstad, O.M., Bovine milk in human nutrition—a review. *Lipids Health Dis.* 2007, *6*, 25.

[85] Calder, P.C., Kew, S., The immune system: A target for functional foods? *Br. J. Nutr.* 2002, *88*(Suppl 2), S165–S177.

[86] Hata, I., Ueda, J., Otani, H., Immunostimulatory action of a commercially available casein phosphopeptides preparation, CPP-III, in cell cultures. *Milchwissenschaft.* 1999, *54*(1), 3–7.

[87] Kawahara, T., Aruga, K., Otani, H., Characterization of casein phosphopeptides from fermented milk products. *J. Nutr. Sci. Vitaminol. (Tokyo).* 2005, *51*, 377–381.

[88] Otani, H., Nakano, K., Kawahara, T., Stimulatory effect of a dietary casein phosphopeptide preparation on the mucosal IgA response of mice to orally ingested lipopolysaccharide from *Salmonella typhimurium. Biosci. Biotechnol. Biochem.* 2003, *67*, 729–735.

[89] Saint-Sauveur, D., Gauthier, S.F., Boutin, Y., Montoni, A., et al., Effect of feeding whey

peptide fractions on the immune response in healthy and *Escherichia coli* infected mice. *Int. Dairy J.* 2009, *19*, 537–544.

[90] Tang, Z., Yin, Y., Zhang, Y., Huang, R., et al., Effects of dietary supplementation with an expressed fusion peptide bovine lactoferricin-lactoferrampin on performance, immune function and intestinal mucosal morphology in piglets weaned at age 21 d. *Br. J. Nutr.* 2009, *101*, 998–1005.

[91] Mahida, Y.R., The key role of macrophages in the immunopathogenesis of inflammatory bowel disease. *Inflamm. Bowel Dis.* 2000, *6*, 21–33.

[92] Meisel, H., Biochemical properties of regulatory peptides derived from milk proteins. *Biopolymers* 1997, *43*, 119–128.

[93] Kitazawa, H., Yonezawa, K., Tohno, M., Shimosato, T., et al., Enzymatic digestion of the milk protein beta-casein releases potent chemotactic peptide(s) for monocytes and macrophages. *Int. Immunopharmacol.* 2007, *7*, 1150–1159.

[94] Otani, H., Hata, I., Inhibition of proliferative responses of mouse spleen lymphocytes and rabbit Peyer's patch cells by bovine milk caseins and their digests. *J. Dairy Res.* 2009, *62*, 339.

[95] Otani, H., Horimoto, Y., Monnai, M., Suppression of interleukin-2 receptor expression on mouse CD4 + T cells by bovine k-caseinoglycopeptide. *Biosci. Biotechnol. Biochem.* 2014, *60*, 1017–1019.

[96] Tellez, A., Corredig, M., Brovko, L.Y., Griffiths, M.W., Characterization of immune-active peptides obtained from milk fermented by *Lactobacillus helveticus*. *J. Dairy Res.* 2010, *77*, 129–136.

[97] de Medina, F.S., Daddaoua, A., Requena, P., Capitán Cañadas, F., et al., New insights into the immunological effects of food bioactive peptides in animal models of intestinal inflammation. *Proc. Nutr. Soc.* 2010, *69*, 454–462.

[98] Vinderola, G., Matar, C., Perdigón, G., Milk fermentation products of *L. helveticus* R389 activate calcineurin as a signal to promote gut mucosal immunity. *BMC Immunol.* 2007, *8*, 19.

[99] Jiehui, Z., Liuliu, M., Haihong, X., Yang, G., et al., Immunomodulating effects of casein-derived peptides QEPVL and QEPV on lymphocytes in vitro and in vivo. *Food Funct.* 2014, *5*, 2061–2069.

[100] Requena, P., Daddaoua, A., Martínez-Plata, E., González, M., et al., Bovine glycomacropeptide ameliorates experimental rat ileitis by mechanisms involving downregulation of interleukin 17. *Br. J. Pharmacol.* 2008, *154*, 825–832.

[101] Requena, P., Daddaoua, A., Guadix, E., Zarzuelo, A., et al., Bovine glycomacropeptide induces cytokine production in human monocytes through the stimulation of the MAPK and the NF-kappaB signal transduction pathways. *Br. J. Pharmacol.* 2009, *157*, 1232–1240.

2

Donkey Milk Proteins: Digestibility and Nutritional Significance

Donata Marletta, Flavio Tidona and
Salvatore Bordonaro

Abstract

Donkey milk is particularly recommended for infant nutrition as substitute of cow milk in case of sensitive neonates (showing cow milk protein allergies). Its protein composition and the ratio between caseins and whey proteins reveals a high similarity with human milk, thus, in the last 10 years, an increasing interest arose to obtain a full characterisation of donkey milk proteins, here acknowledged. Digestibility data, mainly derived *in vitro* with human gastrointestinal enzymes, showed the high digestibility of donkey caseins and major whey proteins, except lysozyme and α-lactalbumin which proved to be quite resistant. The reported antimicrobial properties of donkey milk open concrete possibilities to use donkey milk as natural food preservative. Due to its attractive healthy properties, donkey milk was investigated for useful applications or to develop novel foods characterised by a high nutritional profile.

Keywords: donkey milk, proteins, nutritional value, digestibility, novel food

1. Introduction

A newborn who is fed colostrum (milk of the first 3–5 days of lactation, reinforced with transfer immunity factors) and then mature milk greatly benefits for a healthy growth, fulfilling all nutritional requirements and perinatal passive immunisation. The early feeds should be easily digestible and well tolerated to achieve a suitable intake and reduce the risk of diseases, mainly favoured by the passage of milk proteins and peptides which, after digestion, stimulate the mucosa immune system of the infant [1]. The milk of each species is designed to meet the specific needs of the neonate; however, when breastfeeding is not possible, or after weaning, it becomes important to find an adequate alternative nourishment. This is particularly challenging for the infants affected by cow milk protein allergies (CMPA) that, occasionally, are associated with clinical cross-reactivity between milks of other ruminants [2]. In addition, if multiple milk protein allergies appear, the use of soy-based formulas or extensively hydrolysed protein formulas is not recommended to avoid the risk of cross-sensitisation [3]. Contrarily, equid milk, especially donkey milk, showed to be well tolerated by children with CMPA in terms of clinical tolerability [4, 5], likely associated to the comparable protein fraction composition between donkey and human milk. With the exception of a notably lower fat content, donkey milk is characterised by a gross composition similar to human milk, with a close ratio between casein/whey proteins which is believed to play a crucial role in the sensitisation to cow milk protein fraction, reducing the allergenic capacity [6]. Although further studies, including *in vivo* tests, are desirable to achieve more consistent results on its hypoallergenic properties, an increasing interest is arising around donkey milk. Indeed, recent investigations on the protein and fatty acid profile [7, [8], main mineral composition [9], some vitamin contents, microbiological and hygienic aspects [10, 11], greatly contributed for a wider knowledge on this milk. The majority of surveys on the quality of donkey milk have been conducted in Italy, but some data are also available from the milk of Chinese and Balkan donkey breeds [12]. A distinctive value of donkey milk is especially related to its antimicrobial properties which were more extensively documented [13–15]. Furthermore, numerous evidences reported several health-promoting properties such as antioxidant activity [16], the regulation of immune response in healthy elderly consumers [17] and anti-proliferative and anti-tumour *in vitro* effects on human lung cancer cells [18]. Besides infant diet therapy, the attractive nutritional features shown by donkey milk could also be addressed to other categories (e.g. elderly population) or employed in alternative food formulations.

This chapter will focus on the mainly characterised proteins of donkey milk, showing their nutritional value, the related impact on the human digestive system, and some potential applications in the dairy field.

2. Donkey milk caseins

Total protein content of donkey milk ranges between 15 and 18 g/L and the casein fraction represents about 35–45%, much lower than the milk of ruminants (>70%) but more similar to human milk (<30%). The available knowledge on donkey milk caseins is limited, compared to conventional dairy species and a full characterisation was also complicated by their heterogeneity, partly due to post-translational processes, genetic polymorphism, non-allelic deleted forms [19]. A combination of electrophoretic, chromatographic, and proteomic-based methods allowed the identification of the four casein fractions (α_{s1}-, α_{s2}-, β-, and *k*-casein). Casein (CN) distribution of donkey milk showed β-casein as the predominant one, followed by the α_{s1}-casein whereas α_{s2}-casein was detected as minor component [7, 19]. *k*-casein was only found in traces and is reported to be the most heterogeneous individual casein, likely due to different levels of glycosylation [19]. Donkey CSN2 gene has been sequenced (GenBank FN598778.1) and β-casein's primary structure of donkey milk has been completely characterised [20], using mare's β-CN derived from cDNA, as reference. It is constituted of 226 amino acids and it has a molecular weight of 25,529 Da, containing seven potential phosphorylation sites, together with two additional ones (located at Thr12 and Thr207), analogously to the homologous mare's β-CN. According to Chianese et al. [19], the full-length β-CN and its deleted form were both found equally phosphorylated with 5, 6, and 7 P/mole, together with a novel β-CN variant which showed the same phosphorylation pattern but a higher molecular weight (>28 mass units) than the most common β-casein.

Donkey α_{s1}-CN contains 202 amino acids and has a molecular mass of 24,406 Da, prior to post-translational modifications [21]. The related CSN1S1 gene has been sequenced (GenBank Acc. Num. FN386610) and a rare mutation was reported to be associated (Marletta et al., personal communication) with the apparent absence of this fraction in the milk of a jennet belonging to Ragusano breed [22]. A remarkable heterogeneity of this protein was assigned to either discrete phosphorylation (5, 6, and 7 P/mole) or non-allelic deleted forms, generated by incorrect RNA splicing as already shown in the homologous goat and sheep casein [19]. α_{s1}-CN of donkey and cow shows a low homology and their difference in the amino acid sequence of the IgE-binding linear epitopes may be responsible for the hypoallergenic properties [23]. Regarding the α_{s2}-CN of donkey, the CSN1S2 gene (GenBank Acc. Num. CAX65660.1) has 2 forms (I and II) that differ in structure and encoded protein sequence. The major form (CSN1S2 I) consists of 19 exons and encodes a 221 amino acid protein, whereas the CSN1S2 II form is shorter (16 exons and 168 encoded amino acids). The existence of different splicing isoforms has been also suggested [24, 25], whereas three main phosphorylated components have been described for α_{s2}-CN, each accounting for 10, 11, and 12 P/mole [19]. All these structural variations may influence protein allergenicity [26]. Finally, the CSN3 gene sequence has also been determined in donkey (GenBank Acc. Num. FR822990.1) but only a specific immunostaining was able to detect *k*-casein in donkey milk; it was surely due to the low amount of this fraction and to its great heterogeneity, so that eleven *k*-casein components were found in an individual sample [19].

3. Donkey milk whey proteins

Donkey milk is characterised by a high proportion of whey protein. Most of the alleged nutritional properties of this milk can be attributed to this fraction, that is mainly composed of β-lactoglobulins (β-Lg) α-lactoalbumin (α-La) and lysozyme (Lyz). The other three minor proteins immunoglobulins (Igs), serum albumin (SA), and lactoferrin (Lf) are also present. Even if the whole whey protein fraction is considered to be responsible for the low bacterial count of donkey milk [15, 27], the antimicrobial activity is mainly attributed to Lyz and, to a lesser extent, to Lf [28]. These minor proteins, together with Igs, are believed to work in synergy, for inhibiting microbial growth and reducing the incidence of gastrointestinal infections [1, 29, 30].

Donkey β-Lg, that is the most abundant whey protein, consists of two components: a major β-Lg I of 162 amino acid residues and a minor β-Lg II of 163 residues, which possesses an additional Glycine between the 116th and the 117th residue, as it occurs in mare's milk. β-Lg I (gij125913) represents about 80% of total β-Lg and presents only two variants, A and B: the former is of 18,528 Da, the latter is of 18,514 Da [31]. In contrast, β-Lg II (gij125904) (Mw 18,200 Da), representing only the remaining 20%, has five variants: A, B, C [31, 32], D [33], and E [34] with molecular weights ranging from 18,227 (variant B) to 18,311 (variant D). Very recently, a new variant, with a predicted molecular weight of 18,315, has been identified in Ragusano donkey breed by direct DNA sequencing (Marletta et al., personal communication). β-Lg, absent in human milk, is generally considered to be one of the main causes of CMPA [35] because it can elicit an allergic reaction in sensitive subjects [28]. To this point, the identification of animals producing milk lacking in β-Lg II protein [22] appears promising to solve potential residual cases of reactivity [36]. Interestingly, donkey β-Lg was found to be highly degraded (70%) *in vitro* by human gastric and duodenal juice [37] in comparison to cow counterpart [38]; this feature could enhance the formation of derived bioactive peptides in gut, with potential antimicrobial activity [14]. Donkey α-lactalbumin (α-La) contains 123 amino acid residues and has a molecular weight of 14,215 Da [39]. Only one α-La genetic variant (gij262063) with two isoforms, characterised by different isoelectric points (pI 4.76 and 5.26, respectively), have been identified so far [33, 40]. Even though donkey α-lactalbumin shows a striking sequence homology with C-lysozyme, which is a known powerful antibacterial agent [41], there is no direct evidence of antimicrobial activity of donkey α-La and/or its derived peptides, so far. Lysozyme (Lyz), whose content is particularly high in donkey milk (up to 4 g/L), has two variants (A and B) both containing 129 amino acids (gij126613; gij126614) and a molecular weight of 14,632 Da, which differ in three amino acid substitutions at positions 48, 52, and 61 as previously described [32, 33]. In general, lysozyme has an important role in the intestinal immune response since it acts as a powerful antibacterial protein, splitting the bonds between N-acetylglucosamine and N-acetylmuramic acid of the peptidoglycan [42]; thus, Gram-positive bacteria are more sensible to Lyz than the Gram-negative. However, a synergistic action with lactoferrin is supposed to enhance antibacterial action against also towards some Gram-negative bacteria [15]. These natural preservative properties could be the reason for lengthy shelf-life reported for raw donkey milk [13, 15].

Immunoglobulins (IgGs) of donkey milk show a high content in comparison with human and bovine counterparts [28] since they are supplied to the foal only after parturition to fortify the natural immunopassive system of the neonate. The presence of IgGs in colostrum and then (in lower amount) in mature milk is still a matter of debate for attributing health beneficial effects to a given milk type or for consuming raw milk [43]. Donkey lactoferrin (Lf) is an 80-kg/mol iron-binding multifunctional glycoprotein that exerts several biological activities [40, 44]; it is generally associated with antimicrobial, antiviral, immunomodulatory, and anticarcinogenic activity [[45] for a review], although its content in donkey milk is relatively low compared to lysozyme. Finally, lactoperoxidase (LP), an oxidoreductase enzyme with protective function against microorganism infections, is found at a small concentration in fresh donkey milk [46, 47]. LP is known to be inactivated by high temperature but this enzyme could be of significant nutritional interest in raw-fresh milk, because working in synergy with lactoferrin and lysozyme could contribute to enhance the natural preservative action of donkey milk [15].

4. Remarks on donkey milk digestibility

Digestibility of donkey milk proteins was firstly assessed by Tidona et al. [14] in a simulated gastrointestinal digestive process using human gastric and duodenal juices; donkey caseins proved to be rapidly digested since after 1 hour of digestion only about 7% remained intact. The acid coagulum observed in the acidic conditions of gastric digestion (pH ~2) was very fine and the formation of a soft precipitate was also reported in equine and human milk, which is physiologically more suitable for infant nutrition than the firm coagulum formed by bovine milk [48]. Casein micelle size of donkey milk (about 298 nm) was found to be much larger than the one of human milk (64 nm), as it is inversely related to the *k*-casein content [49]; this condition, together with the relative abundance of β-casein, may be the reason for the high susceptibility to hydrolysis by gastrointestinal enzymes [38]. Certainly the low protein and casein content of donkey milk might favour the fast digestible caseins compared to high casein predominant milk of other species.

Regarding the whey proteins, β-lactoglobulins showed to be quite resistant to gastric enzymes (mainly pepsin) but were highly degraded by human duodenal juice (~30% remained undigested) contrarily to what was reported from cow and goat milk [38]. Digestibility of β-Lgs could be even enhanced in the individual milk of donkeys lacking β-lactoglobulin type II, achieving a higher rate of degradation [37]; this is nutritionally relevant since human milk is typically devoid of β-lactoglobulins. Donkey α-lactalbumin is the most resistant protein, since the 95% was found undigested after 1 hour of *in vitro* digestion [37], so that it reaches the gut relatively intact as already reported for milk of other species [38]. Similarly, lysozyme was quite resistant to human gastrointestinal enzymes, although at a lower extent (~75%) compared to α-lactalbumin, and was also found to be thermal stable after high pasteurisation treatment [27, 50]. Particularly interesting is the high digestibility of Donkey Lf by gastric and duodenal juice [37]. This evidence suggests that lactoferrin might play a further biologic role directly in the gut [51] as well as through its bioactive peptides, called lactoferricin (Lfcin) and lactoferrampin (Lfampin), as already observed in cow milk. The presence of these proteins and the peptides derived (so far unidentified) during digestion could still inhibit sensitive bacteria in the intestine as the antimicrobial activity exhibited by the digested donkey milk was even enhanced [14]. Among the other minor whey proteins (such as immunoglobulins, serum albumin, and lactoperoxidase), weak data are reported given their low concentration in donkey milk, but warrant further investigations to evaluate how the natural immunepassive system may be affected by the digestion process.

5. Nutritional value and potential applications

Since ancient times, queens and empresses like Cleopatra and Poppea used to take a bath in donkey milk, experiencing its smoothing and lenitive properties. Nowadays, donkey milk is still used in cosmetic preparations but the renewed interest shown by the scientific community is mostly oriented for feeding purposes. Namely, donkey milk was defined as "pharma-food" for its nutritional, nutraceutical, and functional properties [16]. To enhance the nutritional value, donkey milk was considered a suitable matrix as carrier of probiotic bacteria, when supplemented in concentrated freeze-dried cells, reporting a high viability of *L. rhamnosus* and *L. paracasei* in donkey milk [52], which could be useful for the prevention and treatment of antibiotic-associated diarrhoea [53].

The peculiar high lysozyme content and the antimicrobial activity shown by native donkey milk were tested by its addition to goat milk, which is often used as hypoallergenic drinkable milk, to evaluate the inhibition of pathogen bacteria under refrigerated storage conditions; an inhibited growth rate was especially evident for *Staphylococcus aureus, Listeria monocytogenes*, and *Campylobacter jejuni* [54]. Moreover, the employment of donkey milk was recently proposed as an alternative to egg lysozyme, in order to prevent blowing defects in long seasoned cheeses [55]: the addition of donkey milk (1.1% v/v), significantly contributed to reduce the presence of coliforms in cheese.

Despite the high lysozyme content, which tends to inhibit several bacteria, some less sensitive lactic acid bacteria were used to ferment donkey milk, which resulted in a non-conventional ingredient to deliver fermented milks of new generation [56]. Recently, a fermented beverage based on donkey milk emulsified with sunflower oil was developed, showing an increased folic acid content, a higher level of polyunsaturated fatty acids, and more favourable lipid quality indexes (atherogenicity and thrombogenicity index) than donkey milk itself [57]. The manufacture process for yogurt production with donkey milk was also attempted with the addition of two adjunct probiotic strains: the resulted product presented a higher antioxidant activity and a lower lactose content, which aims to meet nutritional requirements of certain consumers groups [16]. Besides, rheological aspects related to donkey milk fermented products are limited by the low protein and casein content because the acid-induced coagulation of equine caseins behaves more as a micellar flocculation rather than gelation [28] and a high tendency to syneresis [57]. Indeed, the protein content (low in caseins) and profile (poor in *k*-casein) of donkey milk are not very suitable for transformation as the use of bovine chymosin leads to a very weak gel, without curd formation. Yet, raw donkey milk was surprisingly employed in cheese-making using a new coagulant, a pure camel chymosin which increases the substrate binding and determines a higher milk clotting activity [58]; the result was a soft cheese that allowed merely 3.3% of cheese yield. The manufacture of the cheese was only obtained with fresh donkey milk, since any heat treatment prevented coagulation. Thus the mechanism underlying the enzymatic coagulation of donkey milk needs further research.

Author details

Donata Marletta[1*], Flavio Tidona[2] and Salvatore Bordonaro[1]

*Address all correspondence to: d.marletta@unict.it

1 Department of Agriculture, Food and Environment (Di3A), University of Catania, Italy

2 Council for Research in Agriculture and Agricultural Economy Analysis (CREA), Research Centre for Fodder Production and Dairy Casearie (CREA-FLC), Lodi, Italy

References

[1] Baldi A, Politis I, Pecorini C, Fusi E, Chronopoulou R, Dell'Orto V. Biological effects of milk proteinsand their peptides with emphasis on those related to the gastrointestinal eco system. J Dairy Res. 2005; 72:66–72. DOI: 10.1017/S002202990500110X

[2] Järvinen KM, Chatchatee P. Mammalian milk allergy: clinical suspicion, cross-reactivities and diagnosis. Curr Opin Allergy Clin Immunol. 2009; 9:251–258. DOI: 10.1097/ACI.0b013e32832b3f33

[3] Brill H. Approach to milk protein allergy in infants. Can Fam Physician. 2008; 54:1258–1264. PMCID: PMC2553152

[4] Monti G, Bertino E, Muratore MC, Coscia A, Cresi F, Silvestro L, Fabris C, Fortunato D, Giuffrida MG, Conti A. Efficacy of donkey's milk in treating highly problematic cow's milk allergic children: an in vivo and in vitro study. Pediatr Allergy Immunol. 2007; 18:258–264. DOI: 10.1111/j.1399-3038.2007.00655.x

[5] Swar MO. Donkey milk-based formula: a substitute for patients with cow's milk protein allergy. Sudan J Paediatr. 2011; 11:21–24

[6] Lara-Villoslada F, Olivares M, Xaus J. The balance between caseins and whey proteins in cow's milk determines its allergenicity. J Dairy Sci. 2005; 88: 1654–1660. DOI: 10.3168/jds.S0022-0302(05)72837-X

[7] Polidori P, Vincenzetti S. Protein profile characterization of donkey milk. In: Hurley WL, editor. Milk Protein. In Tech; Rijeka, Croatia 2012. p. 215–232. DOI: 10.5772/45982. ISBN 978-953-51-0743-9.

[8] Martemucc G, D'Alessandro AG. Fat content, energy value and fatty acid profile of donkey milk during lactation and implications for human nutrition. Lipids Health Dis. 2012; 11:113–127. DOI: 10.1186/1476-511X-11-113.

[9] Fantuz F, Ferraro S, Todini L, Piloni R, Mariani P, Salimei E. Donkey milk concentration of calcium, phosphorus, potassium, sodium and magnesium. Int Dairy J. 2012; 24:143–145. DOI: 10.1016/j.idairyj.2011.10.013

[10] Carminati D, Tidona F, Fornasari ME, Rossetti L, Meucci A, Giraffa G. Biotyping of cultivable lactic acid bacteria isolated from donkey milk. Lett Appl Microbiol. 2014; 59:299–305. DOI: 10.1111/lam.12275

[11] Pilla R, Daprà V, Zecconi A,Piccinini R. Hygienic and health characteristics of donkey milk duringa follow-up study. J Dairy Res. 2010; 77:392–397. DOI: 10.1017/S0022029910000221.

[12] Cavallarin L, Giribaldi M, de los Dolores Soto-Del Rio M, Valle E, Barbarino G, Gennero MS, Civera T. A survey on the milk chemical and microbiological quality in dairy

donkey farms located in North Western Italy. Food Control. 2015; 50:230–235. DOI: 10.1016/j.foodcont.2014.08.019

[13] Zhang X, Zhao L, Jiang L, Dong M, Ren F. The antimicrobial activity of donkey milk and its microflora changes during storage. Food Control. 2008; 19:1191–1195. DOI: 10.1016/j.foodcont.2008.01.005

[14] Tidona F, Sekse C, Criscione A, Jacobsen M, Bordonaro S, Marletta D, Vegarud GE. Antimicrobial effect of donkeys' milk digested *in vitro* with human gastrointestinal enzymes. Int Dairy J. 2011; 21:158–165. DOI: 10.1016/j.idairyj.2010.10.008

[15] Šarić L, Šarić BM, Mandić AI, Torbica AM, Tomić JM, Cvetković DD, Okanović DG. Antibacterial properties of domestic Balkan donkeys' milk. Int Dairy J. 2012; 25:142–146. DOI: 10.1016/j.idairyj.2012.03.007

[16] Perna A, Intaglietta I, Simonetti A, Gambacorta E. Donkey milk for manufacture of novel functional fermented beverages. J Food Sci. 2015; 80:352–359. DOI: 10.1111/1750-3841.12862

[17] Jirillo F, Jirillo E, Magrone T. Donkey's and goat's milk consumption and benefits to human health with special reference to the inflammatory status. Curr Pharm Des. 2010; 16:859–863. DOI: 10.2174/138161210790883688

[18] Mao X, Gu J, Sun Y, Xu S, Zhang X, Yang H, Ren F. Anti-proliferative and anti-tumour effect of active components in donkey milk on A549 human lung cancer cells. Int Dairy J. 2009; 19:703–708. DOI: 10.1016/j.idairyj.2009.05.007

[19] Chianese L, Calabrese MG, Ferranti P, Mauriello R, Garro G, De Simone C, Quarto M, Addeo F, Cosenza G, Ramunno L. Proteomic characterization of donkey milk "caseome". J Chromatogr A. 2010; 1217:4834–4840. DOI: 10.1016/j.chroma.2010.05.017

[20] Cunsolo V, Cairone E, Saletti R, Muccilli V, Foti S. Sequence and phosphorylation level determination of two donkey beta-caseins by mass spectrometry. Rapid Commun Mass Spectrom. 2009; 23:1907–1916. DOI: 10.1002/rcm.4087.

[21] Cunsolo V, Cairone E, Fontanini D, Criscione A, Muccilli V, Saletti R, Foti S. Sequence determination of alphas1-casein isoforms from donkey by mass spectrometric methods. J Mass Spectrom. 2009; 44:1742–1753. DOI: 10.1002/jms.1683.

[22] Criscione A, Cunsolo V, Bordonaro S, Guastella AM, Saletti R, Zuccaro A, D'Urso G Marletta D. Donkeys' milk protein fraction investigated by electrophoretic methods and mass spectrometric analysis. Int Dairy J. 2009; 19:190–197. DOI:10.1016/j.idairyj.2008.10.015

[23] Cunsolo M, Muccilli V, Fasoli E, Saletti R, Righetti PG, Foti S. Poppea's bath liquor: the secret proteome of she-donkey's milk. J Proteomics. 2011; 74:2083–2099. DOI: 10.1016/j.jprot.2011.05.036

[24] Cosenza G, Pauciullo A, Annunziata AL, Rando A, Chianese L, Marletta D, Iannolino G, Nicodemo D, Di Berardino D, Ramunno L. Identification and characterization of the

donkey CSN1S2 I and II cDNAs. Ital J Anim Sci. 2010; 9:206–211. DOI: 10.4081/ijas.2010.e40

[25] Saletti R, Muccilli V, Cunsolo V, Fontanini D, Capocchi A, Foti S. MS-based characterization of α(s2)-casein isoforms in donkey's milk. J Mass Spectrom. 2012; 47:1150–1159. DOI: 10.1002/jms.3031

[26] Ballabio C, Chessa S, Rignanese D, Gigliotti C, Pagnacco G, Terracciano L, Fiocchi A, Restani P, Caroli AM. Goat milk allergenicity as a function of α_{s1}-casein genetic polymorphism. J Dairy Sci. 2011; 94:998–1004. DOI: 10.3168/jds.2010-3545

[27] Chiavari C, Coloretti F, Nanni M, Sorrentino E, Grazia L. Use of donkey's milk for a fermented beverage with lactobacilli. Lait. 2005; 85:481–490. DOI: 10.1051/lait:2005031

[28] Uniacke-Lowe T, Huppertz T, Fox PF. Equine milk proteins: chemistry, structure and nutritional significance. Int Dairy J. 2010; 20:609–629. DOI: 10.1016/j.idairyj.2010.02.007

[29] Brumini D, Criscione A, Bordonaro S, Vegarud GE, Marletta D. Whey proteins and their antimicrobial properties in donkey milk: a brief review. Dairy Sci Technol. 2016; 96:1–14. DOI: 10.1007/s13594-015-0246-1

[30] Brumini D, Bø Furlund C, Comi I, Devold TG, Marletta D, Vegarud GE, Monceyr-onJonassen C. Antiviral activity of donkey milk protein fractions on echovirus type 5. Int Dairy J. 2013; 28:109–111. DOI: 10.1016/j.idairyj.2012.08.010

[31] Godovac-Zimmermann J, Conti A, Sheil M, Napolitano L. Covalent structure of the minor monomeric beta-lactoglobulin II component from donkey milk. Biol Chem Hoppe Seyler. 1990; 371:871–879. DOI: 10.1515/bchm3.1990.371.2.871

[32] Herrouin M, Mollè D, Fauquant J, Ballestra F, Maubois JL, Lèonil J. New genetic variants identified in donkey's milk whey proteins. J Protein Chem. 2000; 19:105–115. DOI 10.1023/A:1007078415595

[33] Cunsolo V, Costa A, Saletti R, Muccilli V, Foti S. Detection and sequence determination of a new variant beta-lactoglobulin II from donkey. Rapid Commun Mass Spectrom. 2007; 21:1438–1446. DOI: 10.1002/rcm.2978

[34] Chianese L, De Simone C, Ferranti P, Mauriello R, Costanzo A, Quarto M, Garro G, Picariello G, Mamone G, Ramunno L. Occurrence of qualitative and quantitative polymorphism at donkey beta-Lactoglobulin II locus. Food Res Int. 2013; 54:1273–1279. DOI: 10.1016/j.foodres.2012.11.005

[35] Restani P, Ballabio C, Di Lorenzo C. Molecular aspects of milk allergens and their role in clinical events. Anal Bioanal Chem. 2009; 395:47–56. DOI: 10.1007/s00216-009-2909-3

[36] Wal JM. Cow's milk proteins/allergens. Ann Allergy Asthma Immunol. 2002; 89:3–10. DOI: 10.1016/S1081-1206(10)62115-1

[37] Tidona F, Criscione A, Devold TG, Bordonaro S, Marletta D, Vegarud GE. Protein composition and micelle size of donkey milk with differentprotein patterns: effects on

digestibility. Int Dairy J. 2014; 35:57–62. DOI: 10.1016/j.idairyj.2013.10.018

[38] Inglistad RA, Devold TG, Eriksen EK, Holm H, Jacobsen M, Liland KH. Comparison of the digestion of caseins and whey proteins in equine, bovine, caprine and human milks by human gastrointestinal enzymes. Dairy Sci Technol. 2010; 90:549–563. DOI: 10.1051/dst/2010018

[39] Giuffrida MG, Cantisani A, Napolitano L, Conti A, Godovac-Zimmermann J. The amino acid sequence of two isoforms of alpha-lactalbumin from donkey (*Equus asinus*) milk is identical. Biol Chem Hoppe Seyler. 1992; 373:931–935. DOI: 10.1515/bchm3.1992.373.2.931

[40] Vincenzetti S, Amici A, Pucciarelli S, Vita A, Micozzi D, Carpi FM, Polzonetti V, Natalini P, Polidori P. A proteomic study on donkey milk. Biochem Anal Biochem. 2012; 1:109. DOI: 10.4172/2161-1009.1000109

[41] Lopez-Exposito I, Recio I. Protective effect of milk peptides: antibacterial and antitumor properties. In: Bösze Z editor. Advances in Experimental Medicine and Biology: Bioactive Components of Milk. New York: Springer; 2008. p. 271–294. DOI: 10.1007/978-0-387-74087-4_11

[42] Lopez-Exposito I, Pellegrini A, Amigo L, Recio I. Synergistic effect between different milk-derived peptides and proteins. J Dairy Sci. 2008; 91:2184–2189. DOI: 10.3168/jds.2007-0037

[43] Claeys W, Verraes C, Cardoen S, De Block J, Huyghebaert A, Raes K, Dewettinck K, Herman L. Consumption of raw or heated milk from different species: an evaluation of the nutritional and potential health benefits. Food Control. 2014; 42:188–201. DOI: 10.1016/j.foodcont.2014.01.045.

[44] Yamauchi K, Wakabayashi H, Shin K, Takase M. Bovine lactoferrin: benefits and mechanism of action against infections. Biochem Cell Biol. 2006; 84:291–296. DOI: 10.1139/o06-054

[45] Tomita M, Wakabayashi H, Shin K, Yamauchi K, Yaeshima T, Iwatsuki K. Twenty-five years of research on bovine lactoferrin applications. Biochimie. 2009; 91:52–57. DOI: 10.1016/j.biochi.2008.05.021

[46] Simos Y, Metsios A, Verginadis I, D'Alessandro AG, Loiudice P, Jirillo E, Charalampidis P, Kouimanis V, Boulaka A, Martemucci G, Karkabounas S. Antioxidant and anti-platelet properties of milk from goat, donkey and cow: an *in vitro*, ex vivo and *in vivo* study. Int Dairy J. 2011; 21:901–906. DOI: 10.1016/j.idairyj.2011.05.007

[47] Mariani P. Donkey milk nutraceutical characteristics: a biochemical evaluation of nutritious and clinical properties [PhD thesis]. Camerino: University of Camerino, Italy; 2010. Available at: http://cameprints.unicam.it/284. Accessed 1 April 2015.

[48] Solaroli G, Pagliarini E, Peri C. Composition and nutritional quality of mare's milk. Ital J Food Sci. 1993; 5: 3–10.

[49] Fox PF, McSweeney PLH. Dairy Chemistry and Biochemistry. 1st ed. London: Blackie Academic and Professional Publishers; 1998. 396 p.

[50] Di Cagno R, Tamborrino A, Gallo G, Leone C, De Angelis M, Faccia M. Uses of mares' milk in manufacture of fermented milk. Int Dairy J. 2004; 14:767–775. DOI: 10.1016/j.idairyj.2004.02.005

[51] Brock JH. The physiology of lactoferrin. Biochem Cell Biol. 2002; 80:1–6. DOI: 10.1139/o01-212

[52] Vincenzetti S, Savini M, Cecchini C, Micozzi D, Carpi F, Vita A, Polidori P. Effects of lyophilization and use of probiotics on donkey's milk nutritional characteristics. Int J Food Eng. 2011; 7:1–8. DOI: 10.2202/1556-3758.2032

[53] Salminen S, Isolauri E. Intestinal colonization, microbiota, and probiotics. J Pediatr. 2006; 149: S115–S120. DOI: 10.1016/j.jpeds.2006.06.062

[54] Fratini F, Turchi B, Pedonese F, Pizzurro F, Ragaglini P, Torracca B, Tozzi B, Galiero A, Nuvoloni R. Does the addition of donkey milk inhibit the replication of pathogen microorganisms in goat milk at refrigerated condition? Dairy Sci Technol. 2015; 96:1–8. DOI: 10.1007/s13594-015-0249-y

[55] Cosentino C, Paolino R, Freschi P, Calluso AM. Short communication: Jenny milk as an inhibitor of late blowing in cheese: a preliminary report. J Dairy Sci. 2013; 96:3547–3550. PubMed: 23587377

[56] Coppola R, Salimei E, Succi M, Sorrentino E, Nanni M, Ranieri P, Belli Blanes R, Grazia L. Behaviour of *Lactobacillus rhamnosus* strains in ass's milk. Annals of Microbiol. 2002; 52:55–60

[57] Tidona F, Charfi I, Povolo M, Pelizzola V, Carminati D, Contarini G, Giraffa G. Fermented beverage emulsion based on donkey milk with sunflower oil. Int J Food Sci Technol. 2015; 50: 2644–2652. DOI: 10.1111/ijfs.12936

[58] Iannella G. Donkey cheese made through pure camel chymosin. African J Food Sci. 2015; 9: 421–425. DOI: 10.5897/AJFS2015.1322

3

Bioactive Lactoferrin-Derived Peptides

Adham M. Abdou and Hend A. Elbarbary

Abstract

Lactoferrin (LF) is a member of the transferrin family that is a cationic iron-binding protein. It is an 80-kDa glycoprotein that is found in many secretions in the body and is highly present in milk and colostrums. It exerts antibacterial effects and has a wide range of biological activities. Moreover, it is considered as a precursor of different peptides that have multifunctional bioactivities. During the last decade, several applications of LF and its peptides have been discovered, which has led to its commercial production. Therefore, LF and its peptides can offer a variety of specialized ingredients that can be tailored to meet the needs of natural food preservatives and functional food ingredients.

Keywords: lactoferrin, structure, purification, food preservation, therapeutic activities

1. Introduction

Milk is the first complete functional food devised by nature for the protection and development of newborn mammals. The antimicrobial and bioactive components in milk have received strong attention in recent years. These components have been isolated from a variety of sources including cheese whey and colostrums from hyperimmunized cows [1]. Lactoferrin (LF), lactoperoxidase system, lactoglobulin, and lactolipids are the most common antimicrobial preservatives in dairy industry [2].

The biodefensive properties of mammalian milk proteins have been well described and widely acknowledged for many years. As early as 1930, it was reported that milk possessed active inhibitors [3]. LF was originally discovered in cow's milk in 1939, which was identified as a red iron-binding protein, so it was called LF (lacto=milk; ferrin= iron) [4]. Later, Schade and Caroline [5] isolated an iron-binding protein from human serum "serotransferrin," which was later named transferrin.

Milk is the major source of LF. It is abundantly excreted in colostrums of human milk in a concentration up to 7 g/L and the LF concentration in mature milk declines approximately 7-fold in time during lactation to 1 g/L [6]. LF concentration in milk varies among different mammals. In bovine milk, LF occurs in relatively low amounts from 0.02 to 0.03 g/L, although, in colostrums, this amount is much higher from 2 to 5 g/L [7], whereas, in rats and dogs, no LF has been detected so far [8].

LF, a member of the transferrin family, is a cationic iron-binding protein that is found in many exocrine secretions, including milk, tears, saliva, and serum [9]. It exerts antibacterial effects by limiting iron availability for microorganisms [10]. In addition, it shows some biological activities as immunomodulatory effect and chemoprevention of carcinogenesis [11].

It has been demonstrated that peptides derived from LF by pepsin digestion showed higher antimicrobial activity against both Gram-positive and Gram-negative bacteria than intact LF. However, other proteases such as trypsin, chymotrypsin, and plant and microbial proteases do not generate significant antimicrobial peptides from LF [12]. Antimicrobial peptides purified by high-performance liquid chromatography (HPLC) have been extensively used for the fractionation and identification of peptides on a laboratory scale [13]. This system allows the high resolution of peptides in a short time. Therefore, these peptides are potential food preservatives that can suppress microbial growth. However, liquid chromatography is a relatively expensive system for the large-scale preparation of peptides; moreover, some solvents used in chromatography are harmful to humans and not suitable for food processing [14]. Therefore, there is a large-scale, low-cost, and biocompatible approach to peptide fractionation based on the amphoteric nature of the sample peptides dissolved in water [14]. This approach is referred to as autofocusing that is considered as a potential industrial peptide fractionation process [15]. In the sight of these facts, the concerns of this chapter were

- Structure of LF;
- Methods of production, isolation and purification of peptides derived from LF;
- Stability of LF and its derived peptides; and
- Bioactive properties and applications of LF and its derived peptides.

2. Structure

LF is an iron-binding monomeric glycoprotein with a molecular weight of approximately 80 kDa [7]. Moore et al. [16] demonstrated some physicochemical characteristics of bovine LF. Its carbohydrate content is 11.2%, maximal iron content is 1.4 mg/g, isoelectric point is 8, and absorbance at 465 nm of 1% solution (iron saturation) is 0.58.

LF consists of 692 amino acids. The protein contains a single polypeptide chain folded into two distinct lobes called N- and C-lobe, referring to the N- and C-terminal parts of the molecule, respectively [17]. Both lobes have the same fold, consistent with their high level of sequence identity; in each case, the lobe is subdivided into two sublobes or domains called N1, N2, C1, and C2, respectively. It is separated by a deep interdomain cleft that surfaces an iron-binding site (Fe^{3+}), which is tightly bound in synergistic cooperation with a bicarbonate anion [18]. The use of this anion can affect the metal binding capacity, giving pH control of iron release [19]. Furthermore, LF retains iron to as low as pH 3.5. This gives LF a more potent iron withholding capacity than other transferrins [7]. There is just a single covalent link between the two lobes, which is called a hinge region that forms a short helix conformation with three turns and represent the sequences 334 to 344. In the N-terminal lobe of LF, the sequences start in one domain, move into the second domain, and then back into the first domain before running through a short helix to the C-lobe where the folding pattern is repeated [17].

The two domains fold with a ligand-binding site that is exploited by many binding proteins, including the large family of bacterial periplasmic binding proteins [20]. Moreover, LF is also highly basic due to a unique basic region in the N-terminal of the molecule. One important consequence of this property is that LF can bind with many acid molecules, including heparin and various cell surface molecules [21]. Also, there is a positive charge concentrated in the interlobe region associated with the short helix. This appeals as a likely DNA and many cell types binding region [22].

3. Methods of production, isolation, and purification of peptides derived from LF

3.1. Enzymatic hydrolysis

The antimicrobial sequence of LF was found mainly near its N-terminus, which has an important role in iron chelation. The high proportion of basic residues in the identified domain is allowing its interaction with surface components of microbial cells [23]. Cationic peptide generated upon gastric pepsin cleavage of bovine LF with porcine pepsin, cod pepsin, or acid protease from *Penicillium duporzti* showed strong activity against *Escherichia coli* O111, whereas hydrolysates produced by trypsin, papain, or other neutral proteases were much less active [12]. Bellamy et al. [24] reported that a 25-residue peptide could be released from the N-terminal region of bovine LF with potent bactericidal activity. Interestingly, this reaction is catalyzed at acidic pH by pepsin. After cleavage by this enzyme, the region from amino acids 17 to 41 (FKCRRWQWRMKKLGAPSITCVRRAF) is released as lactoferricin B (Lf-cin B) from bovine LF. Lf-cin B showed broad-spectrum antibacterial activity against both Gram-negative and Gram-positive species, even strains that were resistant to native LF. This indicated that the antibacterial activity of hydrolysate was greater than that of undigested LF with all strains tested approximately 8-fold [24].

3.2. Rennet enzyme and autofocusing technique

LF-cin (RRWQWRMKKLG) as well as other three peptides (KLLSKAQEKFGKNKSRSFQL, APRKNVRWCTISQPEWFKCR, and TRVVWCAVG) was also isolated in a single step from calf rennet hydrolyzed LF by autofocusing technique and were purified by reverse-phase

HPLC. They were characterized by N-terminal Edman sequencing, mass spectrometry, and antibacterial activity determination [25]. The autofocusing is conducted based on the amphoteric nature of the sample using 5 L autofocusing apparatus without using any chemical substances [26]. The basic autofocusing fractions exhibited potent antibacterial activity against *E. coli* and *Bacillus subtilis* in comparison to acidic ones and much higher than the crude LF. In addition, they had greater antibacterial activities than those resulted from autofocusing fractions of LF hydrolyzed by both porcine pepsin and fungal rennet. The most active autofocusing fraction was purified by standard chromatography techniques and assessed for antibacterial activity. The synthetic peptides were found to inhibit the growth of *E. coli* and *B. subtilis* with minimum inhibitory concentration (MIC) ranging from 25 to 100 μg/ml and from 6.25 to 50 μg/ml, respectively. Moreover, the autofocusing can produce stronger antimicrobials than those in crude ones [25].

3.3. Chromatographic techniques

LF has a cationic nature according to its amino acid composition. This gives it suitability to be purified by cation-exchange chromatography such as carboxymethyl-Sephadex [27]. It is considered the most popular procedure for LF peptide purification in most companies. For example, skim milk or cheese-whey is applied to a cation-exchange chromatography column without pH adjustment after its filtration. First, the column is washed with a low concentration (1.6%) NaCl solution in which lactoperoxidase is eluted. Then, LF is eluted with a high concentration (5%) NaCl solution. Ultrafiltration is applied to concentrate LF and is separated from NaCl by dia-filtration. The LF is then freeze- or spray-dried to make a powder [27]. In the same manner, LF can bind transition metals and anionic compounds such as heparin and DNA. These materials have been used to purify LF. Metal (copper)-chelate affinity chromatography followed by gel filtration was reported for the purification of LF from human whey [28]. Heparin-agarose affinity chromatography was used to purify bovine LF and its peptides from the secretions of the involuting mammary gland [29].

4. Stability and solubility of LF and its derived peptides

4.1. Temperature

It is well known that heat treatment of milk and milk protein solutions affect the functional properties of the native proteins. A consideration of heat stability is very important if LF is to be used as a bioactive component of foods [30]. Several studies were recorded and varied about the heat stability of LF. Heating of milk at 65°C for 30 min had no significant effect on LF. However, the entire activity of LF was lost at 85°C/30 min in all kinds of milk. However, camel's milk antimicrobial factors were significantly more heat resistant than cow's and buffalo's milk proteins because it contains a higher concentration of LF [31]. Similarly, Paulsson et al. [32] found that milk LF was unaffected by pasteurization but completely denatured by ultra-high temperature (UHT) treatment and decrease the interaction capacity of LF with bacteria. Luf and Rosner [33] found that heat treatment at 63°C/30 min reduced the native LF content by

40%, whereas high temperature short time treatment of milk had no significant effect on LF denaturation.

LF is denatured more rapidly in its apo-form than in the iron-saturated form. Both apo- and iron-saturated LF are more heat sensitive when treated in milk than in phosphate buffer [34]. Abe et al. [35] reported that LF resisted heating at 80°C/5 min without any significant loss of the iron-binding capability. After heating at 100°C/5 min, LF was still able to bind approximately 85% of the amount of iron bound by the unheated LF. All iron-binding capability was lost when LF was heated to 120°C in phosphate buffer.

The thermostability of LF was dependent on both pH and ionic strength. LF was thermostable at acidic pH than neutral pH [35]. However, under more extreme conditions (120°C, pH 2), LF is acid hydrolyzed; however, it retained antimicrobial activity independent of iron binding [36]. Kawakami et al. [37] reported that, at pH 3.5, LF was resistant to heating at ionic strength 0.37 or below, but turbidity and precipitation occurred at ionic strength above 0.47. LF and its derived peptides' antibacterial activity are varied with the growth temperature. Decreased temperature resulted in a corresponding decrease in MIC and minimum bactericidal concentration (MBC). This was most pronounced for *E. coli*, which increased its sensitivity for LF with decreased growth temperature [38].

4.2. pH

pH is an important factor affecting the denaturation of LF. The rate of LF denaturation was highest at pH above 5.2 in which it mainly affects the rate of the unfolding and aggregation stages of denaturation [39]. In the same aspect, Saito et al. [40] reported that LF and its hydrolysates were denatured to an insoluble state by heat treatment under neutral or alkaline conditions above pH 6. In contrast, they remain soluble after heat treatment under acidic conditions at pH 2 to 5.

LF retained their functional activities at pH ranging from 2 to 7.4 [41]. LF is very stable at pH 4 and high temperature. It resisted heating at 90°C for 5 min at pH 4 without any loss of iron-binding capacity, antigenic activity, or antibacterial activity. LF treated at pH 2 or 3 and 100°C or 120°C for 5 min was apparently degraded, but antibacterial activity was equal to or stronger than that of unheated one [35].

LF was demonstrated to bind to the surface of some pathogenic microorganisms associated with intramammary infections. This binding is optimal at acidic pH with time-dependent binding surface varied according to bacterial species [42]. Griffith and Humphreys [43] concluded that LF was active near neutral pH only in the presence of bicarbonate ions. Because bicarbonate is secreted in the lumen of the intestine, the conditions are favorable for the antimicrobial activity of LF [44].

In the combination between pH and temperature, Murdock and Mathews [45] reported that, under low pH and refrigeration conditions, LF and its pepsin-digested peptides could limit the growth or reduce the population of pathogenic bacteria in dairy products. Moreover, Troost et al. [46] concluded that LF resisted stomach pH and survive gastric transit after oral administration.

4.3. Solubility

The iron-binding capability of bovine LF affects mainly its solubility. Ferrous iron was not stable in solution and was easily changed to insoluble ferric state, but the solubility of the ferrous iron was stabilized in solution in the presence of LF [47].

At acidic pH 2 to 5, LF hydrolysate in the heated samples at 80°C to 120°C remained soluble and the solutions were clear. On the other aspect, at neutral and alkaline pH 6 to 10, turbidity and gel formation occurred, which is markedly increased with temperature rise. At pH 11, the LF hydrolysate in the heated samples remained soluble, but the color of the samples became dark [35].

4.4. Salts

Reiter et al. [48] suggested that citrate reduces the antibacterial activity of LF by competing with the iron-binding proteins. In contrast, bicarbonate enhances the iron-binding properties of LF.

Salanlah and al-Obaidi [49] studied the effect of pH, temperature, magnesium, and calcium on the bactericidal activity of LF against *Yersinia pseudotuberculosis*. The bactericidal activity of LF was higher at acid pH. However, it was not efficient at 4, 15, and 25°C, but it was effective at 37°C. The activity of LF-derived peptides was time and concentration dependent. Calcium did not affect their activity up to 60 mM, whereas magnesium reduced the activity of LF only. Furthermore, Branen and Davidson [50] reported that the addition of EDTA enhanced the activity of hydrolyzed LF in tryptic soy broth against Enterohemorrahgic *E. coli*, *Listeria monocytogenes*, and *Salmonella enteritidis*. The EDTA chelate the excess of cations in broth that reduces the activity of hydrolyzed LF.

4.5. Proteases

Holo-LF is more resistant to proteolysis than apo-LF. The five different fragments with molecular weight that ranged from 25 to 53 kDa were generated from trypsin digest of LF. However, β-lactoglobulin has the ability to reduce the susceptibility of LF to tryptic digestion, suggesting that complex formation is not a mechanism for protecting LF against intestinal degradation [51].

4.6. Interaction with other antimicrobial agents

The anti-*Candida* activity of LF and its peptides or Lf-cin in combination with clotrimazole was shown synergistic by checkerboard analysis. These results indicate that LF-related substances function cooperatively with azole antifungal agents against *Candida albicans*. The concentration of Lf-cin required for inhibiting the growth of *C. albicans* decreased in the presence of relatively low concentrations of clotrimazole [52].

The concomitant use of LF with antibiotic cefodoxime proxetil resulted in a synergistic activity against *Staphylococcus aureus*, *E. coli*, *Klebsiella pneumoniae*, and *Pseudomonas aeruginosa* and an additive activity against *E. coli* strain. The antibiotic MIC in the presence of LF was reduced to <1/64 with an efficacy rate of 53/57 (92.9%) in a patient group with infections [53].

Soukka et al. [54] reported that both LF or its peptides and lactoperoxidase system showed bactericidal activity against *Streptococcus mutans* at low pH. LF hydrolysate enhanced lactoperoxidase enzyme activity against *S. mutans* but decreased the yield of antimicrobial components. Denisova et al. [55] reported that the combined administration of LF, lactoperoxidase, and lactoglobulin obtained from cow milk's as well as LF obtained from human milk contributed to the elimination of *Shigella sonnei* from the lungs of infected mice. They prevented the death of the animals. On the contrary, each one has no protective action. López-Expósito and Recio [56] reported that there is a synergistic effect when Lf-cin B was combined with bovine LF against both *E. coli* and *Staphylococcus epidermidis*. Bovine LF increased its antimicrobial activity when it was assayed with bovine αs2-casein f (183-207).

5. Bioactive properties and applications of LF and its derived peptides

5.1. Antimicrobial activity

The role of LF in innate defense enhances the broad spectrum of antimicrobial activity. Accordingly, various modes of antimicrobial effects have been reported for LF [2].

5.1.1. Stasis effect

The bacteriostatic mechanism is related to the high iron-binding affinity of the protein that deprives iron-necessary for microbial growth. Because the bacteriostatic properties of LF are due to its iron-binding ability, the protein is capable of retarding the growth of a broad range of Gram-negative and Gram-positive bacteria (*E. coli, Salmonella typhimurium, L. monocytogenes, B. subtilis, Bacillus stearothermophilus,* and *Shigella dysenteriae*) and certain yeasts [57]. However, bacteriostasis is often temporary because some Gram-negative bacteria adapt to iron-restrictive conditions by synthesizing low molecular weight iron chelators (siderophores) that can remove iron from LF [58]. In the same way, Bishop et al. [59] stated that 0.02 mg/ml apo-LF showed marked inhibition of growth of coliforms. In addition, it can inhibit the growth of *Klebsiella* spp. and *Aerobucter aerogenes* at concentrations of 0.2 and 2 mg/ml, respectively.

5.1.2. Cidal effect

The bactericidal effect of Lf-cin against some Gram-negative and Gram-positive bacteria has not any relation simple iron deprivation. It causes a rapid loss of bacterial viability [60]. It was observed that Lf-cin could bind to the outer membrane of Gram-negative bacteria causing alterations in microbial permeation that lead to bacterial death [61]. Moreover, it has the ability to interfere with the carbohydrate metabolism of invading bacteria or LF's ribonuclease activity, which may interfere with microbial protein synthesis [10].

5.1.3. Cationic effect

Current opinion on the mechanism of action of basic antibacterial peptides is focused on their interaction with the negatively charged elements in the membranes of susceptible bacteria causing an increase in cell permeability [62, 63]. These elements are lipopolysaccharide in Gram-negative bacteria and lipotechoic acid in Gram-positive bacteria. This is in agreement with the study that demonstrates that only basic autofocusing fractions have antibacterial activity [25]. Notably, 6 of the 20 amino acid residues in peptide (KLLSKA-QEKFGKNKSRSFQL) and 5 of 11 amino acid residues of peptide (RRWQWRMKKLG) (Lf-cin B fragment) revealed from rennet digestion of LF are basic residues and showed antibacterial activity and this feature probably has an important role in determining their potent bactericidal properties [25]. In addition, Hoek et al. [64] reported that the first 10-amino acid residues of Lf-cin B had more potent antibacterial activity than the rest of 25-amino acid residues) FKCRRWQWRMKKLGAPSITCVRRAF)against *E. coli L361*. In the same way, antimicrobial domain was identified in the N1-domain of LF, designated as lactoferrampin (WKLLSKAQEKFGKNKSR), corresponding to the amino acids 268 to 284 of LF. This peptide exhibited candidacidal activity, which was substantially higher than the activity of LF. Furthermore, lactoferrampin was active against *B. subtilis, E. coli,* and *P. aeruginosa* [65].

Morten et al. [66] reported that the activity increases with an increased amount of tryptophan (Trp) residues, especially against *B. subtilis*. Trp is thought to function as an anchor in membrane proteins. The aromaticity of Trp has been suggested to interact with both hydrophobic and hydrophilic constituents of the membrane due to its amphipathic nature [67]. Trp is also suggested to act as a needle that pulls α-helices across phospholipid membranes. In cationic peptides, the insertion of Trp residues into bacterial cell membranes disturbs the packing of the phospholipids and causes a weakening of the membrane [68]. The presence of several Trp residues in the 11-residue Lf-cin B peptide (RRWQWRMKKLG) clearly leads to a more effective disruption of the membrane [69]. Moreover, the presence of Trp in Lf-cin B fragment revealed from rennet digest may be responsible for the antibacterial activity by this mechanism [25].

The hydrophobicity, amphipathicity, and net positive charge have been described as important characteristics of antimicrobial peptides [70]. The penetrating ability of antimicrobial peptides to the surface membrane is known as the carpet model [71, 72]. In this model, the peptides first bind to the phospholipid head groups at the surface of the membrane. The hydrophobic core portion of the peptide is then able to interact with the lipid bilayer when hydrophobicity threshold concentration is above 3, causing a disruption of membrane permeability [72]. However, some basic and hydrophobic peptides generated from rennet digested LF have no antibacterial activity [25, 73]. These facts suggest that not only the basic and hydrophobic nature but also other factors may contribute to the antibacterial activity.

5.1.4. Adhesion-blockade effect

Bacterial adherence to target epithelia is an important step in the pathogenesis of the disease. There are many forms of adhesion sites according to the nature of microorganisms.

5.1.4.1. Adhesion-blockade of enteric pathogens

Several carbohydrates, such as 0.1% fructose or 0.5% glucose, strongly inhibit the adherence of shigellae to guinea pig colonic cells. Fructose-containing peptides from LF also inhibit the adhesion of *Shigella flexneri* to colonic epithelial cells [74]. LF-derived peptides have the ability to inhibit the hemagglutination activity of type 1 fimbriated *E. coli* [75]. The agglutination reaction was specifically inhibited by glycopeptides derived from LF or α-methyl-D-mannoside. These observations indicate that the glycans of LF could serve as receptors for type 1 fimbrial lectin of *E. coli*. Furthermore, more than 3 log CFU/g *E. coli* reduction in feces resulted from the oral administration of LF (20 mg/ml in 20% sucrose solution) [75]. Bovine LF inhibited the binding of fibronectin, fibrinogen, collagen type I, collagen type IV, and laminin to bacteria. Electron microscopy revealed the loss of type 1, CFA-I and CFA-I1 fimbria of *E. coli* grown in broth containing 10 μM LF. These data suggest a strong influence of LF on adhesion-colonization properties of *E. coli* [2].

5.1.4.2. Adhesion-blockade of oral pathogens

LF-derived peptides have the ability to adsorb hydroxyapatite, which resulted in interfering with the attachment of *S. mutans* to hydroxyapatite [76]. Also, LF-derived peptides inhibit the growth of *Prevotella intermedia* by interfering with the binding of this bacteria to fibronectin, collagen type I and type IV, and laminin [77]. LF-dependent adhesion-inhibition of *Actinomyces comitans* and *P. intermedia* to fibroblasts and Matrigel could involve the binding of LF to both the bacteria and substrata. The decreased adhesion may be due to the blocking of both specific adhesin-ligand and nonspecific charge-dependent interactions [78].

5.2. Food preservation

LF possesses an intrinsic bactericidal activity that is unrelated to its capacity to bind iron [62]. From this point, there has been an increased interest in the food industry for using LF and its peptides as a preservative in a wide variety of food products. Furthermore, LF has been used as a functional food ingredient in some products such as in infant formula, supplemental tablets, yoghurt, skim milk, drinks, and pet foods. In addition, these products have many therapeutic effects, including anti-infection, improvement of gastrointestinal microflora, immunomodulation, anti-inflammation, and antioxidation [52].

A diverse range of Gram-positive and Gram-negative bacteria was found to be susceptible to the inhibition and inactivation by Lf-cin B. It had the ability to reduce the bacterial population by approximately 3 $\log_{10}$ CFU/ml for both *E. coli* and *B. subtilis* at concentration 31 μg/ml [64]. In a study reported by Shin et al. [79] about the activity of bovine LF and its peptides, they demonstrated that MIC against *E. coli* O157:H7 was 3000 μg/ml for LF, 100 to 200 μg/ml for LF hydrolysate, and 8 to 10 μg/ml for Lf-cin B in 1% bactopeptone broth. In addition, Lf-cin B killed these bacteria at a concentration above 10 μg/ml.

In the meat industry, LF-cin B was primarily tested in ground beef at concentrations of 50 or 100 μg/ml, where it was found to cause a maximum of 2 log CFU/g reductions at 4°C or 10°C [80]. The LF-derived peptides can also be sprayed onto carcasses at a concentration of

3.26 ml spray/kg beef or 65.2 mg/kg beef for preventing bacterial contamination during processing or can be applied to finished beef surface before final packaging to extend the shelf life [81]. In addition, it has been used in the preservation of poultry, pork, fish, and other seafood [82].

The addition of EDTA has shown a positive effect on the antimicrobial activity of Lf-cin. Lf-cin B was unable to inhibit the growth of *E. coli* at 37°C at a concentration of 1600 μg/ml, whereas the addition of 100 or 400 μg/ml EDTA, depending on the strain of *E. coli*, totally prevented the growth [50]. However, Murdock and Mathews [45] reported that there was no effect for EDTA on the antimicrobial action of Lf-cin B in UHT milk even at 4000 μg/ml Lf-cin and 10 mg/ml EDTA.

On the contrary, LF and its peptides have been used in the preservation of raw, pasteurized, and UHT milk, butter, and cheese due to its antimicrobial activity against pathogenic and spoilage bacteria [83]. Moreover, Hugunin [84] reported that LF should be added after pasteurization and fermentation of yoghurt to extend the shelf life of yoghurt and stimulate the growth of its starter cultures. Chen and Allen [85] reported that LF preserves milk with a concentration of 2 mg/ml. Suzuki et al. [41] recorded that LF at a concentration of 1 mg/ml inhibited the formation of hydroperioxides by 48%, whereas the application of 11 mg/ml completely inhibited both the formation of hydroperioxides and several microorganisms. In 2009, Elbarbary reported that the 11-residue Lf-cin B fragment (RRWQWRMKKLG) generated from calf rennet digest of LF has a significant activity in reduction of approximately 3 $\log_{10}$ CFU/ml of *E. coli* and *B. subtilis* populations in milk.

Nowadays, the surface of vegetables, fruit, and carrot juice was coated with Lf-cin B to extend their shelf life. It is more significant when it was added to juice filtrate after its dialysis. As the concentration of cations in filtrate decreases, the antimicrobial efficacy of LF-cin B increases [9].

Moreover, LF can be incorporated into edible films. This is applied for dual purposes that enhance food safety and preservation, as this film acts as a physical barrier and prevents lipid oxidation as well as an antimicrobial substance [86].

Emerging preservation technologies, such as the combination of high hydrostatic pressure with LF, have proven promising in the effort to overcome the limitations in LF and its peptide application [120]. The use of LF and LF-cin B in combination with high pressure (155–400 mPa) enhanced the bactericidal activity against *E. coli*, *S. enteritidis*, *S. typhimurium*, *S. sonnei*, *S. flexneri*, *Pseudomonas fluorescens*, and *S. aureus* in potassium phosphate buffer at 20°C and showed more marked bacterial reduction than the application of LF or Lf-cin B alone [87].

On the contrary, a novel LF derivative [activated LF (ALF)] can be applied as alternative to LF. Naidu [82] reported that ALF has more potent bacteriostatic efficacy against *E. coli* in contaminated beef steak, and the MIC of ALF and LF was 62 and >1000 μg/ml, respectively. In the same manner, Ransom et al. [88] recorded that ALF inhibited the growth of *L. monocytogenes*, *S. typhimurium*, and *E. coli* O157:H7 in beef carcass adipose tissue after its dipping into ALF followed by dipping into 2% lactic acid in comparison to LF, which had no effect on the growth of bacterial population.

5.3. Antimicrobial effect in human therapy

5.3.1. *Antibacterial effect*

The antibacterial function of LF has been substantiated by both *in vitro* [89] and *in vivo* [90]. It appears that two different mechanisms involving two separate domains of the protein contribute to the antibacterial function of LF, which are its bacteriostatic and bactericidal mechanisms [89].

The ingestion of milk containing LF at 40 mg/ml will result in the formation of LF-cin B at a molar concentration corresponding to 4.5% of ingested LF. The highest concentration of Lf-cin B was found mainly in the stomach then in the upper small intestine then the lower one. Not all ingested LF was digested and some LF were detected in lower gastrointestinal tract [91].

Feeding the infants with LF or its peptides-enriched formula at a concentration of 1 mg/ml for 2 weeks leads to the increase and establishment of bifidobacterium-predominant flora in infant [92], whereas the ratios of Enterobacteriaceae, *Streptococcus,* and *Clostridium* showed a tendency to decrease [93].

The oral administration of LF hydrolysate prepared by porcine pepsin in mice showed an inhibitory effect on the proliferation of *Clostridium ramosum* [94]. Furthermore, supplementation of the milk diet with LF or a pepsin-generated hydrolysate of LF resulted in a significant suppression of bacterial translocation from the intestines to the mesenteric lymph nodes, and the bacteria involved were mainly members of the family Enterobacteriaceae. This ability of LF to inhibit bacterial translocation may be due to its suppression of bacterial overgrowth in the guts of milk-fed mice [95].

Lf-cin B was identified as the most active antimicrobial peptide generated from pepsin digest. It has significant bacteriostatic and bactericidal activities against multiple strains of *Helicobacter pylori* [96], which is mainly associated with chronic gastritis, peptic ulcer, and gastric cancer. There are numerous possible mechanisms by which Lf-cin B alone may suppress *H. pylori* infection. First, it possesses a direct antimicrobial activity against *Helicobacter* species [97]. Second, Lf-cin B may interfere with *H. pylori* adhesion to the gastric surface [98]. Third and last, Lf-cin B may suppress the production of proinflammatory cytokines, tumor necrotic factor, as a part of a regulatory role in the host immune response [99]. There is also evidence that Lf-cin B can exert a direct bactericidal effect on certain *S. mutans* and *Vibrio cholerae* [100].

5.3.2. *Antiviral activity*

In addition to the antibacterial activity of LF and its peptides, they display antiviral activity against a wide range of human and animal viruses, both DNA and RNA viruses, including rotavirus, respiratory syncytial virus, herpes virus, HIV, hepatitis C virus (HCV), and poliovirus. The antiviral effect of LF lies in the early phase of infection [101]. The antiviral mechanism of LF against viral infection depends on preventing the entry of virus in the host cell either by interfering with adsorption of the virus to the target cell, as LF is capable of binding to viral particles or its envelope protein [102], or by interfering with the viralcoreceptor interaction [103]. The N-terminal region of LF proved to be essential for its antiviral activity that is responsible for binding to the receptor specific to viral infection, which in turn act as a binding site for the initial interaction of virus with host cells [104] .

Another mechanism for the antiviral effect of LF is the ability to block viral replication in target cells [105]. Further studies indicated that the protective effect of LF was due to the up-regulation of natural killer (NK) cells, monocytes, and granulocytes, which eliminated the infection [104]. Tanaka et al. [106] reported that patients with chronic viral infection, such as HCV, when given LF orally at a dose 1.8 or 3.6 mg/day for 8 weeks showed a decrease in their serum alanine transaminase and viral concentration. In another study, 25 patients with chronic hepatitis C genotype 1b received a 6-month course of LF. The serum level of HCV RNA significantly decreased in patients given LF during the 6 months of treatment [107].

Fujihara and Hayashi [108] reported that LF and its peptides could inhibit the infection in mouse with herpes simplex virus type-1 (HSV-1) by preventing HSV-1 plaque formation after a topical application of 1% LF. However, it did not inhibit the propagation of the virus.

The administration of LF (1 mg/g body weight) protected mice from death due to infection with murine cytomegalovirus (MCMV) through the augmentation of NK cell activity. LF-treated mice showed a significant increase in the NK cell activity but not of the cytolytic T lymphocytes that recognize an MCMV-derived peptide [109].

Bovine LF and its peptides could also inhibit the integrin-mediated internalization of adenovirus into host cells through its binding to viral polypeptides III and IIIa [110].

5.3.3. Antifungal activity

The antifungal activity of LF has been studied. It is mainly through the ability of LF to change the structure of the cell wall of fungi, inactivate the enzyme, and deprive fungi from iron [111].

C. albicans and other *Candida* species are common pathogens that frequently cause oral infections in immunocompromised individuals due to the suppression of local as well as systemic defense mechanisms [112]. LF with concentrations ranging from 0.5 to 100 mg/ml is able to inhibit the growth of several *Candida* species [113]. Moreover, Takakura et al. [114] showed the protective activities of LF administered orally against oral candidiasis. LF at a dose of more than 0.5 g/kg/day exerts therapeutic activity and facilitated the recovery from oral candidiasis within 5 to 7 days. Furthermore, Masci [115] reported that a mouthwash containing LF and lysozyme was effective against oral candidiasis in immunocompromised patients rather than oral administration.

LF has also been shown to improve inflammatory diseases such as dermatophytosis caused by the fungus *Trichophyton mentagrophytes* and intractable stomatitis [116].

5.4. Antiparasitic activity

The role of LF in parasitic diseases is not well defined and may involve multiple mechanisms. Preincubation of *Toxoplasma gondii* and *Eimeria stiedae* sporozoites with LF peptides reduced their infectivity in animal models [117]. Mice challenged with *T. gondii* survived for 35 days as a result of oral administration of 5.0 mg Lf-cin. Furthermore, mice challenged with *T. gondii* cysts were still alive after an intraperitoneal injection of 0.1 mg Lf-cin. In contrast, 80% of untreated mice died due to acute toxoplasmosis within 14 days [118]. This suggests an effect of basic peptides on parasite membrane integrity and/or interaction with host tissues. Moreover, LF inhibits some parasites via the stimulation of the process of phagocytosis where immune cells engulf and digest foreign organisms [118]. Also, LF has the ability to reduce the binding of *Plasmodium berghei* to the surface of the target cells [119]. Other reported antiparasitic activities appear to involve interference with parasite iron acquisition, such as *Pneumocystis carinii* [120]. In other parasites, such as *Trichomonas foetus*, LF appears to act as a specific iron donor and could thus be expected to enhance infection [121].

5.5. Enhancer of iron absorption

The role of LF as an enhancer of iron absorption was proposed because LF binds a majority of iron in breast milk and therefore could facilitate the uptake of iron in the small intestine of infants [122]. The notion that LF is involved in iron absorption was supported by studies that showed the presence of specific LF receptor on brush membranes from human intestine [123]. Lönnerdal and Bryant [124] showed that iron is equally well absorbed from LF (whether heat-treated or untreated) and ferrous sulfate. Thus, iron provided by dietary LF is likely to be well utilized in human adults. The ability of LF to bind iron naturally support its function in intestinal iron absorption [123].

5.6. Anticancer activity

LF has shown anticancer activity in experimental lung, bladder, tongue, colon, and liver carcinogenesis on rats [125]. Also, in another study, LF suppressed the incidence of oral cancer by 50% [126]. *In vivo* studies have indicated that LF may induce apoptosis in cancer cells; in addition, the immunomodulatory activity of LF is critical in preventing cancer growth by enhancing the activity of NK cells [127]. Therefore, LF offers promise as a potential chemopreventive agent for oral cancer. Currently, LF is used as an ingredient in yogurt, chewing gums, infant formulas, and cosmetics [126].

5.7. Immunomodulatory activity

The immunomodulatory activity of LF is attributed to its ability to up-regulate T-cell proliferation, boost NK cell number, promote lymphocyte maturation, and also play a role in the myelopoietic process [128]. Moreover, LF released from neutrophils forms an integral part of the innate immunity protecting against infection through its antimicrobial activity [129].

6. Conclusion

Several studies proved that LF and its peptides have great potential for use as natural antimicrobials for food preservation and other health benefits in addition to the well-known iron-binding activity. Because LF is commercially available from bovine milk (~90 MT/year worldwide) and considered as a highly safe food additive, it may be an excellent agent for administration to humans. The enzymatic hydrolysis of LF results in the elaboration of functional peptides that exert, in some cases, greater functionality than the native LF. These peptides could serve as an attractive model for the development of new natural food preservatives, functional food ingredients, food supplements, and drug discovery.

Author details

Adham M. Abdou* and Hend A. Elbarbary

*Address all correspondence to: dradham@yahoo.com

Food Control Department, Faculty Veterinary Medicine, Benha University, Moshtohor, Kaliobiya, Egypt

References

[1] Farrell HM, Jamenez-Flores R, Bleck GT, Brown EM, Butler JE, Creamer LK, Hicks CL, Hollar CM, Ngkwai-Hang KF, Swaisgood HE: Nomenclature of the proteins of cow's milk—sixth revision. J Dairy Sci 2004; 87: 1641–1647.

[2] Naidu AS. Natural Food Antimicrobial Systems. CRC Press, Boca Raton; 2000.

[3] Lyer S, Lonnerdal B: Lactoferrin, lactoferrin receptors and iron metabolism. Eur J Clin Nutr 1993; 47: 232–241.

[4] Sorensen M, Sorensen MPL: The proteins in whey. C R Trav Lab Carlsberg 1939; 23: 55–59.

[5] Schade AL, Caroline L: An iron-binding component in human blood plasma. Science 1946; 104: 340–341.

[6] Hennart PF, Brasseur DJ, Delogne-Desnoeck JB, Dramax MM, Robyn CE: Lysozyme, lactoferrin and secretory immunoglobulin A content in breast milk: Influence of duration of lactation, nutrition status, prolactin status and parity of mother. Am J Clin Nutr 1991; 53: 32–39.

[7] Baker EN, Baker HM, Kid RD: Lactoferrin and transferrin: Functional variations on a common structural framework. Biochem Cell Biol 2002; 80: 27–34.

[8] Masson L, Heremans JF: Lactoferrin in milk from different species. Comp Biochem Physiol B 1971; 39: 119–129.

[9] Chantaysakorn P, Richter RL: Antimicrobial properties of pepsin-digested lactoferrin added to carrot juice and filtrates of carrot juice. J Food Protect 2000; 63: 376–380.

[10] Ye XY, Wang HX, Liu F, Ng TB: Ribonuclease cell free translation-inhibitory and superoxide radical scavenging activities of the iron-binding protein lactoferrin from bovine milk. Int J Biochem 2000; 32: 235–241.

[11] Tomita M, Wakabayashi H, Yamauchi K, Teraguchi S, Hayasawa H: Bovine lactoferrin and lactoferricin derived from milk: Production and applications. Biochem Cell Biol 2002; 80: 109–112.

[12] Tomita M, Bellamy W, Takase M, Yamauchi K, Wakabayashi H, Kawase K: Potent antibacterial peptides generated by pepsin digestion of bovine lactoferrin. J Dairy Sci 1991; 74: 4137–4142.

[13] Mant CT, Zhou NE, Hodges RS: Amino acids and peptides. In: Chromatography Part B: Applications. Heftman E, editor. Elsevier Science Publishers: Amsterdam, The Netherlands, 1992; pp. 75–150.

[14] Yata M, Sato K, Ohtsuki K, Kawabata M: Fractionation of peptides in protease digests of proteins by preparative isoelectric focusing in the absence of added ampholyte: A biocompatible and low-cost approach referred to as autofocusing. J Agric Food Chem 1996; 44: 76–79.

[15] Hashimoto K, Sato K, Nakamura Y, Ohtsuki K: Development of continuous type apparatus for ampholyte-free isoelectric focusing (autofocusing) of peptides in protein hydrolysates. J Agric Food Chem 2006; 54: 650–655.

[16] Moore SA, Anderson BF, Groom GR, Haridas M, Baker EN: Three dimensional structure of bovine lactoferrin at 2.8 Å resolution. J Mol Biol 1997; 274: 222–236.

[17] Powell MJ, Ogden JE: Nucleotide sequence of human lactoferrin DNA. Nucleic Acids Res 1990; 18: 4013–4013.

[18] Vorland LH: Lactoferrin: a multifunctional glycoprotein. Acta Pathol Microbiol Immunol Scand 1999; 107: 971–981.

[19] Baker HM, Masson AB, He QY, Baker EN: Ligand variation in the transferring family: The crystal structure of the H249Q mutant of the human transferring N-lobe as a model for iron binding in insect transferrins. Biochemistry 2001; 40: 11670–11675.

[20] Quiocho FA, Ledvina PS: Atomic structure and specificity of bacterial periplasmic receptors for active transport and chemotaxis: Variation of common themes. Mol Microbiol 1996; 20: 17–25.

[21] Zakharova ET, Shavlovski MM, Bass MG, Gridasova AA, Pulina MO, De Filippis V, Beltramini M, Di Muro P, Salvato B, Fontana A, Vasilyev VB, Gaistskhoki VS: Interaction of lactoferrin with ceruloplasmin. Arch Biochem Biophys 2000; 374: 222–228.

[22] Wu HF, Monroe DM, Church FC: Characterization of the glycosaminoglycan binding region of lactoferrin. Arch Biochem Biophys 1995; 317: 85–92.

[23] Tomita M. Active peptides of lactoferrin. In: Proceedings of the International Dairy Federation Seminar. Indigenous Antimicrobial Agents of Milk: Recent Developments; 1994. pp. 7–13.

[24] Bellamy W, Takase M, Yamaauchi K, Wakabayashi H, Kawase K, Tomita M: Identification of bactericidal domain of lactoferrin. Biochim Biophys Acta 1992; 1121: 130–136.

[25] Elbarbary HA, Abdou AM, Park EY, Nakamura Y, Mohamed HA, Sato K: Novel antibacterial lactoferrin peptides generated by rennet digestion and autofocusing technique. Int Diary J 2010; 20: 646–651.

[26] [26]Hashimoto K, Sato K, Nakamura Y, Ohtsuki K: Development of a large-scale (50 L) apparatus for ampholyte-free isoelectric focusing (autofocusing) of peptides in enzymatic hydrolysates of food proteins. J Agric Food Chem 2005; 53: 3801–3806.

[27] Yoshida S, Ye-Xiuyun Y: Isolation of lactoperoxidase and lactoferrins from bovine milk acid whey by carboxynlethyl cation exchange chromatography. J Dairy Sci 1991; 74: 1439–1444.

[28] Lonnerdal B, Carlsson J, Porath J: Isolation of lactoferrin from human milk by metal-chelate affinity chromatography. FEBS Lett 1977; 75: 89–92.

[29] Rejman JJ, Hegarty HM, Hurley WL: Purification and characterization of bovine lactoferrin from secretions of the involuting mammary gland: Identification of multiple molecular forms. Comp Biochem Physiol 1989; 93B: 929–934.

[30] Mata L, Sanchez L, Headon DR, Calvo M: Thermal denaturation of human lactoferrin and its effect on the ability to bind iron. J Agric Food Chem 1999; 46: 3964–3970.

[31] Elagamy EI: Effect of heat treatment on camel proteins with respect to antimicrobial factors: A comparison with cow's and buffalo's milk proteins. Food Chem 2000; 68: 227–232.

[32] Paulsson MA, Svensson U, Kishore AR, Naidu AS: Thermal behaviour of bovine lactoferrin in relation to bacterial interaction and antibacterial activity.Bull Int Dairy Fed 1994; 215.

[33] Luf W, Rosner E: Thermal stability of lactoferrin in bovine milk. Wien Tierarztl Monatsschr 1997; 84: 70–73.

[34] Sanchez L, Peiro JM, Castillo H, Perez MD, Ena JM, Calvo M: Kinetic parameters for denaturation of bovine milk lactoferrin. J Food Sci 1992; 57: 873–879.

[35] Abe H, Saito H, Miyakawa H, Tamura Y, Shimamura S, Nagao E, Tomita M: Heat stability of bovine lactoferrin at acidic p H. J Dairy Sci 1991; 74: 65–71.

[36] Saito H, Miyakawa H, Tamura Y, Shimamura S, Tomita M: Potent bactericidal activity of bovine lactoferrin hydrolysate produced by heat treatment at acidic p H. J Dairy Sci 1991; 74: 3724–3730.

[37] Kawakami H, Tanaka M, Tatsumi K, Dosako S: Effect of ionic strength and p H on the thermostability of lactoferrin. Int Diary J 1992; 2: 287–298.

[38] Vorland LH, Ulvaten H, Andersen J, Haukland HH, Rekdal Q, Svendsen JS, Gutteberg TJ: Antibacterial effects of lactoferricin B. Scand J Infect Dis 1999; 31: 179–184.

[39] Law A J, Leaver J: Effect of pH on the thermal denaturation of whey proteins in milk. J Agric Food Chem 2000; 48: 672–679.

[40] Saito H, Takase M, Tamura Y, Shimamura S, Tomita. M: Physicochemical and antibacterial properties of lactoferrin and its hydrolysate produced by heat treatment at acidic p H. Adv Exp Med Biol 1994; 357: 219–226.

[41] Suzuki YA, Kelleher SL, Yalda D, Wu L, Huang J, Huang N, Lönnerdal BO: Expression, characterization, and biologic activity of recombinant human lactoferrin in rice. J Pediatr Gastroenterol Nutr 2003; 36: 190–199.

[42] Naidu AS, Miedzobrodzki J, Andersson M, Nilsson LE, Forsgren A, Watts JL: Bovine lactoferrin binding to sex coagulase- negative Staphylococci isolated from bovine intramammary infections. J Clin Microbiol 1990; 28: 2313–2319.

[43] Griffith E, Humphreys J: Bacteriostatic effect of human milk and bovine colostrums on *Escherichia coli*: Importance of bicarbonate. Infect Immun 1977; 15: 396–401.

[44] Reiter B: Review of the progress of dairy science: Antimicrobial systems in milk. J Dairy Res 1978; 45: 131–147.

[45] Murdock CA, Mathews KR: Antibacterial activity of pepsin-digested lactoferrin on foodborne pathogens in buffered broth systems and ultra-high temperature milk with EDTA. J Appl Microbiol 2002; 93: 850–856.

[46] Troost FJ, Steijns J, Saris WHM, Brummer RJM: Gastric digestion of bovine lactoferrin *in vivo* in adults. J Nutr 2001; 131: 2101–2104.

[47] Nagasako Y, Shimamuro S, Tomita M, Saito H, Tamura Y: Iron-binding properties of bovine lactoferrin in iron rich solution. J Dairy Sci 1993; 76: 1876–1881.

[48] Reiter B, Brock JH, Steel ED: Inhibition of *Escherichia coli* by bovine colostrum and postcolostral milk. II. The bacteriostatic effect of lactoferrin on a serum susceptible and serum resistant strain of *E. coli*. Immunology 1975; 28: 83–95.

[49] Salanlah AA, al-Obaidi AS: Effect of some physical and chemical factors on the bactericidal activity of human lactoferrin and transferrin against *Yersinia pseudotuberculosis*. N Miaobiol 1995; 18: 275–281.

[50] Branen J, Davidson PM: Activity of hydrolysed lactoferrin against foodborne patho-

genic bacteria in growth media: The effect of EDTA. Lett Appl Microbiol 2000; 30: 233–237.

[51] Brock JH, Pineiro A, Lampreave F: The effect of trypsin and chymotrypsin on the antibacterial activity of complement, antibodies and lactoferrin and transferrin m bovine colostrum. Ann Vet Res 1978; 9: 287–294.

[52] Wakabayashi H, Abe S, Okutomi T, Tansho S, Kawase K, Yamaguchi H: Cooperative anti-*Candida* effects of lactoferrin or its peptides in combination with azole antifungal agents. Microbiol Immunol 1996; 40: 821–825.

[53] Chimura T, Hirayama T, Nakahara M: *In vitro* antimicrobial activities of lactoferrin, its concomitant use with cefpodoxime proxetil and clinical effect of cefpodoxime proxetil. Jpn J Antibiot 1993; 46: 482–485.

[54] Soukka T, Lumikari M, Tenovuo J: Combined inhibitory effect of lactoferrin and lactoperoxidase system on the viability of *Streptococcus mutans*, serotype c. Scand J Dental Res 1991; 99: 390–396.

[55] Denisova II, Gennadeva TI, Zhdanova, TV, Khristova M, Khazenson LB: The action of lactoperoxidase, lactoferrin and lactoglobulin on *Shigella sonnei* in an *in vivo* experiment. Z Mikrobiol Epidemiol Immunobiol 1996; 1: 66–69.

[56] López-Expósito I, Recio I: Protective effect of milk peptides: antibacterial and antitumor properties. Adv Exp Med Biol 2008; 606: 271–293.

[57] Pakkanen R, Aalto J: Growth factors and antimicrobial factors of bovine colostrum. Int Diary J 1997; 7: 285–297.

[58] Crosa JH: Genetics and molecular biology of siderophore-mediated iron transport in bacteria. Microbiol Mol Biol Rev 1989; 53: 517–530.

[59] Bishop JG, Schanbacher FL, Fergnson LC, Smith KL: *In vitro* growth inhibition of mastitis- causing coliform bacteria by bovine apo-lactoferrin and reversal of inhibition by citrate and high concentrations of apo lactoferrin. Infect Immun 1976; 14: 911–918.

[60] Arnold RR, Russell JE, Champion WJ, Gauthier JJ: Bactericidal activity of human lactoferrin: influence of physical conditions and metabolic state of the target microorganism. Infect Immun 1981; 32: 655–660.

[61] Chapple DS, Mason DJ, Joannou CL, Odell EW, Gant V, Evans RW: Structure-function relationship of antibacterial synthetic peptides homologous to a helical surface region on human lactoferrin against *Escherichia coli* serotype O111. Infect Immun 1998; 66: 2434–2440.

[62] Epand RM, Vogel HJ: Diversity of antimicrobial peptides and their mechanisms of action. Biochim Biophys Acta 1999; 1462: 11–28.

[63] Ulvatne H, Haukland HH, Olsvik O, Vorland LH: Lactoferricin B causes depolarization of the cytoplasmic membrane of *Escherichia coli* ATCC 25922 and fusion of negatively charged liposome. FEBS Lett 2001; 492: 62–65.

[64] Hoek KS, Milne JM, Grieve PA, Dionysius DA, Smith R: Antibacterial activity of bovine lactoferricin-derived peptides. Antimicrob Agents Chemother 1997; 41: 54–59.

[65] Van der Kraan MIA, Groenink J, Nazmi K, Veerman ECI, Bolscher JGM, Amerongen AVN. Lactoferrampin: A novel antimicrobial peptide in the N1-domain of bovine lactoferrin. Peptides. 2004; 25: 177–183.

[66] Morten BS, Haug BE, Rekdal Q, Skar ML, Stensen W, Svendsen JS: Important structural features of 15-residue lactoferricin derivatives and methods for improvement of antimicrobial activity. Biochem Cell Biol 2002; 80: 65–74.

[67] SchifferM, Chang CH, Stevens FJ: The functions of tryptophan residues in membrane proteins. Protein Eng 1992; 5: 213–214.

[68] Wimley WC, White SH: Determining the membrane topology of peptides by fluorescence quenching. Biochemistry 2000; 39: 161–170.

[69] Schibli DJ, Hwang PM, Vogel HJ: The structure of the antimicrobial active center of lactoferricin B bound to sodium dodecyl sulfate micelles. FEBS Lett 1999; 446: 213–217.

[70] Hong J, Oren Z, Shai Y: Structure and organization of hemolytic and nonhemolytic diastereomers of antimicrobial peptides in membranes. Biochemistry 1999; 38: 16963–16973.

[71] Shai Y: Mechanism of the binding, insertion and destabilization of phospholipid bilayer membranes by alpha-helical antimicrobial and cell nonselective membrane-lytic peptides. Biochim Biophys Acta 1999; 1462: 55–70.

[72] Stark M, Liu L, Deber CM: cationic hydrophobic peptides with antimicrobial activity. Antimicrob Agents Chemother 2002; 46: 3585–3590.

[73] Elbarbary HA. Emerging opportunities of natural antimicrobials in the dairy industry (Ph.D. thesis). Egypt: Faculty of Veterinary Medicine, Benha University; 2009.

[74] Izhar M, Nuchanlowitz Y, Mlrelman D: Adherence of *Shigella flexneri* to guinea pig intestinal cells is mediated by a mucosal adhesin. Infect Immun 1982; 35: 1110–1118.

[75] Teraguchi S, Shin K, Fukuwatari Y, Shimanlura S: Glycans of bovine lactoferrin function as receptors for the type 1 fimbrial lectin of *Escherichia coli*. Infect Immun 1996; 64: 1075–1077.

[76] Visca P, Berlutti F, Vittorioso P, Dalmastri C, Thaller MC, Valenti P: Growth and adsorption of *Streptococcus mutans* 6715-13 to hydroxyapatite in the presence of lactoferrin. Med Microbiol Immunol (Berl) 1989; 178: 69–79.

[77] Alugupalli KR, Kalfas S, Edwardsson S, Forsgren A, Arnold RR, Naidu AS: Effect of

lactoferrin on interaction of *Prevotella intermedia* with plasma and subepithelial matrix proteins. Oral Microbiol Immunol 1994; 9: 174–179.

[78] Alugupalli KR, Kalfas S: Characterization of the lactoferrin-dependent inhibition of the adhesion of actinomycetemcomitan, *Prevotella intermedia* and Prevotella nigrescens to fibroblasts and to a reconstituted basement membrane. APMIS 1997; 105: 680–688.

[79] Shin K, Yamauchi K, Teraguchi S, Hayasawa H, Tomita M, Otsuka Y, Yamazaki S: Antibacterial activity of bovine lactoferrin and its peptides against enterohemorrhagic *Escherichia coli* O157:H7. Lett Appl Microbiol 1998; 26: 407–411.

[80] Venkitanarayanan KS, Zhao T, Doyle MP: Antibacterial effect of lactoferrin B on *Escherichia coli* O157:H7 in ground beef. J Food Protect 1999; 62: 747–750.

[81] Food and Drug Administration (FDA). Center for Food Safety and Applied Nutrition. Agency Response Letter. GRAS Notice no. GRN 000067. http://www.foodsafety.gov. 2001.

[82] Naidu AS: Activated lactoferrin—A new approach to meat safety. Food Technol 2002; 56: 40–45.

[83] Champagne CP, Laing RR, Roy D, Mafu AA: Psychrotrophs in dairy products: Their effects and their control. Crit Rev Food Sci Nutr 1994; 34: 1–30.

[84] Hugunin A. Whey Products in Yoghurt and Fermented Dairy Products. U.S. Dairy Export Council: California, USA; 1999. pp. 1–8.

[85] Chen HY, Allen JC: Human milk antibacterial factors: the effect of temperature on defense systems. Adv Exp Med Biol 2001; 501: 341–348.

[86] Medina I, Tombo I, Satué-Gracia MT, German JB, Frankel EN: Effects of natural phenolic compounds on the antioxidant activity of lactoferrin in liposomes and oil-in-water emulsions. J Agric Food Chem 2002; 50: 2392–2399.

[87] Masschalck B, van Houdt R, Michiels CW: High pressure increases bactericidal activity and spectrum of lactoferrin, lactoferricin and nisin. Int J Food Microbiol 2001; 64: 325–332.

[88] Ransom JR, Sofos JN, Belk KE, Scanga JA, Hossner KL, Childs KD, Yemm RS, Smith GC. Susceptibility of *Escherichia coli* O157:H7, *Salmonella typhimurium*, and *Listeria monocytogenes*, inoculated onto beef tissues, steaks and RTE products, to lactic acid, lactoferrin, and activated lactoferrin. Final Report to National Cattlemen's Beef Association, Centennial, CO. Center for Red Meat Safety, Colorado State University: Fort Collins, CO; 2003. pp. 1–41.

[89] Brock JH: The physiology of lactoferrin. Biochem Cell Biol 2002; 80: 1–6.

[90] Van Hooijdonk ACM, Kussendrager KD, Steijnsa JM: *In vivo* antimicrobial and antiviral activity of components in bovine milk and colostrum involved in non-specific defence. Br J Nutr 2000; 84: 127–134.

[91] Kuwata H, Yip TT, Yamauchi K, Teraguchi S, Hayasawa H, Tomita M, et al.: The survival of ingested lactoferrin in the gastrointestinal tract of adult mice. Biochem J 1998; 334: 321–323.

[92] Roberts AK, Chierici R, Hill MJ, Volpata S, Vigi V: Supplementation of an adapted formula with bovine lactoferrin: I. Effect on the infant fecal flora. Acta Paediatr 1992; 81: 119–124.

[93] Kawaguchi S, Hayashi T, Masano J, Okuyama K, Suzuki T, Kawase K: A study concerning the effect of lactoferrin enriched infant formula on low birth weight infants with special reference to changes in intestinal flora. Perinat Med 1989; 19: 557–562 (in Japanese).

[94] Teraguchi S, Slun K, Ogata T, Kingaku M, Kaino A, Miyauchi H, Fukuwatari Y, Shimamura S: Orally administered bovine lactoferrin inhibits bacterial translocation in mice fed bovine milk. Appl Environ Microbiol 1995; 61:4131–4134.

[95] Teraguchi S, Shin K, Ozawa K, Nakamura S, Fukuwatari Y, Tsuyuki S, Namihira H, Shimamura S: Bacteriostatic effect of orally administered bovine lactoferrin on proliferation of *Clostridium* species in the gut of mice fed bovine milk. Appl Environ Microbiol 1995; 61: 501–506.

[96] Dial EJ, Lichtenberger LM: Effect of lactoferrin on *Helicobacter felis* induced gastritis. Biochem Cell Biol 2002; 80: 113–117.

[97] Appelmelk BJ, An YQ, Geerts M, Thijs BG, De Boer HA, Maclaren DM, De Graaf J, Nuijens JH: Lactoferrin is a lipid A-binding protein. Infect Immun 1994; 62: 2628–2632.

[98] Wada T, Aiba Y, Shimizu K, Takagi A, Miwa T, Koga Y: The therapeutic effect of bovine lactoferrin in the host infected with *Helicobacter pylori*. Scand J Gastroenterol 1999; 34: 238–243.

[99] Shimtzu N, Ikehara Y, Nozaki K, Kaminishi M, Tsuda H, Sugiyama A, Katsuyama T, Tatematsu, M. Effects of lactoferrin administration in *Helicobacter pylori* infection animal model. In: Lactoferrin: structure, function and applications. Shimazaki K, Tsuda H, Tomita M, Kuwata T, Perraudin JP, editors. Elsevier Science Publishers B.V.: Amsterdam, The Netherlands; 2000. pp. 209–215.

[100] Losnedahl KJ, Wang H, Aslam M, Sixiang Z, Hurely WL. Antimicrobial proteins in milk. Illinois Dairy Report, Department of Animal Sciences, University of Illinois; 1996.

[101] Van der Strate BWA, Beljaars L, Molema G, Harmsen MC, Meijer DKF: Antiviral activities of lactoferrin. Antiviral Res 2001; 52: 225–239.

[102] Superti F, Ammendolia MG, Valenti P, Seganti L: Antirotaviral activity of milk proteins: lactoferrin prevents rotavirus infection in the enterocyte-like cell line HT-29. Med Microbiol Immunol (Berl) 1997; 186: 83–91.

[103] Berkhout B, van Wamel JLB, Beljaars L, Meijer DKF, Visser S, Floris R: Characteriza-

tion of the anti-HIV effects of native lactoferrin and other milk proteins and protein-derived peptides. Antiviral Res 2002; 55: 341–355.

[104] Swart PJ, Kuipers ME, Smit C, Pauwels R, DE Bethune MP, De Clercq E, Huisman H, Meijer DKF: Antiviral effects of milk proteins: acylation results in polyanionic compounds with potent activity against human immunodeficiency virus type1 and 2 *in vitro*. AIDS Res Hum Retroviruses 1996; 12: 769–775.

[105] Groot F, Geijtenbeek TBH, Sanders RW, Baldwin CE, Sanchez-Hernandez M, Floris R, Van Kooyk Y, de Jong E, Berkhout B: Lactoferrin prevents dendritic cell-mediated human immunodeficiency virus type 1 transmission by blocking the DC-SIGN-gp 120 interaction. J Virol 2005; 79: 3009–3015.

[106] Tanaka K, Ikeda M, Nozaki A, Kato N, Tsuda H, Saito S, Sekihara H. Lactoferrin inhibits hepatitis C virus viremia in patients with chronic hepatitis C: a pilot study. Jpn J Cancer Res 1999; 90: 367–371.

[107] Iwasa M, Kaito M, Ikoma J, Takeo M, Imoto, I, Adachi Y, et al: Lactoferrin inhibits hepatitis C virus viremia in chronic hepatitis C patients with high viral loads and HGV genotype 1b. Am J Gastroenterol 2002; 97: 766–767.

[108] Fujihara T, Hayashi K: Lactoferrin inhibits herpes simplex virus type-1 (HSV-1) infection to mouse cornea. Arch Virol 1995; 140: 1469–1472.

[109] Shimizu K, Matsuzawa H, Okada K, Tazume S, Dosako S, Kawasaki Y, Hashimoto K, Koga Y: Lactoferrin-mediated protection of the host from murine cytomegalovirus infection by a T-cell dependent augmentation of natural killer cell activity. Arch Virol 1996; 141: 1875–1889.

[110] Pietrantoni A, Di Biase AM, Tinari A, Marchetti M, Valenti P, Seganti L, Superti F: Bovine lactoferrin inhibits adenovirus infection by interacting with viral structural polypeptides. Antimicrob Agents Chemother 2002; 47: 2688–2691.

[111] Nikawa H, Samaranayake LP, Tenovuo J, Pang KM, Hamada T: The fungicidal effect of human lactoferrin on *Candida albicans* and *Candida krusei*. Arch Oral Biol 1993; 38: 1057–1063.

[112] Samaranayake LP. In: Host factors and oral candidosis. Mac Farlane TW, editor. Wright: London, United Kingdom; 1990. pp. 66–103.

[113] Kuipers ME, de Vries HG, Eikelboom MC, Meijer DKF, Swar PJ: Synergistic fungistatic effects of lactoferrin in combination with antifungal drugs against clinical *Candida* isolates. Antimicrob Agents Chemother 1999; 43: 2635–3641.

[114] Takakura N, Wakabayashi H, Ishibashi H, Teraguchi S, Tamura Y, Yamaguchi H, Abe S: Oral lactoferrin treatment of experimental oral candidasis in mice. Antimicrob Agents Chemother 2003; 47: 2619–2623.

[115] Masci JR: Complete response of sever, refractory oral candidiasis to mouthwash containing lactoferrin and lysozyme. AIDS 2000; 14: 2403–2404.

[116] Sato R, Inanami O, Tanaka Y, Takase M, Naito Y: Oral administration of bovine lactoferrin for treatment of intractable stomatitis in feline immunodeficiency virus (FIV)-positive and FIV-negative cats. Am J Vet Res 1996; 57: 1443–1446.

[117] Omata Y, Satake M, Maeda R, Saito A, Shimazaki K, Yamauchi K, Uzuka Y, Tanabe S, Sarashina T, Mikami T: Reduction of the infectivity of *Toxoplasma gondii* and *Eimeria stiedai* sporozoites by treatment with bovine lactoferricin. J Vet Med Sci 2002; 63: 187–190.

[118] Isamida T, Tanaka T, Omata Y, Yamauchi K, Shimazaki K, Saito A: Protective effect of lactoferricin against *Toxoplasma gondii* infection in mice. J Vet Med Sci 1998; 60: 241–244.

[119] Shakibaei M, Frevert U: Dual interaction of the malaria circumsporozoite protein with the low density lipoprotein receptor-related protein (LRP) and heparin sulfate proteoglycans. J Exp Med 1996; 184: 1699–1711.

[120] Cirioni O, Giacometti A, Barchiesi F, Scalise G: Inhibition of growth of pneumocystis carinii by lactiferrins alone and in combination with pyrimethamine, clarithromycin and minocyclin. J Antimicrob Chemother 2000; 46: 577–582.

[121] Tachezy J, Kulda J, Bahnikova I, Suchan P, Razga J, Schrevet J: *Trichomonas foetus*: iron acquisition from lactoferrin and transferring. Exp Parasitol 1996; 83: 216–228.

[122] Iyer S, Lönnerdal B: Lactoferrin, lactoferrin receptors and iron metabolism. Eur J Clin Nutr 1993; 47: 232–241.

[123] Davidson LA, Lönnerdal B: Fe-saturation and proteolysis of human lactoferrin: Effect on brush-border receptor-mediated uptake of Fe and Mn. Am J Physiol 1989; 257: G930–G934.

[124] Lönnerdal B, Bryant A: Absorption of iron from recombinant human lactoferrin in young US women. Am J Clin Nutr 2006; 83: 2305–2309.

[125] Tsuda H, Sekine K, Fujita K, Ligo M: Cancer prevention by bovine lactoferrin and underlying mechanisms—A review of experimental and clinical studies. Biochem Cell Biol 2002; 80: 131–136.

[126] Chandra Mohan KV, Kumaraguruparan R, Prathiba D, Nagini S: Modulation of xenobiotic-metabolizing enzymes and redox status during chemoprevention of hamster buccal carcinogenesis by bovine lactoferrin. Nutrition 2006; 22: 940–946.

[127] Lönnerdal B, Iyer S: Lactoferrin: molecular structure and biological function. Annu Rev Nutr 1995; 15: 93–110.

[128] Actor JK, Hwang SA, Kruzel ML: Lactoferrin as a natural immune modulator. Curr Pharm Des 2009; 15: 1956–1973.

4

Allergenic Milk Proteins: Friend or Foe Nutritional Proteins?

Guillermo Docena, Paola Smaldini,
Renata Curciarello and Angela María Candreva

Abstract

Allergies are complex inflammatory diseases with a not fully understood etiology. Several factors, including genetic, environmental, age of exposition, diet, etc., are associated with the induction of these diseases. The incidence of allergies has increased during the last decades and constitutes the most common immune-based disease worldwide. According to the hygiene hypothesis, a lower exposure to pathogens and commensal microbes that reside in the intestinal lumen is responsible for the rapid rise of the prevalence of atopic and allergic disorders, specifically food allergy. To overcome this tendency, the immunological mechanisms underlying this pathology should be better understood, which will undoubtedly impact the development of novel therapies. A large body of evidence demonstrates that immunotherapies constitute corrective treatments of the impaired regulation of the immune system in allergic patients.

The aim of this chapter is to present an overview about allergic diseases and food allergies, mechanisms involved, differences with toxic reactions and food intolerances, a brief description of the main milk allergenic proteins, and the new therapeutic strategies derived from the old immunotherapies to restore oral tolerance, and finally, to describe different pathologies associated with milk allergy.

Keywords: milk allergens, food allergy, IgE, mucosal immunology, immunotherapy

1. Introduction to allergic diseases

1.1. Overview of allergic responses

The significant contribution of milk to the nutritional intake and health is well recognized. With a unique nutrient profile, it constitutes a relevant source of essential nutrients (minerals, vitamins, riboflavin, amino acids, proteins) and energy. This makes milk and dairy products essential components in the diet and is considered a protective food. For this reason, it has been included as an integral component of different programs in several countries. In the United States, there is a federal child nutrition program, which includes hundreds of millions of half-pints of milk. In particular, the importance of cow's milk proteins (CMP) resides in its significant amount, but also in its exceptionally high quality. Fluid milks contain approximately 3.5% proteins, being caseins 82% of the total milk proteins and whey proteins the remaining 18%. Regarding the high quality of these components, it provides all of the amino acids required by humans, in a distribution that resembles that needed by the human body [1]. In addition, milk contains other biological factors that are active in humans: growth factors (the human active insulin growth factor-1), cytokines (transforming growth factor-β), and immunoglobulins (IgM, IgG, and IgA).

Although milk is one of the most nutritious foods with health-promoting properties, in a restricted proportion of the population, it may have adverse effects. It has been described that dairy intake may increase the risk of prostate and ovarian cancer, type 1 diabetes, multiple sclerosis, acne, increased cholesterol levels, and probably atherosclerosis, lactose intolerance in a high proportion of individuals, and finally, is the most common food allergen in the world [2–5].

Cow's milk allergy (CMA) accounts for most diagnosed food allergies, mainly in the first year of life (6–8% at 1 year of age) [6], it is rare in adults (1–3%) and it can cause severe anaphylaxis [4, 7]. The natural history of food allergy indicates that it is lost over time. However, this process of outgrowing a food allergy does not mean a cure for the disease and is highly dependent on the food and individual susceptibility. For some foods, the intestinal mucosa "learns" how to manage dietary antigens during childhood, while for other foods sensitivity persists in adulthood.

Allergy is a chronic condition involving an abnormal reaction of the body to substances in the environment that are harmless for most people. These substances that act as immunogens are called *allergens* and were previously exposed to patients by inhalation, ingestion, injection, or skin contact. The immune system considers the allergen as a potential harmful antigen or threat, and triggers a misguided, inappropriate, and exaggerated reaction that comprises complex circuits of molecular and cellular components that promote inflammation. These mechanisms are called hypersensitivity mechanisms and any of the four types of hypersensitivity reactions described by Gell and Coombs [8] can be involved in an allergic disease. Since different organs can be targeted (nose, eyes, skin, lungs, gastrointestinal tract), allergic disorders are not a unique clinical entity. Allergy is actually a syndrome, with a spectrum of disorders with a heterogeneous and variable clinical presentation. It may include hay fever, food allergies, atopic dermatitis, allergic asthma, and anaphylaxis; and symptoms may include itchy eyes, runny nose, difficult breathing, congestion, wheezing, skin rash, eczema, urticaria, bronchoconstriction, or diarrhea. Most allergic reactions are mild to moderate and may cause irritation and discomfort. However, a small number of people may experience a life-threatening reaction called anaphylaxis, which is a severe condition that requires immediate life-saving medication (epinephrine).

Another term that should be defined is *atopy*. It is the genetic predisposition to develop allergic diseases and is mainly mediated by IgE-dependent or immediate Type I mechanisms. The allergic reaction mediated by IgE antibodies is called atopic allergy and patients are atopic. The role of the Type 2 T helper (Th2) cell-mediated immune response against innocuous environmental antigens in the immune pathogenesis of allergic atopy is well documented by an extended body of experimental evidences. The critical role of Th2 cytokines, such us interleukin (IL) 4, IL-5, IL-9, and IL-13, in the initiation, maintenance, and amplification of human allergic inflammation has been thoroughly characterized, while the tissue damage produced by the sustained inflammation is mainly due to eosinophils, neutrophil, etc (type IV hypersensitivity).

Allergens are innocuous, environmental, and widely distributed antigens that upon frequent exposure to genetically predisposed individuals (atopic subjects) can lead to immunologic sensitization (sensitization phase of the allergic response) (**Figure 1**) that involves the production of IgE antibodies by plasma cells. Upon re-exposure to allergens, immediate and delayed (late-phase) responses may occur and tissue inflammation arises (effector phase). If

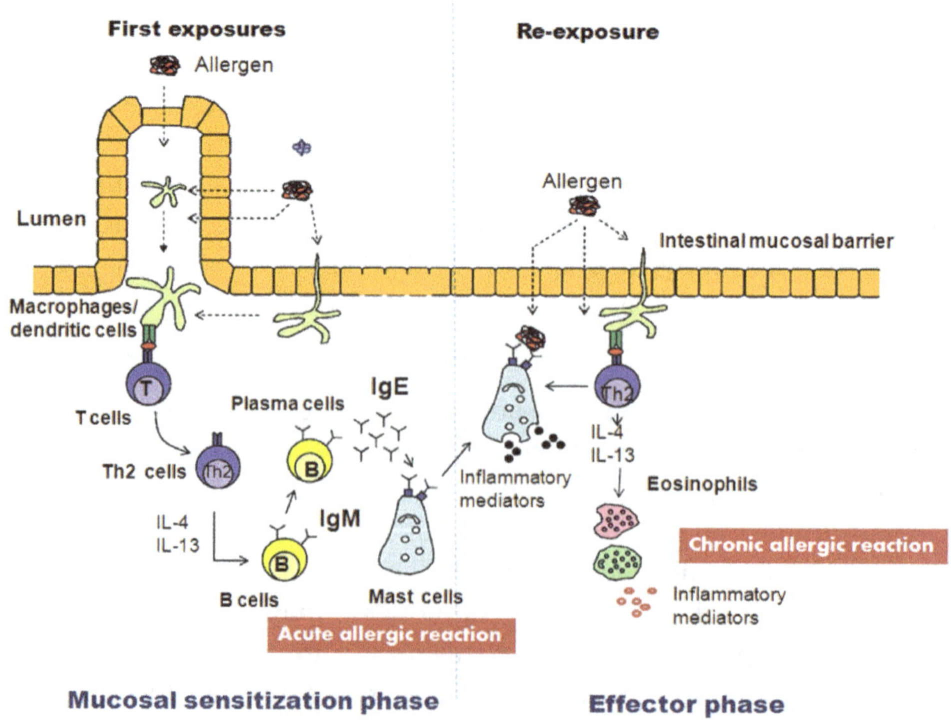

Figure 1. Mechanisms involved in an acute and chronic inflammation of an allergic reaction.

this mechanism is repeatedly triggered, different cells are attracted to the inflamed area and the condition may progress to a clinically detectable disease that is indicated by reversible or even irreversible tissue remodeling. In this state, a myriad of soluble mediators that are released by inflammatory cells under different specific or unspecific triggers (allergens, viruses, tobacco smoke, air pollutants, etc) exacerbate the severity of the disease.

During the sensitization phase, which occurs in the first exposures to environmental allergens, IgE antibodies are produced by plasma cells and secreted. Soluble antibodies are rapidly bound to high affinity surface receptors for IgE (FcεRI) on tissue mast cells, circulating basophils, and bone marrow-released eosinophils. Cells with surface-bound IgE antibodies are called sensitized cells and are prepared to rapidly respond to the specific antigen through the membrane-bound IgE antibodies. After activation, cells secrete numerous vasoactive and proinflammatory mediators, such as histamine, accumulated in cytoplasmic granules, and induce the synthesis of other pro-inflammatory mediators (leukotrienes, prostaglandins, growth factors, etc). Continuous re-exposure to allergen may occur, sensitized cells are activated, and these soluble mediators maintain tissue inflammation.

The term *allergen* refers to the immunogen (the substance that induce the immune response) or antigen (the substance that reacts with the induced immunological elements) involved in an allergenic reaction. Most of them are carbohydrates, proteins, and glycoproteins with common sequential, conformational, structural, and evolutionary features [9]. Most of the allergenic proteins are concentrated in 70 protein families, out of the 10,000 described (2–5% of all known structural protein families), and food allergens (500 allergens) are distributed in approximately 20 families and 4 super-families (prolamins, prophilins, cupins, and Bet v1 homologues).

The most common natural sources of allergens are comprised in household dust mites, pollens, animal dander, insect stings, moulds, some drugs (antibiotics), and certain foods. Despite the high number of food allergens described, a relatively small number of allergens cause a high proportion of food allergies. More than 170 foods have been reported to be allergenic. However, allergy to certain foods appears to be especially common. The "Big Eight" is referred to the eight foods (milk, egg, soy, fish, shellfish, peanuts, wheat, and tree nuts) that produce the 80–90% of food allergies, milk being the most allergenic food worldwide. In order of prevalence, and depending on the population studied, the most common food allergens are milk, egg, peanut, tree nuts, crustaceans, shellfish, fish, wheat, and soy.

1.2. Definition of the different adverse reactions caused by foods

To unify the terms used by the medical community and the common people, which is sometimes influenced by social and cultural perceptions creating confusion, and to avoid misdiagnosing, it is useful to define the different conditions that involve the exposure to foods and may produce symptoms (**Figure 2**). There are still too many situations where people, and even certain professionals, do not recognize that food allergy is a medical condition, not a food preference. The clinician should be aware that adverse reactions to food can be categorized based on whether the immune system is involved or not.

The food-related *adverse reactions* comprise any abnormal reaction that is produced by a food or food component. They include both toxic and non-toxic reactions. The former are only dependent on the food and maybe caused by the presence of histamine (chocolate, strawberry, etc), histamine-releasing factors (pineapple), or most commonly, contaminated foods (bacteria toxins from *Staphylococcus, Clostridium*, etc.). The latter is dependent on individual susceptibility and involves *food intolerances*, which are independent of immune activation, and the immune-mediated *food allergy*. In addition, allergic reactions to food components can be classified as IgE-mediated (immediate gastrointestinal hypersensitivity, eosinophilic esophagitis, anaphylaxis, etc) or non-IgE-mediated (food-protein-induced enteropathy, allergic colitis, and protocolitis, etc.). Therefore, food intolerances (lactose intolerance, anatomical abnormalities of the gastrointestinal tract, etc.) and food poisoning are separate conditions, they are not food allergies. Similar clinical reactions can occur to some chemicals and food additives (urticaria); however, if they do not involve the immune system, they are known as adverse reactions rather than allergy (toxic adverse reactions).

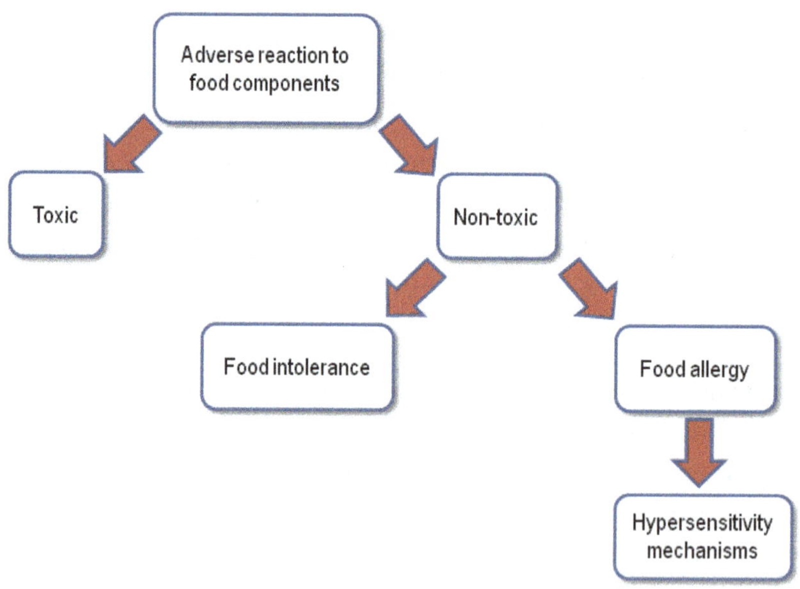

Figure 2. Classification of adverse reactions to food components.

In conclusion, *food allergy* is defined as an adverse reaction elicited on exposure to a given food and is mediated by a specific immune response that occurs reproducibly on time, and might be either IgE-mediated and/or non-IgE-mediated.

The literature reflects that it is very common the overdiagnosis of food allergies due to the difference between food allergy and food intolerance, which is sometimes confusing [10, 11]. For diagnosis, once an adverse reaction is suspected, it should be demonstrated that the immune system is involved and the offending food should be identified to avoid unnecessary restriction diets that may affect growth, nutritional deficiencies, or impair the quality of life. Many children are unnecessarily placed on restrictive food diets on the basis of serum food-specific IgE testing or skin-prick testing, which are the most common complementary assays to clinical history. The presence of specific IgE does not necessarily imply a clinical allergy [11]. Therefore, the gold standard assay for confirmation of food allergy, the oral food challenge, should be done. Nevertheless, in most places, it is complicated and rather impracticable to perform as a standardized challenge, and hence food allergy is commonly not confirmed, and overdiagnosis is possible.

1.3. Main causes of allergic diseases

It is widely accepted that although allergies can develop at any age, the risk of developing allergies is genetically determined. If neither parent is allergic, the chance for allergies is about 15%. If one parent is allergic, the risk increases to 30%, and if both are allergic, the risk is greater than 60%. Several polymorphic genes in almost all chromosomes have been identified in different allergic populations [12, 13]; however, all of them only determine susceptibility. Nevertheless, the expression of the allergic phenotype is dependent on the gene-environment interaction. A large body of experimental evidences has identified during the last 5–6 decades environmental factors as the main inducers of allergy in susceptible individuals. This makes allergy inheritance a complex polygenic disorder and is considered a multifactorial disease.

There are clear evidences that in Western societies, these disorders are increasing in frequency, with reports showing that prevalence has doubled over 10-year periods. It was initially seen in UK, other countries of Europe, and USA, that 30–35% of people are affected of asthma and rhinitis. But more recently, food allergy has emerged over the last 10–15 years as a "second wave" of the allergy epidemic, affecting 10% of infants in Australia. As changes in genes take many hundreds of years in humans, the genetic basis alone cannot account for the current increase in allergy seen over the past decades. New studies suggest that environmental factors are responsible for these changes. Epidemiological studies have clearly shown that modification of the pattern of microbial exposure of children represents a key factor to understand the changes in severity and prevalence of atopic diseases. The hygiene hypothesis, formerly expressed by Strachan [14] and later modified by Bach [15], is nowadays the most reasonable explanation for the striking increase observed from the twentieth century in the incidence of many chronic inflammatory disorders, including allergies and autoimmune disorders. The modern urbanization and lifestyle have generated immunoregulatory problems attributable to depletion of exposure to organisms that are present in the urban environment with which mammals co-evolved and shaped the immune system. The immune system needs to come into contact with a variety of microorganisms, while it is developing at the infant stage, in order that it responds appropriately later in life. We now live in an environment with less contact with pathogens and commensal microorganisms since we use cleaners containing anti-microbial agents, antibiotics, better vaccines, and compulsive vaccination programs, more hygienic food preparations, etc. While children living in farms were directly exposed to animals, and their environment contained a range of microbial agents and plant-derived agents, most of people in westernized societies now live in cities with minimal exposure to animals. The literature reflects that people with a rural lifestyle have a lower incidence of allergy. Inadequate exposure to environmental microorganisms may therefore result in the immune system of atopic children developing a tendency toward allergy. The development of a new generation of antibiotics and vaccines have protected people from many infections that previously killed large numbers of the population and have proven to be essential to protect the health of the population. However, the cost of this may be the reduction in the pressure placed on the immune system to mount a strong immune response against infectious agents. This may be one of the factors to explain why the immune system in a restricted, but increasing, percentage of the population is now reacting to innocuous allergens. In other words, the initial interpretation by Strachan [14] proposed a missing immune deviation of allergen-specific responses from Th2 to Th1 immune profile, mainly as consequence of the reduced production of IL-12, interferon-α, and interferon-γ by innate cells (mainly dendritic cells, NK cells, and innate lymphocytes), which are stimulated by prokaryotic products. The reduced contact with pathogens that prime Th1 responses in early life can result in a stronger and compensatory induction of the contra-regulatory Th2 response. Notwithstanding, this postulate could not explain the contemporary close rise in the prevalence observed for the Th1-mediated autoimmune diseases (type 1 diabetes, multiple sclerosis, and inflammatory bowel diseases) (**Figure 3B**) in developed countries, and the epidemiological observations reporting a low prevalence of allergy in geographical regions characterized by chronic helminth infections (stronger inducers of Th2 as well as suppressive cytokines). Therefore, an alternative view has emerged, which suggests the importance of a reduced immune suppression rather than missing immune deviation. The modified and unifying hypothesis later suggested by Bach postulated that a lower microbial burden may favor increased prevalence of allergy by inducing a lower activity of regulatory cells (dendritic and T cells). Epidemiological and experimental findings suggest that both mechanisms, missing immune deviation, and reduced immune suppression, may be independently involved in allergic and autoimmune disorders [16], and both, allergic diseases and autoimmunity, can independently and simultaneously increase. Therefore, the expanded hygiene hypothesis may provide a better understanding of these epidemiological changes observed in immune-mediated inflammatory disorders.

Since it has recently been demonstrated the importance of the intestinal microbiota in shaping the immune system, changes in the composition of the microbiota is suggested as a critical factor to generate impaired mucosal and systemic regulatory circuits. The gut is always in a state of controlled inflammation, and regulatory cells are abundant in this tissue. Defects in Tregs promoted by the "unhealthy microbiota" or "bad microbiota" undoubtedly impact in the regulation of several immune mechanisms, including those that protect from allergy and autoimmune diseases. In conclusion, the modern hygiene hypothesis postulates that the restricted exposure to pathogenic and commensal microorganisms may be implicated in the increased observed in several immunological disorders.

This hypothesis not only provides the theoretical framework to explain the rise in the prevalence of immune disorders but also has therapeutic implications, as it will be discussed later in this chapter. Allergen-specific immunotherapy has proven to be the unique disease-modifying therapy for allergy. It has been demonstrated in treated patient the suppressive role for the therapy-induced Tregs and secreted IL-10.

2. Allergenic proteins

2.1. Milk allergens

2.1.1. Allergens of cow's milk

Acute reactions in CMA patients can be caused by several proteins in cow's milk. Cow's milk contains approximately 30–35 g/L of CMP. Milk protein fractions can easily be obtained using either chymosin (rennin) or acid precipitation (pH 4.6). The insoluble coagulum containing the whole casein fraction constitutes approximately 82% of the CMP, while lactoserum or whey proteins, which comprise approximately the remaining 18% of the CMP, constitute the soluble fraction. Both fractions contain allergens that could elicit allergic symptoms in susceptible individuals. The casein fraction or Bos d 8 (from *Bos domesticus*) that contains αS1-, αS2-, β-, and κ-casein (32%, 10%, 28%, and 10% of total CMP, respectively) constitutes the major allergens of milk. The whey major allergens are α-lactalbumin (Bos d 4) and β-lactoglobulin (Bos d 5) (5% and 10% of total CMP, respectively), while less common allergens are serum albumin (Bos d 6) and immunoglobulins (Bos d 7) [17, 18].

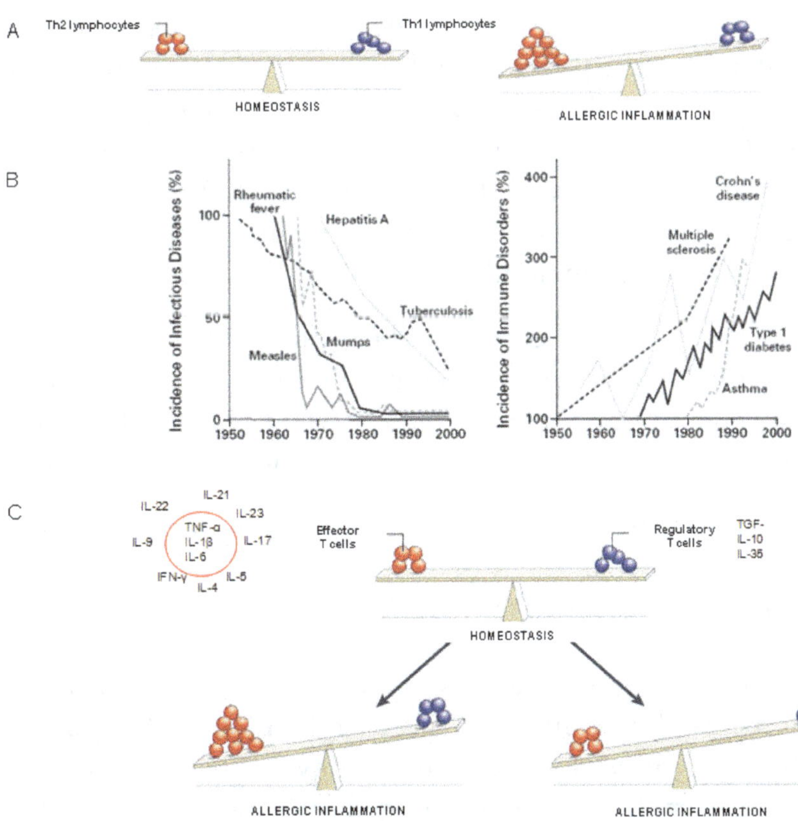

Figure 3. The hygiene hypothesis originally defined by Strachan and re-formulated by Bach. *A*, Strachan postulated that the balance between Th1 and Th2 cells is impaired in allergy. *B*, Infectious diseases were controlled in the last decades while Th1- and Th2-immune-mediated diseases increased in prevalence; C, Bach postulated that an impaired balance between Treg and effector T cells promotes immune-mediated inflammatory disorders. Part of figure was taken with the permission from Bach [15], NEJM 2002.

2.1.1.1. Caseins

The allergen Bos d 8 refers to a heterogeneous mixture of proteins. The casein fraction of milk proteins contains components that belong to two unrelated protein families, one family comprising αS1-, αS2-, and β-caseins (the αs2-casein gene is not expressed in humans), while κ-casein constitutes the other family. Even within the α-/β-casein family, sequence identities are below 15%. Hence, the entry Bos d 8 was demerged into four separate allergens: Bos d 9.0101 (αS1-casein), Bos d 10.0101 (αS2-casein), Bos d 11.0101 (β-casein), and Bos d 12.0101 (κ-casein) in a proportion of 37/37/13/13%, respectively. The name Bos d 8, which is widely established and has been used in numerous publications and names of commercial diagnostic tests, was kept and designates the whole casein fraction.

There is considerable similarity in the caseins from different mammalian milks used for human consumption, which explains their IgE cross-reactivity [19]. The analysis of sequences showed that bovine caseins had more than 80% of sequence homology with goat and sheep caseins, while bovine β-casein has 50% of sequence homology with human β-casein.

2.1.1.2. Whey proteins

Whey contains essentially globular proteins. The major allergens of lactoserum are β-lactoglobulin and α-lactalbumin. The β-lactoglobulin (Bos d 5) is the only lipocalin that acts as a food allergen and occurs naturally as a dimer. It has no homologous counterpart in human milk, which explains its high immunogenicity. The relative resistance of β-lactoglobulin to acid hydrolysis and gut proteases determines that this protein can be absorbed as a native antigen in the intestinal mucosa. Lipocalins have a high allergenic potential, and several allergens of animal origin belong to this protein family.

Alpha-lactalbumin (Bos d 4), the other major whey protein allergen, is a monomeric protein stabilized by four disulfide bonds with a calcium-binding domain. The complete amino acid sequence of bovine α-lactalbumin shows extensive homology with hen's egg white lysozyme but also with human α-lactalbumin; however, it has been described as a milk allergen.

Other minor whey allergens are bovine serum albumin, immunoglobulins, and lactoferrin.

2.1.2. Milk allergens from other mammals

Since cow's milk is the most consumed milk, the incidence of CMA is higher than for other milk sources. Nevertheless, goat's milk and sheep's milk represent important sources of milk allergens. Due to the high degree of sequence homology and hence cross-reactivity with cow's milk, goat's and sheep's milk are potent allergens for most cow's milk allergic patients. Notwithstanding, and for unknown reasons, some patients with CMA can tolerate these alternative milks as dairy substitutes during treatment. As it can be observed in **Table 1**, bovine casein sequences showed more than 80% of sequence homology with goat and sheep caseins, and the major whey allergens showed a sequence homology higher than 90% [20]. In addition, milks of other mammals such as horse, donkey, or camel are also used as dairy substitutes.

Protein	Concentration g/L	Allergen name	Molecular weight kDa	pI	Sequence homology % amino acid identity
Whole casein fraction (80%)	29.5	Bos d 8			αS1 and αS2-CAS from cow milk: 22.5%αS1-CAS from cow milk and sheep/goat milk: 87/89%αS2-CAS from cow milk and sheep/goat milk: 87/89%
αS1-casein	12–15	Bos d 9.0101	32.4	4.9–5	
αS2-casein	3–4	Bos d 10.0101		5.2–5.4	
β-casein	9–11	Bos d 11.0101	26.6	5.1–5.4	β-CAS from cow milk and sheep/goat milk: 91%β-CAS from cow milk and human milk: 50%
κ-casein	3–4	Bos d 12.0101	19	5.4–5.6	κ-CAS from cow milk and sheep/goat milk: 84%
Whey fraction (20%)	6.3				
α-lactalbumin	1–1.5	Bos d 4	14.2	4.8	α-LA from cow milk and water buffalo: 99% α-LA from cow milk and sheep/goat milk: 94%/95% α-LA from cow milk and human milk: 78%
β-lactoglobulin	3–4	Bos d 5	18	5.3	β-LG from cow milk and water buffalo/mouflon: 98%/95% β-LG from cow milk and sheep/goat milk: 93%/94% cockroach allergen Bla g 4: about 20%
Bovine serum albumin	0.1–0.4	Bos d 6	67.0	4.7–4.95	BSA from sheep milk: 92% BSA from cow milk and pig, cat, human, rhesus macaque, horse milk: between 74 and 79%
Lactoferrin	0.09	Lactoferrin	80	8.7	Lactoferrin from cow milk and human: 69%
Immunoglobulins	0.6–1	Bos d 7	150	–	–

Table 1. Main characteristics of the major cow's milk proteins and sequence homology with other mammal milks.

2.2. Methods employed for characterization and prediction of allergenicity

2.2.1. Methods used for detection of allergens

Different methods can be used to study the allergenicity of milk proteins. Polypeptide masses usually range between 5 and 70 kDa; however, many allergens are oligomers with molecular masses greater than 200 kDa. Allergenicity assessment is generally performed using human IgE-containing sera, polyclonal antisera (rabbit, rat, mouse, goat, sheep, camelid, chicken, etc), or experimental animals (pig, mice, rabbits, etc). Protein-based methods employed to identify and characterize allergens usually involve immunochemical detection protocols such as the radio-allergosorbent test (RAST), enzyme allergosorbent test (EAST), dot blot, immunoblotting, and enzyme-linked immunosorbent assay (ELISA). These tests can be developed as qualitative, semi-quantitative, or quantitative assays using polyclonal antisera, human sera, or monoclonal antibodies. Furthermore, the ELISA technique is the most widely used assay to routinely detect and quantify allergens in foods due to its high precision, simple handling, and potential for standardization. Another emerging technology applied for allergens analysis is the use of biosensors. Biosensor instruments, based on the interaction between allergens and specific antibodies, can monitor the presence of a protein in real time. The immobilization of an antibody to a sensor chip surface allows the on-line recognition of a protein based on surface plasmon resonance. It has been applied for the detection of a few potentially allergenic foods like peanut hazelnut, egg, and milk in the food processing industry.

In addition, cell-based assays are used for characterization or quantification of allergens. The basophil activation and histamine release test can be used to confirm the allergenicity of a protein (two surface IgE epitopes on the allergen are needed to activate basophils). Sensitized basophils are used and cell activation can be assessed by ELISA (detection of histamine released in the supernatant), flow cytometry (translocation of CD63 and CD203c from the membrane of cytoplasmic granules to the plasmatic membrane), or electrochemically (modification of impedance in rat basophilic leukemia RBL-2H3 mast cells transfected with FCεRI upon interaction of IgE with the allergen).

In order to characterize the presence of epitopes and further investigate its antigenicity (B epitopes) or allergenicity (IgE epitopes), epitope mapping is performed. This point is critical for developing protein- or peptide-based immunotherapies. Linear or conformational, B or T epitopes can be assessed using different strategies, and synthetic or recombinant peptides are commonly used. However, the main drawback in the development of these techniques is often the high cost and time-consuming. For this reason, functional proteomics or immunoproteomics constitute nowadays a high-throughput technology to identify epitopes. These methods employ mass spectrometry (MS) to study the interactions of peptides with the major histocompatibility complex or antibodies. Given the high sensitivity and high accuracy in determining molecular mass and high capacity for analysis, MS techniques currently figure prominently in identifying B-cell or T-cell epitope targets for vaccine and immunotherapy development [21].

Furthermore, epitope mapping is important in diagnosis (the so called "component-resolved diagnosis," which is based on individual IgE pattern reactivity according to recombinant allergen recognition) and treatment (taylor-made immunotherapy and vaccine development).

2.2.2. Methods used for prediction of allergenicity

In addition to the different immunochemical techniques employed to assess the allergenicity of new proteins or to predict cross-reactive allergens, several online databases and computational methods have been developed to effectively determine potential allergens. Bioinformatics comparison of food proteins provides a mechanism to identify proteins that can lead an increased risk of food allergic reactions. The use of bioinformatics tools may help to identify new allergenic or similar proteins to an allergen that might induce cross-allergic reactions. This point will undoubtedly impact on allergy diagnosis, prognosis of potential reactions, and therapy. Several allergen databases and allergen prediction web tools are available on the Internet and can be easily operated by common users: IUIS Allergen Nomenclature Allergen, Database for Food Safety, Allergome, The Immune Epitope Database (IEDB), Structural Database of Allergenic Proteins (SDAP), Allermatch, AllerTool and AllergenPro, and others.

3. Clinical implications

3.1. Epidemiology and triggering factors

The perception of milk allergy is far more frequent than confirmed CMA, due to the few surveys of secular and geographical trends in food allergy in adults and children. Reports of CMA prevalence range between 1 and 17.5% among preschoolers, between 1 and 13.5 % at 5–16 years of age, and among 1–4% of adults [22, 23]. Clinical symptoms of CMA commonly appear during the first months of life, usually within days or weeks after feeding with CM-based formulas have been started or may sometimes be seen in exclusively breast-fed infants. With such an early age of onset, symptoms of an erythematous rash or hives shortly after intake of CM formula are suggestive of food allergy. Urticaria, exanthema, or both are more common in IgE-mediated CMA, whereas children with non-IgE-mediated CMA more often have atopic eczema and diarrhea. Tolerance to CM develops in children with non-IgE-mediated CMA significantly earlier than in the IgE-positive group (5 years old vs. 8 years old or more). At age of 8 years, children with IgE-positive CMA are more frequently sensitized to birch pollen, animal dander, and foods than those with IgE-negative CMA. Furthermore, it has been reported that CMA increased the risk for sensitization to inhalant allergens [24, 25].

Patients with IgE-mediated CMA develop gastrointestinal symptoms in 32–60% of cases, skin symptoms in 5–90%, and anaphylaxis in 0.8–9 % of cases. This frequency of anaphylaxis is the main concern pointed out in many CMA studies [22].

A variety of risk factors are proposed to influence food allergy or sensitization: sex (male sex in children), race/ethnicity (increased among Asian and black children compared with white children), genetics (familial associations, HLA, and specific genes), atopy (co-morbid atopic dermatitis), vitamin D insufficiency, dietary fat (reduced consumption of omega-3-polyunsaturated fatty acids), reduced consumption of antioxidants, increased use of antacids (reducing digestion of allergens), breastfeeding, obesity (an inflammatory state), increased hygiene, and the timing and route of exposure for foods. Although it is not clear which factors are key in the increase in the incidence of food allergies observed in the last decades, the aforementioned hygiene hypothesis provides a convincing interpretation. Nevertheless, other non-considered factors should be taken into consideration. Changes in the diet might have a causal link with allergy development. A decreased intake of fruits and vegetables, changes in the type of fat included in the diet, the timing of feeding events during infancy, nutrients and micronutrients in the diet, such as long-chain polyunsaturated fatty acids, vitamin D, folic acid, etc, may likely affect food allergy development. However, a number of nutritional and dietary variables might be interfering food allergy manifestation in infants, instead of considering only one nutrient or dietary characteristic, which might be an oversimplification of the complex interactions taking place.

Although several theories have been proposed to explain the protective effect of breastfeeding, pros and cons are still controversial. It has been proposed that the effect of exclusive breastfeeding delays the introduction of cow's milk, thus preventing early sensitization. Nevertheless, it contradicts the fact that the administration of very small amounts of the allergen benefits tolerance induction and allergy control. Therefore, it is considered that the intake of CMP in the mother's diet during lactation could be relevant. It has been widely probed that mother's dietary native antigens are present in breast milk. Furthermore, and not less important, immunomodulatory components are present in breast milk, such as IgA, IgG, TGF-β, macrophages, and dendritic cells.

The most striking component that may be affected by lifestyle and diet is the composition of the microbiota. It has been demonstrated that relatively harmless microorganisms (helminths, saprophytic mycobacteria, lactobacilli, etc) that have been present throughout mammalian evolution can drive maturation of regulatory immune cells and prime immunoregulation by release of IL-10 and TGF-β. Developed countries with modern lifestyle have a diminished contact with these microbes, while they were kept intact in rural life. Additionally, it has been demonstrated that allergic patients have less frequency of lactobacilli and increased coliforms in the microbiota compared with microbiota of healthy individuals [26]. This means that a modification on the microbial community may induce a change in the background of the bystander suppression that is continuously induced in the gut. Compelling data in atopic individuals supports that different factors may induce a dysbiosis and be detrimental for the host [27].

In conclusion, a suboptimal microbiome depleted of beneficial bacteria can alter the host homeostasis in the gut, thereby increasing an aberrant immune response to innocuous allergens and autoantigens.

The evidence that link the composition of the microbioma and the immune system provides new insights into the causes of the increase in several immunopathologies and suggests potential new therapeutic targets.

3.2. Current treatments

CMA is a heterogeneous disorder with no single immunologic mechanism involved. Once food allergy to CMP is suspected, avoidance of the allergen is the only available treatment for infants. For 6-month-old or younger babies, milk should be substituted and the recommended formulae are currently extensively hydrolyzed proteins or amino acid-based formulae as the best choice for infants with high-risk of anaphylaxis [28]. As mentioned before, milks of alternative sources are also employed, with different consequences regarding its clinical tolerance. Mammalian milks, rice-, and soy-based formulae are the most used alternatives. As previously mentioned, most of the patients do not tolerate animal milks, although, and, for not fully understood reasons, a restricted proportion of patients can perfectly tolerate goat's milk and sheep's milk. Soy-based formulae are also frequently used in children older than 6 months old [29]. Nonetheless, soy intolerance is observed in some IgE-mediated or non-IgE-mediated milk allergic patients. This phenomenon can be explained by either co-sensitization or cross-allergenicity between soy and milk allergens. We have used a food allergy mouse model to CMP and characterized several soy allergens that cross-react with caseins. We identified B and T epitopes shared between the main soy allergens and bovine caseins [30–32]. Again, only a restricted proportion of milk allergic patients react against soy proteins during restriction diet. It probably depends on the individual pattern of reactivity, affinity of IgE antibodies, and on some unknown factor [33]. Therefore, this point adds a new restriction to the choice of a dairy substitute. In addition, the nutritional inadequacy of soy-based formulae makes them not recommendable for patients younger than 6 months old.

Therefore, even though a restriction diet seems to be the easiest and most efficient treatment for milk allergic patents, several drawbacks arise. Finally, patient and family education is crucial to prevent accidental reactions

3.3. Adverse reactions and related causes

Approximately 50% of patients with food allergy suffer accidental reactions upon contact with food and non-food containing the allergenic component during the restriction diet. This can be explained by several factors: contamination of food with the allergen (utensils at restaurant, cross-contamination during the food manufacturing or processing, during cooking, etc.), ignorance of the composition of the food (caseinate means the presence of milk proteins), mislabeling of food composition (omission), presence of cross-reactive components (soy, peanut, fish, nuts, etc), etc. Most of these accidental reactions are often severe and life-threatening for patients, who should be aware of self-use of pre-loaded pen injections (epinephrine) [34, 35]. In addition, and to make the situation even more complicated, there is an increasing tendency of CMA to persist with time.

Therefore, the current standard of care for the management of IgE-mediated food allergy involves the identification of causative foods and avoidance of the allergens while always having self-injectable epinephrine available [36]. The avoidance of the immunologic stimulus seems to be efficient; however, it is not a corrective procedure of the impaired immune response and accidental reactions are very common for patients. Considering all of these complications, there is a need in the medical community to develop disease-modifying treatments.

3.4. New insights on immunotherapy and relevance of mouse models for experimental allergy

Immunotherapy is nowadays a promising treatment for food allergies. It has been successfully used in different immunopathologies (cancer, autoimmunity, etc) to induce, enhance, or suppress an immune response. In allergy, it is nowadays accepted that it has the potential for disease modification [37]. Although allergen-specific immunotherapy has been used for the treatment of IgE-mediated allergy longer than a hundred years, the first randomized clinical trial of oral immunotherapy for food allergy was done in 2008 [38]. The controlled and stepwise administration of milk to patients with IgE-mediated CMA rendered a clinical tolerance to 200-fold higher amounts of milk than placebo-treated patients. Nevertheless, the development of long-term tolerance was unlikely. It should be mentioned that all clinical trials report the presence of adverse reactions during immunotherapy. For this reason, there are no approved therapies for food allergy in the clinical practice, and still remains as an experimental therapy.

Tolerance is the state in which a person can consume a food without any allergic symptom in weeks, months, or years after cessation of regular and therapeutic exposure to the food antigen. In this state, long-term clinical unresponsiveness is maintained [39, 40]. This mechanism contrasts with desensitization, which depends on the regular ingestion or exposure to the food allergen, once tolerance has been achieved, to sustain unresponsiveness. The immunologic mechanisms underlying the development of tolerance are yet not fully understood, but regulatory T cells are likely involved.

In recent years, the controlled oral or sublingual administration of the allergen has gained increasing attention. It has been demonstrated that a substantial number of allergic patients can tolerate gradually increasing amounts of the food allergen, which can probably assure that patients are protected against accidental natural exposure to the allergen-containing food. However, most of these clinical trials showed limitations [41, 42].

The development of animal models for food allergy holds great potential as powerful biological tools to investigate the underlying mechanisms involved in the allergic pathway and for developing and testing novel treatments to restore tolerance. Due to the ethical concerns and the chance of fatal anaphylactic reactions in humans, great interest has arisen in the use of animal models that resemble the pathology in man [43].

Mice are the predominant laboratory animals used to study many diseases for several reasons. However, the main limitation is that allergy is not spontaneously elicited in mice, due to its genetic background. Therefore, it should be artificially induced using pro-Th2 adjuvants (hydroxide aluminum, cholera toxin, Staphylococcus enterotoxin B, etc.)[44–46]. Since the first models developed, there has been significant progress in their optimization. The use of mucosal adjuvants, such as cholera toxin, promotes the induction of the inflammatory process in the intestinal mucosa. These models closely resemble the pathology in man. For this reason, food allergy mouse models have become a useful tool as a pre-clinical assay.

The characterization of new mucosal adjuvant (molecular patterns from microorganisms) and new routes of administration (oral, nasal, or sublingual) will undoubtedly impact in the development of novel disease-modifying therapies using the offending allergen in a controlled and step-wise administration strategy. Immunomodulatory and tolerogenic immunotherapies that restore the immunoregulatory networks are promising treatments for food allergy.

3.5. Other pathologies associated to food allergies

Many studies in the last decades linked the exposure to foods with disorders non-related with food allergy and might be considered food intolerances. The appearance of some disturbances in the central nervous system such as migraine, epilepsy, and hyperkinetic syndrome has been reported, or schizophrenia in patients undergoing a gluten-free diet. However, in this case, genetic and immunological components might be involved in symptoms associated with psychiatric disturbances [47]. It has been hypothesized that defects in the intestinal barrier allow the passage of neuroactive peptides of food origin and interfere with the central nervous system. This hypothesis has been proven in CMA children, in whom a slight improvement in autistic symptoms was achieved after milk restriction diet. This observation, along with the presence of high levels of IgA antibodies specific for the incriminated foods, was interpreted as a consequence of the increased absorption of protein fragments by the intestinal mucosa due to a peptidase defect detected in autistic patients. High prevalence of immune-mediated conditions, including asthma, allergic rhinitis, atopic dermatitis, urticaria, type 1 diabetes, and inflammatory bowel disease, has been reported in children with autism spectrum disorders [48–51]. In addition, peptides derived from casein gastrointestinal digestion may have an opiate-like effect on brain cells. It has been demonstrated in experimental animals that the slow digestion of caseins promotes natural morphine-like substances known as casomorphins. These peptides enter the bloodstream and reach the brain where they act through specific neuronal receptors and cause addictions to dairy products.

4. Conclusions

Milk and dairy products are excellent sources of high-quality proteins, peptides, and amino acids with a range of potential health benefits and functional properties. Hence, they have attracted the interest of researchers and the food industry. Since no severe detrimental properties have been described in its components, symptoms attributed to milk intake are mainly associated to individual susceptibility. These reactions are sorted into food intolerances and food allergies, being the latter immune-mediated disorders which can be caused by any of the four types of hypersensitivity mechanisms. Strikingly, these inflammatory diseases have shown a rise in the last decades which was attributable to lifestyle changes in modern societies of developed countries. Nevertheless, the controlled administration of the "offending" proteins to food allergic patients can reverse the impaired immune response. Nowadays, immunotherapy is the only disease-modifying treatment for allergic patients.

In conclusion, the evolutive and traditional culture of milk consumption by man through thousands of years indicates its importance as a natural nutrient.

Author details

Guillermo Docena*, Paola Smaldini, Renata Curciarello and Angela María Candreva

*Address all correspondence to: guidoc@biol.unlp.edu.ar

Institute of Immunologic and Physiopathologic Studies - IIFP, School of Sciences, The National University of La Plata, La Plata, Argentina

References

[1] Hambraeus L. The significance of mother's milk and breast-feeding for development and later life. Bibliotheca Nutrion et Dieta. 1982;31:1–16.

[2] Dahl-Jorgensen K, Joner G, Hanssen K. Relationship between cows' milk consumption and incidence of IDDM in childhood. Diabetes Care. 1991;14(11):1081–1083.

[3] Malosse D, Perron H, Sasco A, Seigneurin J. Correlation between milk and dairy product consumption and multiple sclerosis prevalence: a worldwide study. Neuroepidemiology. 1992;11(4):304–312.

[4] Rona RJ, Keil T, Summers C, Gislason D, Zuidmeer L, Sodergren E, et al. The prevalence of food allergy: a meta-analysis. The Journal of Allergy and Clinical Immunology. 2007;120(3):638–646.

[5] Lanou AJ. Should dairy be recommended as part of a healthy vegetarian diet? The American Journal of Clinical Nutrition. 2009;89:1638–1642.

[6] Fiocchi A, Brozek J, Schünemann H, Bahna SL, von Berg A, Beyer K, et al. World Allergy Organization (WAO) Diagnosis and Rationale for Action against Cow's Milk Allergy (DRACMA) guidelines. Pediatric Allergy and Immunology: Official Publication of the European Society of Pediatric Allergy and Immunology. 2010;21 Suppl 2:1–125.

[7] Ludman S, Shah N. Managing cow's milk allergy in children. BMJ 2013;5424. 1–8

[8] Gell PGH, Coombs RRA. Clinical Aspects of Immunology. 1st ed. Oxford, England: Blackwell; 1963.

[9] Fitzsimmons CM, Dunne DW. Survival of the fittest: allergology or parasitology? Trends in Parasitology. 2009;25(10):447–451.

[10] Scheider Chafen J, Newberry SJ, Riedl MA, Bravata DM, Maglione M, Suttorp MJ, et al. Diagnosing and managing common food allergies a systematic review. The Journal of American Medical Association. 2015;303(18):1848–1856.

[11] Fleischer DM, Bock SA, Spears GC, Wilson CG, Miyazawa NK, Gleason MC, et al. Oral food challenges in children with a diagnosis of food allergy. The Journal of Pediatrics. 2011;158(4):578–583.e1.

[12] Amoli MM, Hand S, Hajeer AH, Jones KP, Rolf S, Sting C, et al. Polymorphism in the STAT6 gene encodes risk for nut allergy. Genes and Immunity. 2002;3(4):220–224.

[13] Wang D. Risk factors of allergic rhinitis: genetic or environmental? Therapeutics and Clinical Risk Management. 2005;1(2):115–123.

[14] Strachan DP. Hay fever, hygiene, and household size. BMJ (Clinical Research Ed.) [Internet]. 1989;299(6710):1259–1260.

[15] Bach J-F. The effect of infections on susceptibility to autoimmune and allergic diseases. The New England Journal of Medicine. 2002;347(12):911–920.

[16] Romagnani S. The increased prevalence of allergy and the hygiene hypothesis: missing immune deviation, reduced immune suppression, or both? Immunology. 2004;112(3): 352–363.

[17] Wal J-M. Bovine milk allergenicity. Annals of Allergy, Asthma & Immunology 2004;93(5):S2–S11.

[18] James J, Burks W, Eigenmann P. Food Allergy. British Elsevier Inc.; 2012. 1-309 p. ISBN-13: 9781437719925

[19] Docena G, Rozenfeld P, Fernández R, Fossati CA. Evaluation of the residual antigenicity and allergenicity of cow's milk substitutes by in vitro tests. Allergy. 2002;57(2): 83–91.

[20] Genève PE. Cow's Milk. Internet Symposium on Food allergens 2:2000.http:// www.food-allergens.de

[21] Poms RE, Klein CL, Anklam E. Methods for allergen analysis in food: a review. Food Additives and Contaminants. 2004.21:1–31.

[22] Steinke M, Fiocchi A, Kirchlechner V, Ballmer-Weber B, Brockow K, Hischenhuber C, et al. Perceived food allergy in children in 10 European nations. A randomised telephone survey. International Archives of Allergy and Immunology. 2007;143(4):290–295.

[23] Kilgallen I, Gibney MJ. Parental perception of food allergy or intolerance in children under 4 years of age. Journal of Human Nutrition and Dietetics. 1996;9:473–478.

[24] Saarinen KM, Pelkonen AS, Mäkelä MJ, Savilahti E. Clinical course and prognosis of cow's milk allergy are dependent on milk-specific IgE status. The Journal of Allergy and Clinical Immunology. 2005;116(4):869–875.

[25] Kattan JD, Cocco RR, Järvinen KM. Milk and soy allergy. Pediatric Clinics of North America. 2011;58(2):407–426.

[26] Björkstén B, Naaber P, Sepp E, Mikelsaar M, Bjo B. The intestinal microflora in allergic Estonian and Swedish. Clinical & Experimental Allergy. 1999;29:342–346.

[27] Feehley T, Stefka AT, Cao S, Nagler CR. Microbial regulation of allergic responses to food. Seminar in Immunopathology. 2013;34(5):1–27.

[28] Koletzko S, Niggemann B, Arato A, Dias J, Heuschkel R, Husby S, et al. Diagnostic approach and management of cow's-milk protein allergy in infants and children: ESPGHAN GI Committee practical guidelines. Journal of Pediatric Gastroenterology and Nutrition. 2012;55(2):221–229.

[29] Niggemann B, Binder C, Dupont C, Hadji S, Arvola T, Isolauri E. Prospective, controlled, multi-center study on the effect of an amino-acid-based formula in infants with cow's milk allergy/intolerance and atopic dermatitis. Pediatric Allergy and Immunology. 2001;(2):78–82.

[30] Candreva AM, Smaldini PL, Curciarello R, Cauerhff A, Fossati CA, Docena H, et al. Cross-reactivity between the soybean protein P34 and bovine caseins. Allergy, Asthma & Immunology Research. 2015;7:60–68.

[31] Curciarello R, Smaldini PL, Candreva AM, González V, Parisi G, Cauerhff A, et al. Targeting a cross-reactive Gly m 5 soy peptide as responsible for hypersensitivity reactions in a milk allergy mouse model. PloS One. 2014;9(1):e82341.

[32] Smaldini P, Curciarello R, Candreva A, Rey MA, Fossati CA, Petruccelli S, et al. In vivo evidence of cross-reactivity between cow's milk and soybean proteins in a mouse model of food allergy. International Archives of Allergy and Immunology. 2012;158(4):335–346.

[33] van Ree R. Clinical importance of cross-reactivity in food allergy. Current Opinion in Allergy and Clinical Immunology. 2004;4(3):235–240.

[34] Allen KJ, Remington BC, Baumert JL, Crevel RWR, Houben GF, Brooke-Taylor S, et al. Allergen reference doses for precautionary labeling (VITAL 2.0): clinical implications. The Journal of Allergy and Clinical Immunology. 2014;133(1):156–164.

[35] Thang CL, Boye JI, Zhao X. Low doses of allergen and probiotic supplementation separately or in combination alleviate allergic reactions to cow b-lactoglobulin in mice. J Nutr. 2013;143:136-41.

[36] Boyce JA, Assaád A, Burks W, Jones SM, Sampson HA, Wood RA, et al. Guidelines for the diagnosis and management of food allergy in the United States. The Journal of Allergy and Clinical Immunology. 2014;126(6 0):301–402.

[37] Bousquet J, Lockey R, Malling HJ. Allergen immunotherapy: therapeutic vaccines for allergic diseases. A WHO position paper. The Journal of Allergy and Clinical Immunology. 1998;102(4 Pt 1):558–562.

[38] Skripak JM, Nash SD, Rowley H, Brereton NH, Oh S, Hamilton RG, et al. A randomized, double-blind, placebo-controlled study of milk oral immunotherapy for cow's milk allergy. The Journal of Allergy and Clinical Immunology. 2008;122(6):1154–1160.

[39] Staden U, Rolinck-Werninghaus C, Brewe F, Wahn U, Niggemann B, Beyer K. Specific oral tolerance induction in food allergy in children: efficacy and clinical patterns of reaction. Allergy. 2007;62(11):1261–1269.

[40] Passalacqua G, Landi M, Pajno GB. Oral immunotherapy for cow's milk allergy. Current Opinion in Allergy and Clinical Immunology. 2012;12(3):271–277.

[41] Keet CA, Seopaul S, Knorr S, Skripak JM, Wood R. Long-term follow-up of oral immunotherapy for cow's milk allergy. The Journal of Allergy and Clinical Immunology. 2014;132(3):737–739.

[42] Keet CA, Frischmeyer-Guerrerio PA, Thyagarajan A, Schroeder JT, Hamilton RG, Boden S, et al. The safety and efficacy of sublingual and oral immunotherapy for milk allergy. The Journal of Allergy and Clinical Immunology. 2012; 129(2):448–455.

[43] Takeda K, Gelfand EW. Mouse models of allergic diseases. Current Opinion in Immunology. 2009;21(6):660–665.

[44] Ganeshan K, Nelsen C, Hadsaitong A, Schleimer R, Luo X, Bryce PJ. Impairing oral tolerance promotes allergy and anaphylaxis: a new murine food allergy model. Journal of Allergy and Clinical Immunology. 2009;123(1):231–238.

[45] Smaldini PL, Ibañez AE, Fossati CA, Cassataro J, Docena GH. Oral delivery of Brucella spp. recombinant protein U-Omp16 abrogates the IgE-mediated milk allergy. Human Vaccines & Immunotherapeutics. 2014;10(7):1–9.

[46] Li X, Schofield B, Huang C, Kleiner G, Sampson H. A murine model of IgE-mediated cow's milk hypersensitivity. The Journal of Allergy and Clinical Immunology. 1999;103:206–214.

[47] Egger J. Food allergy and disorders of the central nervous system in childhood. Food Allergy in Infancy and Childhood. Springer: Berlin, Heidelberg; 1989. p. 107–124.

[48] Mostafa GA, Reda SM, Abd El-Aziz MM, Ahmed SA. Sputum neurokinin A in Egyptian asthmatic children and adolescents: relation to exacerbation severity. Allergy. 2008;63(9):1244–1247.

[49] Magalhães ES, Pinto-Mariz F, Bastos-Pinto S, Pontes AT, Prado EA, deAzevedo LC. Immune allergic response in Asperger syndrome. Journal of Neuroimmunology. 2009;216(1–2):108–112.

[50] Chaidez V, Hansen R, Hertz-Picciotto I. Gastrointestinal problems in children with autism, developmental delays or typical development. Journal of Autism and Developmental Disorders. 2015;44(5):1117–1127.

[51] Gurney JG, Mcpheeters ML, Davis MM. Parental report of health conditions and health care use among children with and without autism. Archives of Pediatrics & Adolescent Medicine. 2015;160:825–830.

5

Bioactive Peptides from Milk

R. Anusha and O.S. Bindhu

Abstract

Milk is a major source of dietary energy, protein and fat. Due to their specific biological properties leading to health benefits, bioactive peptides (BAPs) derived from milk proteins have been a subject of intensive research during past two decades. These peptide sequences, encrypted within proteins, are liberated *in vivo* during gastrointestinal digestion or *in vitro* by fermentation with proteolytic starter cultures or using proteases. BAP generally comprises 2–20 amino acid (AA) residues. Isolation and characterization of BAP of different bioactivities from milk protein hydrolysates of cow, buffalo, camel, goat, mare, sheep, donkey and yak milks have been reported. Bioactivities of BAP, which depend on constituent AAs and the sequence, include mineral binding, opioid, angiotensin-converting enzyme (ACE) inhibition, immunomodulatory, cytotoxicity, antibacterial and antithrombotic. This chapter focuses on the methodologies adopted to produce BAPs and their prospective role in health enhancing nutraceuticals/pharmaceuticals.

Keywords: bioactive peptides, bioactivities, casein, milk proteins, nutraceuticals, whey

1. Introduction

One of the achievements of mankind over a century is the remarkable progress in the field of healthcare including management of communicable diseases leading to increase in life expectancy of population around the world. However, there is a tremendous upsurge in non-communicable diseases (NCDs) such as cardiovascular diseases (CVDs), diabetes, obesity, etc., which are associated more to life style changes and eating habits than to hereditary [1]. About two-thirds of the 57 million deaths that occurred globally in 2008 were due to NCD [2]. Further one-fourth of these deaths occurred before the age of 60. One among the four major behavioural risk factors identified for NCD is unhealthy diet. Healthy diet and physical activity were an integral part of the WHO's 2008–2013 action plan of the global strategy for the prevention and control of NCDs. Essential nutrients, standards and guidelines for diet and food preparation were considered prime for nutrition in the twentieth century [3]. New challenges now arise due to increased life expectancy, drastic life style modifications and elevated costs of healthcare. This has directed to new nutritional concepts which target maximum well-being and good health along with minimum disease risk throughout the lifespan. Nutraceuticals and functional foods are an attempt towards resolving these health issues and to create a healthy society [4]. The idea of nutraceuticals/functional foods is more often mentioned as an emerging field. However, the concept has been well acknowledged in ancient Indian Vedic texts and Chinese traditional medicine where it quotes "food and medicine have common source." Egyptians and Sumerians are just a few other civilizations that show evidence of food being used as medicine and in disease prevention [5].

There is no agreement on the definition of nutraceutical and functional foods. The term nutraceutical was first mentioned by Defelice in 1989 [6]. It was defined by Zeisel as "those diet supplements that deliver a concentrated form of a presumed bioactive agent from a food, presented in a non-food matrix and used to enhance health in dosages that exceed those that could be obtained from normal foods" [7]. European Commission Concerted Action on Functional Food Science in Europe (FUFOSE) originally proposed that a food can be considered "functional" if it provides positive impact upon one or more physiological functions in addition to its nutritional effects and improves health via reducing disease. Nutraceutical is also defined as an isolated or purified product from food that is generally in medicinal form and not generally related to food where as a functional food is defined as foods that offer more than basic nutrition by providing additional physiological benefit [8]. The biological active substances in functional foods (designer foods/medicinal or therapeutic foods) can either be an essential macro/micronutrient, a non-nutritive component or a component whose nutritive value is not listed as essential [9]. Functional foods are generally grouped into conventional foods containing natural bioactive substances, enriched or modified foods with bioactive substances and synthesised food. These foods as alone or in combination are considered functional foods. Foods/food product could be prepared using different approaches to make it functional: (a) by removing a component that may be identified to cause harmful effect when consumed, (b) elevating a constituent naturally present to induce targeted effect, (c) introducing a micro/macro nutrient for its positive effect and (d) substituting a constituent usually with macronutrient or increasing bioavailability of a food component [5]. Japan was the first country to emerge in production of functional foods to cope with accelerating cost of healthcare in 1980, mainly to improve the health condition of the aging population [5, 10]. Scientific studies have catered to the growing awareness on the relationship between nutrition and health issues that have led the end users' attention towards healthy foods. Functional or biologically active molecules from food have been widely studied and assessed for moderating these conditions as well as to restore their normal physiological functioning.

Biopeptides from dietary proteins have exhibited potential benefits to intervene these abnormal biophysiological conditions [11]. Bioactive peptides (BAPs) have been defined as protein fragments that have a progressive impact on body functions or conditions ultimately influencing health [12]. When BAPs are encrypted within their parent proteins, they are described as "cryptide" or a hidden peptide and are released by gastric digestion of food proteins or by exogenous/endogenous proteolysis by plant, animal or microbial proteases (especially during fermentation) [13]. These peptides demonstrate diverse physiological functions including metabolic functions, immunomodulatory, microbicidal, thrombolytic and pre/probiotic functions in human system [14]. Existing knowledge about milk-protein-derived BAPs as pharmaceutical constituent for potent drug or dietary supplement in the form of formulations of nutraceuticals is reviewed in detail in this chapter.

2. Sources of BAPs

Food proteins are not only a source of nutrients for maintenance of proper body functions but are also considered as a source of BAPs which can promote health and prevent diseases. BAPs are released from food proteins by endogenous proteolysis during gastric digestion or by exogenous hydrolysis using physical (heat), chemical (acid/alkali) or plant/animal/microbial proteases during food processing [15, 16]. Unlike physical/chemical treatments, enzymatic hydrolysis has an advantage of producing intact BAPs without any residual/toxic chemicals in the end products [17]. The size of active sequences may vary from 2 to 20 amino acid (AA) residues and generally have molecular weight <6000 Da [18]. BAP from different sources are found to possess common structural properties with hydrophobic AA residues in addition to lysine, proline or arginine [19]. Low molecular weights (dimer–heptamer) of BAPs are resilient to gastrointestinal (GI) tract enzymes and are easily absorbable in an intact form into the blood stream, hence suitable for therapeutic formulations or as functional foods. Higher molecular weight BAPs, in contrast, are possibly degraded during their passage through GI tract making them either inactive or less active. However, in few cases they may also become more active and thus more absorbable to be transported to target organs. Although milk proteins are considered as the most important source for BAPs, animal, plant and marine proteins also contain potential bioactive sequences [13, 20, 21].

2.1. Milk as a source of BAP

Milk is a characteristic secretion by female mammals with an array of bioactive substances to meet whole nutritional requirements, defensive and physiological functions of the young ones of that particular species. It is considered as one of the nearly complete single foods available in nature for maintaining human health and growth. Although the major mammalian milk that is consumed by humans is that from bovine, milk from sheep, goats, yak, horses and camels is also used. Constituents of milk are similar among species but differ significantly between species [22]. Major constituents of milk include carbohydrates, proteins, lipids, micronutrients and traces of various other organic and inorganic compounds. Lactose is the key carbohydrate present in milk, which is a disaccharide formed from glucose and galactose with β 1-4 glycosidic linkage. Fat content of cow's milk is ~3.25%. Bovine milk fat composition is distinct from other sources due to their diet and the presence of a rumen.

Normal bovine milk comprises about 3.5% of protein, 80% and 20% of which are constituted by casein and whey proteins, respectively [23]. Sub-fractions of casein and whey possess specific biological properties.

Milk from different animal sources, such as bovine, goat, yak and donkey, has been extensively studied for production of BAPs and for their potential positive impact on health, and physiological benefits have been experimentally investigated [24]. Among these, cow's milk has proven to be the best source. To effectively produce biofunctional peptides, a combination of such processes is also preferred [15]. Expression of dormant bioactivities of BAPs, encrypted within the sequence of parent proteins, depends on the site of proteolysis, i.e., in mammary gland or at a GI location as well as the specificity of proteases. Milk-protein-derived BAPs have claimed to be health enhancing components that can be used to reduce the risk of disease or to enhance a certain physiological function, such as potential antimicrobial, antihypertensive, antioxidative, cytomodulary, immune stimulating, opiate effects, etc., in humans. BAPs derived from milk protein may function as exogenous regulatory substances at diverse intestinal and peripheral sites of mammals. They may be absorbed via carrier-mediated or paracellular transport to regulate activities of target organs [12].

The primary physiological role of casein was considered to be a source of AAs required for neonatal nutrition. However, subsequent studies have shed light on its diverse physiological roles including prevention of pathological calcification of mammary gland by micellar caseins [25]. It is a phosphoprotein that precipitates at 20°C from raw skimmed milk at pH 4.6. The electrophoretic separation and AA sequence homology evaluations of different casein fractions revealed its heterogeneous nature with four major families of casein: α_1, α_2, β and κ [26]. Individual casein differs in the phosphate content and calcium-binding properties. Generally, α_1-, α_2- and β-caseins bind calcium strongly and precipitate at relatively low calcium concentrations and hence referred as "calcium-receptive caseins" having concentrated phosphoserine and proline regions. In contrast, κ-casein is "non-receptive casein" (not sensitive to these calcium concentrations), which can stabilise up to 10 times its mass of the calcium-sensitive caseins and possesses one phosphoserine [27]. Caseins α_1 and β have potential to liberate more than 20,000 biofunctional peptides each. Biopeptide from casein was the first identified food-derived BAP that boosted Vitamin D-independent bone calcification in rachitic infants in 1950 [28]. Few milk-derived peptides reveal multi-functional properties, i.e., specific peptide sequences having two or more different biological activities, as casein's primary structure possesses overlapping peptides or "strategic zones" in certain regions. These zones are generally protected from proteases [29]. For example, a strategic zone containing immuno-stimulating and opioid peptides has been reported to reside between residues 51–63 and 60–70 of human and bovine β-casein, respectively [30, 31].

Whey is the chief by-product or even considered as a co-product of cheese and other dairy industries. Concentration of protein in whey depends on various factors, such as milk source, time of the year, type of animal feed, lactation stage, processing quality and whey type [32]. Whey proteins constitute 20% of the whole milk protein, which are heterogeneous in nature. Discovery of whey dates back to 3000 years ago. Seventeenth and eighteenth centuries observed whey being used for its medicinal purpose [33]. Whey proteins are now considered as the highest quality natural protein available [34]. In recent past, there has been an exponential research on whey proteins due to the presence of nutrients that are found to have an active role in the betterment of human health [35, 36]. Whey proteins constitute 20% of whole milk protein which are heterogeneous in nature. All whey proteins are soluble, perform functions like carrier ligands and are involved in various biological activities. Elevated nutritional values of whey proteins are attributed to their diverse AA composition [15]. Globular whey proteins contain acidic/basic and hydrophilic/hydrophobic AAs along their polypeptide chains, well balanced throughout α-helix motifs [35]. With increased magnitude of dairy industry, enormous quantity of whey is produced across the globe. A substantial portion of it is wasted or underutilised as animal feed, in turn increasing the cost of its disposal. Use of whey as a source of BAP provides economic health potential for global population [37, 38]. Under controlled conditions, hydrolysis of whey releases peptides that could be a part of preventive or therapeutic applications [39]. Recent literature reports some biologically important whey peptides: α- and β-lactorphin, lactoferricin B, glycomacropeptide (GMP) to name a few. Commercial formulations with whey as an ingredient are also available, making it a value-added product [40].

3. Production and purification of milk BAPs

Selection of appropriate food protein followed by its enzymatic hydrolysis is the general process used for BAP production. BAPs can be released by *in vivo* or *in vitro* hydrolysis of the parent proteins. They can be produced *in vitro* by (1) enzymatic hydrolysis with digestive enzymes, (2) fermentation of milk with proteolytic starter cultures and (3) proteolysis by enzymes derived from animals, plants or microorganisms [14]. Once the AA sequence is known, it is also possible to synthesise BAPs through the chemical methods, enzymatic synthesis and/or by recombinant DNA technology [15].

3.1. Protein hydrolysis under simulated gastrointestinal tract conditions

In vivo digestion of protein, although possible, is rarely used for the production of BAPs as it requires removal of intestinal contents from animals after feeding protein diet. However, it can be simulated. The protein of interest is first treated with pepsin for a few hours to stimulate digestion, followed by treatment with pancreatin or with a combination of trypsin and chymotrypsin mimicking digestion in stomach at specific pH. This method also enables estimation of BAPs released from food proteins though the stimulation of intestinal proteases after food intake is not taken into consideration [12]. BAPs produced via *in vitro* stimulation of gastric enzymes from milk are mainly ACE inhibitory peptides. Digestive enzymes in combinations with other proteases from bacterial/fungal sources are also used in the generation of BAPs from various food proteins [22].

3.2. Fermentation with proteolytic starter cultures

The presence of lactic acid bacillus (LAB) in milk fermentation can be either as spontaneous or as inoculated starter cultures. Although under spontaneous fermentations the growth of LAB cannot be predicted or controlled, this procedure has been practiced and carried out traditionally for years [41]. The fermentation process by lactic acid bacteria starter cultures, used in the dairy industry, involves the hydrolysis of milk proteins (particularly casein) into peptides and AAs which are used as nitrogen sources necessary for their growth. Several lactic acid bacteria are widely used and their role can be divided into starter and non-starter cultures. The proteolytic system of lactic acid bacteria consists of a cell wall-bound protease and other intracellular proteases [42]. Different starter cultures and fermentation conditions affect the composition and nature of released BAPs. Studies have exhibited that LAB proteases especially from *Lactococcus* sp. and *Lactobacillus* sp. could hydrolyse ≥40% of peptide bonds of α_1 and β caseins resulting in the release of various oligopeptides that are further hydrolysed by complex peptidases [43]. Pure cultures as well as mixed cultures are used for fermentation. Apart from lactic acid bacteria, yeast and other microflora may also be present in a few commercial starter cultures. Peptide profiling of bovine kefir, which is a fermented milk beverage, revealed 236 unique peptides that were released from caseins during its production by kefir grains [44]. BAPs with physiological benefits, such as ACE inhibition, immunomodulation, anti-oxidative, antimicrobial, etc., have been identified from microbial fermented products [45–47]. An endopeptidase from *Lactobacillus helveticus* improved the production of casein-derived ACE inhibitory peptides, VPP (Val-Pro-Pro) and IPP(Ile-Pro-Pro) via carboxy terminal processing mechanism [46]. Even with insufficient evidence to establish the cause–effect relationship between the maintenance of blood pressure (BP) and the consumption of fermentation-derived peptides, food products fortified with such BAPs could prove valuable as nutritional therapeutics at least in certain sub-population [42].

3.3. Enzymatic hydrolysis by proteases

Plant, animal and microbial proteases have been used for the production of BAPs from the protein source. Enzymatic hydrolysis is more suitable for the production of BAPs, especially in food and pharmaceutical industries, due to the lack of residual organic solvents or toxic chemicals in the end product. Further hydrolysis takes place under mild controlled conditions and yield-predictable end product. The widely used enzymes of animal and plant origin include pepsin, trypsin, chymotrypsin, bromelain, papain or ficin. One of the best and cheap sources of animal origin proteinases is pancreases, by-products of the meat industry. Recently, proteases from plant sources are gaining attention. ACE-inhibitory peptides and anti-ulcerative peptides were studied from peptides synthesised from whey protein hydrolysis by *Cynara cardunculus* [48, 49]. Proteases from extracts of celery, fennel, parsley and ficin latex have shown potential in releasing peptides from skim and whole milk with ACE inhibitory and antioxidant properties [50, 51]. However, these preliminary observations need further validation through clinical studies involving human volunteers to confirm such promising *in vitro*/animal study results [49]. Microbial proteases widely used are those obtainable from the *Bacillus* sp., *Bifidobacterium*, proteases from the LAB and few fungi [14, 17]. Furthermore, LAB and their products are considered safe (generally recognized as safe, GRAS). Microorganisms are a relatively cheap source of proteases. Microbial proteases, especially from bacteria, provide few advantages over proteases from other sources. A recent study showed that *Bifidobacterium bifidum MF 20/5, Arsukibacterium ikkense, L. helveticus, Bacilus thermoproteolyticus*, etc., could release peptides from milk and milk products that were potent ACE inhibitors, antimicrobial and antioxidant agents [52–55].

3.4. Chemical synthesis and recombinant DNA technology

Enzymatic production of peptides generally does not produce adequate quantities and is not always economical. Chemical and recombinant technologies are used in such conditions to produce peptides in large quantities. However, they require specialised equipment, skilled labours and incur high cost. Further these techniques can be employed only when the peptide sequence is known. The size of peptide determines the kind of technology to be used. Chemical synthesis is preferred for producing medium and small peptides (5–80 AAs) and recombinant technology for larger peptides and proteins [56]. Numerous such peptides have been approved to be used as drugs in the past decade. This application greatly depends on chemical modifications of peptides to improve membrane permeability, receptor affinity and stability along with decrease in hepato-renal clearance of the peptide. It is also proven that bioavailability of peptides improves with chemical modification, such as bond replacement and conformational modifications including peptide cyclisation. Chemical synthesis has proved that large-scale peptide synthesis at reasonable cost can be achievable through solid phase [57]. Easy peptide separation from other products and impurities remain its major advantage. However, requirement of side-chain protection, racemisation, decrease in yield during removal of protective chains and toxic nature of solvents used are a few drawbacks associated with this method. Chemical synthesis is still considered as one of the best suited procedures to synthesise medium-sized peptides of therapeutic significance [16].

Recombinant DNA technology involves gene expression in microorganisms using modern techniques of cloning to synthesise one or more peptides concurrently. This method requires high cost involving research; however once established it will lead to their economic yield. The *Escherichia coli* based bacterial expression system is widely preferred. Human and bovine β-casomorphins were produced through this technique [15]. Particularly, this technology finds superiority in synthesis of large peptides and proteins [16]. Additionally, it comparatively offers effective alternative for bulk production of peptides [58].

3.5. Purification and characterisation

Purification of peptides after its production from any of the above-mentioned process is prerequisite for its industrial use. Peptides with more than 95% purity is required for NMR studies, enzyme studies, monoclonal antibody production, in treatment of disease, clinical research and structure–function relationship studies. Immunological studies and PAGE analysis require purity of >70–80% [58]. Ultra-filtration is performed using series of membrane filters with different molecular weight cut-off resulting in minimising non-peptide concentration in the hydrolysate. Purification procedure often used is RP-HPLC. This enables rapid purification and detection of peptides in a mixture. Ion exchange, affinity, size-exclusion chromatography and capillary electrophoresis are also being preferred [12]. Peptide characteristics are evaluated through mass spectrometry [58].

4. Absorption of BAPs

BAPs produced after their release from parent proteins by GI enzymes at times exert their activity locally or after absorption at the peripheral organs [14]. Some peptides may undergo further hydrolysis by peptidases in the course of transport of brush border and cytoplasm. Intestinal absorption plays an important role in the effectiveness of their biological functions. Even with little unambiguous *in vivo* evidence, the absorption of BAPs is suggested to be possibly operating through any of the following pathways: carrier-mediated transporter system, passive diffusion and transcytosis, paracellular pathways via tight junctions and endocytosis [21].

4.1. Carrier-mediated transport system

Carrier-mediated transport system actively transports bi- and tripeptides. Cytoplasmic peptidases hydrolyse peptides to AAs in the intestinal epithelial cells; however, some peptides resist this hydrolysis and are transported across the membrane [59]. PepT1 (transmembrane protein) is an intestinal proton-dependent peptide transporter with a wide range of substrate specificity for more than 400–800 di- and tripeptides. L-AA-containing peptides are preferred over D-AA/D-stereoisomers for transporting across membranes [60]. It is a bidirectional transporter where membrane potential and proton gradient decide the rate and direction of absorption. About 70–80% of peptides are absorbed through PepT1 transport system due to its high transporting capacity and intestinal expression [61].

4.2. Passive diffusion and transcytosis

Maternal colostrum-derived γ-globulin permeability associated with passive immunisation by mammalian neonatal small intestine has long been established. Lipid-soluble peptides easily enter enterocytes and are hydrolysed by the action of cytosolic peptidases releasing AAs for intercellular metabolism. Some AAs from enterocytes are moved into portal circulation through transporters on basolateral membrane [62]. Role of this system in absorption is considered small, as peptides are most likely to be hydrolysed by cytoplasmic peptidases [63]. Large polar proteins (>600 Da) are hydrolysed before they enter into enterocytes. Vesicles are formed by invagination of apical membranes that capture the large peptides. These vesicles fuse with lysosomes to form phagolysomes where enzymatic digestion of the macromolecule occurs. Proteins that resist this hydrolysis could be drawn into enterocytes by means of receptor-mediated, absorptive or fluid-phase transcytosis to be secreted into basolateral membrane. Di- and tripeptides are generally hydrolysed by cytosolic peptidases; however, those that resist are hydrolysed by vascular endothelial tissue/plasma peptidases [63–65].

4.3. Paracellular pathway through tight junctions

Paracellular pathway is considered important in BAP absorption in intact form. Tight junctions are formed by adhesive membrane proteins (claudins, occludin, tricellulin and junctional adhesion molecule A) between intestinal epithelial cells creating small pores to enable passive diffusion of peptides. Diffusion through tight junctions depends on permeability and elasticity of these junctions that are controlled by these proteins [65, 66]. Tight junctions generally allow diffusion of cations and inert small molecules (<600 Da) and also larger oligopeptides [67].

4.4. Endocytosis

Intestinal epithelium of adult mammals could absorb proteins in small quantities by endocytosis even though such absorption is much smaller (<0.1%) in comparison with other absorption system; however, they are biologically significant. Larger peptides are generally internalised by endocytosis [fluid phase and/or absorptive (receptor)]. Fluid phase (pinocytosis) is non-specific endocytosis that begins with plasma membrane "pinching off" to vesicle that contains dissolved peptides in extracellular (EC) fluid. Vesicle or pinosome drifts inwardly to perinuclear region to bind with lysosome. Peptides are hydrolysed during this and vesicle is recycled back to plasma membrane. In absorptive or receptor endocytosis, the membrane binding site is specific for the peptide to be transported. Receptor–ligand binding results in clustering of this complex into clathrin-coated regions of plasma membrane followed by endocytosis. These complexes are transported to endosomes where low pH of the endosome disassociates receptor–ligand complex and receptor is either recycled to plasma membrane or may be degraded [67, 68].

5. Bioactivities of BAPs from milk

Naturally occurring bioactive molecules are preferred in comparison with chemosynthetic components for functional food preparation. Protein-derived BAPs are presently gaining attention due to their safe and economic benefits [69]. Food technologists and scientists have been concentrating in understanding relationship between casein- and whey-derived BAPs with their bioactivities. Deciphering their AA sequences along with an understanding on their bioavailability and stability *in vivo* have helped in the development of health-promoting milk protein [70]. **Figures 1** and **2** illustrate various bioactivities of milk-derived peptides. However, relationships between their structural properties and functional activities have not been completely elucidated. Many BAPs from milk proteins are relatively small (e.g., 2–9 AAs), possessing hydrophobic AA residues in addition to proline, lysine or arginine groups. Several *in silico* approaches including quantitative structure activity relationship (QSAR) as an application of chemometric and bioinformatics methods have been used in an attempt to predict the activity of these peptides. Structure–function relationship elucidating studies highlighted the significant contribution by specific AA residues, such as tyrosine (Y), leucine (L), proline (P) and tryptophan (W), present in BAPs for diverse physiological activities [71]. Recent studies have shown that many tryptophan(W)-containing peptides originate from milk proteins, which have been shown *in vitro* to display a wide range of bioactivities such as ACE inhibition along with antioxidant, antidiabetic and satiating-related properties [71].

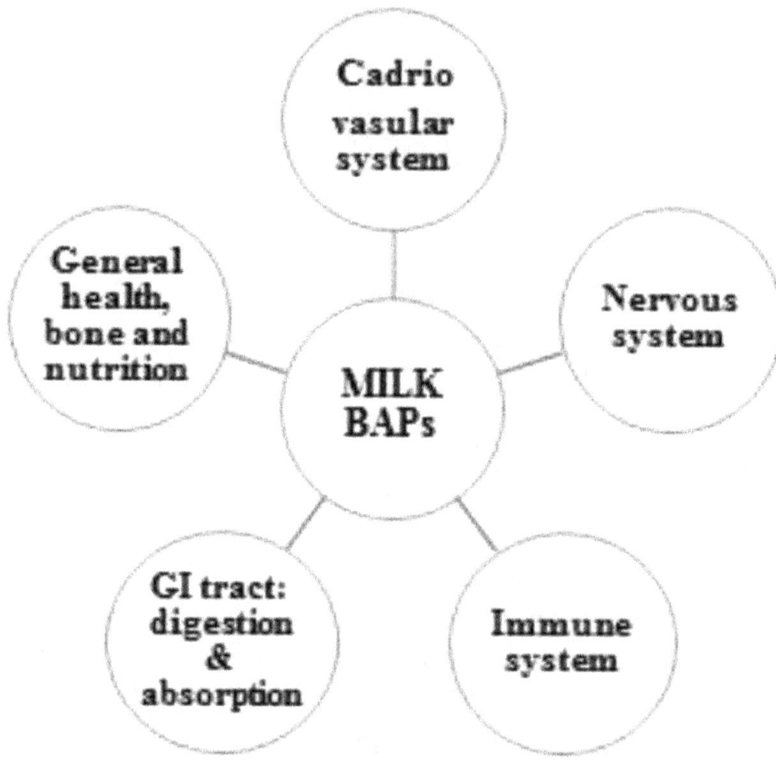

Figure 1. Physiological role of milk bioactive peptides [72, 26].

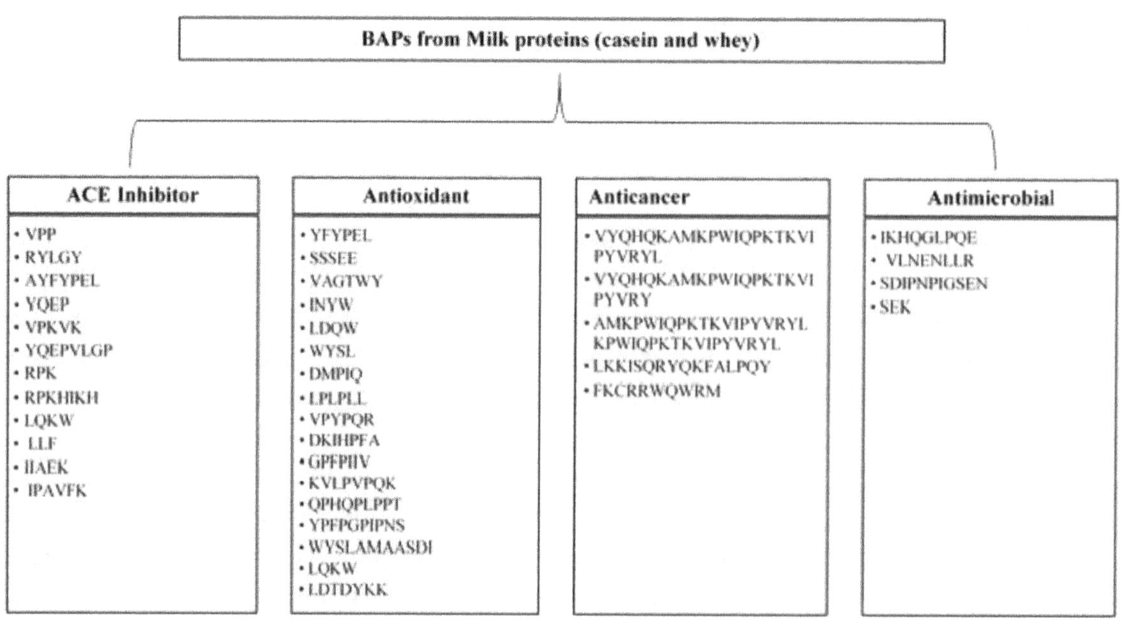

Figure 2. Representative BAPs from milk proteins and their prominent bioactivities [40, 73, 74]. *AA represented by one letter code.

5.1. Functionalities of casein and whey peptides

5.1.1. Mineral binding property

Normal bone development in humans requires adequate calcium supply. Low ingestion of calcium levels are prevalent among infants (mostly pre-term), young and post-menopausal women, elderly who have low calcium absorption and those in treatment or prevention of osteoporosis and children with rickets [75]. Enhancement of Vitamin D-independent bone calcification was the suggested role of casein-derived phosphorylated peptides in rachitic infants [14]. Specific caseinophosphopeptides (CPPs) that are phosphorylated regions released form α- and β-caseins can form soluble organophosphate salts functioning as mineral carriers, especially for calcium. CPPs were the first BAPs to be mentioned by Mellander [28] in 1950. CPPs have been shown to possess mineral binding properties thereby improving bioavailability of metal ions [76]. CPP-ACP (amorphous calcium phosphate) exhibited anti-plaque, remineralisation and demineralisation prevention of teeth [77]. Increased bioavailability of iron in the presence of CPP has also been reported [78]. Whey peptides also exhibited mineral-binding properties [79]. Tryptic hydrolysates of whey proteins inhibited calcium phosphate formation thereby favouring calcium absorption. [80]. Significant elevation in mice calcium absorption in the presence of tryptic hydrolysate of whey (Ser-Thr-Glu-Tyr-Gly) suggested their possible utility as functional food or dietary supplement to prevent calcium deficiency [81].

5.1.2. Antioxidant activity

Aerobic metabolism inherently releases free radicals which are scavenged by antioxidants in the biological system that protects against the damage caused by reactive oxygen species. Free-radical-mediated reactions play a major role in cellular damage and aging via lipid oxidation and production of secondary lipid products contributing to major degenerative diseases, such as CVD, diabetes and even in neurological disorders [82]. A number of milk BAPs have been shown to possess antioxidant property. This ability of peptides to bind/interact with radical species and/or inhibit oxidation could be a boon to boost human health [83]. Casein peptide, derived through pepsin hydrolysis, bearing the sequence Tyr-Phe-Tyr-Pro-Glu-Leu (YFYPEL) was found to show superoxide anion scavenging activity [84]. Further they identified that the preferred sequences were EL>YFYPEL>FYPEL>YPEL>PEL, suggesting that Glu-Leu in the sequence is important for this activity. Casein-derived peptides exhibited antioxidant property via inhibition of lipid peroxidation. It was further observed that proteolysis or dephosphorylation of casein did not impair its antioxidative potential [85]. Peptide with SerP-SerP-SerP-Glu-Glu domain has proven to possess free radical scavenging, hydroxyl and metal chelating properties [19, 86]. A recent study on peptides derived from sodium caseinate by proteases (papain, trypsin and pancreatin) demonstrated the contribution of proteolytic conditions on the extent of antioxidant activity [87]. Antioxidative potential of whey protein hydrolysates (WPHs) was associated with both low and high molecular mass fractions [88, 89]. α-Lactalbumin (α-LA) hydrolysed by thermolysin-yielded peptides with antioxidative properties. Peptides with Trp or Tyr at their extremities were found to possess higher antioxidative activity in comparison with gallic acid and Trolox [90]. Trp-Tyr-Ser-Leu purified from whey protein displayed high radical scavenging and superoxide radical scavenging activities among peptides obtained through other enzymes [91]. Many peptides from whey have been shown to be multifunctional and possess antioxidant property along with ACE inhibition and antimicrobial properties [92, 93]. IIAEK and IPAVFK peptides, derived from tryptic hydrolysate of β-lactoglobulin (β-LG), possessed potent ACE inhibitor potential while VAGTWY was found to be an effective antioxidant having 1.7-fold higher activity in comparison with commercial antioxidant [94]. Hydrolysates derived from sheep cheese whey through hydrolysis by protease from *Bacillus* species P7 were effective as antioxidant and antihypertensive agents. They exhibited 3.2-fold increases in radical scavenging activity and could potentially be engaged in retarding oxidation in food products or as a nutraceutical [95].

Lipid peroxidation leads to spoilage, reduced shelf life along with release of free radicals that possibly decrease the nutritional value of food. Independent or combined antioxidant or antimicrobial roles by whey BAPs have extended their utility as natural food preservation agents. Ability of lactoferrin (Lf) to retard growth of microbes or limit lipid oxidation finds its application in food preservation as it acts as antimicrobial and physical barrier between food and microbes. It can also be used as spray or directly applied to beef carcasses improving its shelf life [96]. Peptide sequence Trp-Tyr-Ser-Leu-Ala-Met-Ala-Ala-Ser-Asp-Ile from α-LA and β-LG hydrolysed by commercial proteases was found to possess radical scavenging activity higher than butylated hydroxyl anisole indicating its utility as antioxidant in food industry [97]. Alcalase-treated whey protein isolate (WPI) hydrolysates proved to possess antioxidant property by acting as metal ion chelator, radical stabiliser and hydrogen donor thereby inhibiting lipid oxidation ascertaining their potential to replace synthetic antioxidants [98]. Radical scavenging activity of fragments LQKW and LDTDYKK of β-LG-enriched WPC were identified [99]. Antioxidant peptides and their application have been reviewed by Power et al. [18]. Critical contribution of thermal processing contributing to maintenance of milk protein bioactivity was delineated. Lf and β_2 transforming growth factor were significantly high in low-heat-treated WPC mainly from acid whey that could induce immune response via proliferation and cytokine responses of intestinal epithelial cells. These findings could be vital in maintaining optimal bioactivity infant's formula thereby influencing maturation and immune modulation of developing intestine [100].

5.1.3. *Angiotensin-converting enzyme inhibitors*

CVD is the leading cause of mortality and morbidity globally. BAPs from milk have gained prominence in protection against CVD owing primarily to their antihypertensive effects along with hypocholesterolaemic, anti-inflammatory and anti-oxidant potentials [101]. Multifunctional ACE regulates blood pressure via catalytic conversion of angiotensin 1 to angiotensin 2 which is a potent vasoconstrictor. ACE inhibition is known to be the major strategy in BP treatment using pharmacological drugs. Due to their side effects, there is continuous search for safe and possibly natural inhibitors [102]. Extensive research has been focussed on ACE inhibitory biopeptides. Casokinins that are casein-derived peptides, found in α_{1}-, β- and κ-caseins, are very effective in inhibiting ACE [72]. Casein peptide, VPP, is a potential inhibitor of ACE. Research suggests that monocyte adhesion to inflamed endothelia is well moderated by VPP that may prove vital in the prevention of atherosclerosis [103]. Pepsin hydrolysed bovine casein (HBC) with molecular mass <3000 were studied for their antihypertensive property. HBC exhibited nearly 10 times decrease in DBP substantiating its potential as ACE inhibitor and an antihypertensive agent. Pepsin hydrolysate of casein has been commercialised and patented as potential hypotensive agents containing α_1 peptides [99, 104]. Structure–activity studies have shown that three residues in the C-terminal region of BAPs seem to interact with active centre of ACE. This interaction may enhance the ACE inhibitory activity if hydrophobic/aromatic AAs (Trp, Tyr and Phe) or imino acid Pro are situated in this location. Additionally, Arg and Lys residue's positive charge may increase the inhibition [105]. Further C-terminal region's importance in ACE inhibition was delineated through QSAR model [106, 107]. A recent study on α-casein-derived peptides identified a potent ACE inhibitor using endoproteinase from *Aspergillus niger* [108]. Enzymatic hydrolysis of β-LG obtained from chemical hydrolysis of ovine and caprine whey exhibited consistently high ACE activity. It was postulated that end product of thermolysin hydrolysates, LQK from LQKW and LL from LLF, possibly possessed ACE inhibitory action [109, 110]. Stage 1 hypertensive young adults treated with dietary intake of WPC beverage daily for 6 weeks resulted in a significant decline in diastolic and systolic blood pressure [111]. Hydrolysates of camel milk, colostrum and colostral whey proteins by pepsin and pancreatin to mimic intestinal digestion presented increased ACE inhibition along with multifunctional effects like radical scavenging and antimicrobial effect. Biopeptides α-LA and immunoglobulin (IgG) exhibited multifunctional activity including ACE inhibition [92]. Recently, optimisation studies to produce whey hydrolysate from WPC containing 70% protein with commercial enzyme through response surface methodology demonstrated up to ~65% inhibition of ACE and 50% antioxidant property [93].

5.1.4. Opioid activity

Opioid peptides are short sequence AAs required by the brain for important functions and these mimic the effect of opiates in the brain. Opioid peptides are peptides that exert affinity towards opiate receptors. Some of the functions of these peptides are to increase analgesic action, moderate social behaviour, stimulate endocrine secretions, increase GI transient time and thereby inhibiting intestinal peristalsis, motility, etc. [112]. Activity of opioid agonist peptides is inhibited by the opioid antagonist naloxone both *in vitro* and *in vivo*. All opioid receptor ligands are proved to possess some characteristic features, such as N-terminal Tyr residue, another Phe, Tyr or another aromatic AA at the third or fourth position from the N-terminal end. These features ensure appropriate fit between the peptide and receptor for the specific target activity [113]. One of the earliest discovered BAPs from milk was casein opioid peptide, β-casomorphins. Water buffalo, sheep and human milk also contain analogues of β-casomorphins. Exorphins derived from α-casein of milk corresponding to bovine α_1-casein and casoxins from κ-casein mimic as opioid antagonists [114]. Recent research on casein-derived β-casomorphins-7 has shown to have positive influence on neurogenesis through redox-based epigenetic effects [115]. Enzymatic hydrolysis of whey by GI enzymes individually or in combination was used *in vitro* to release an agonist peptide, serorphin, from bovine serum albumin (BSA). It also led to the production of antagonist peptides, lactoferroxin A, B and C from Lf, α- lactorphin from α- LA and β-lactorphin from β-LG [116]. α-Lactorphin was shown to upregulate mucin gene (MUC5AC) expression and secretion in human colonic goblet-like cells [117]. Trypsin hydrolysates of β-LG and β-lactorphin exhibit similar results possibly operating via opioid pathway. Results confirm that mucin secretion could be induced via opioid receptor and whey peptides as promising gastrointestinal protective agents [118].

5.1.5. Antithrombotic activity

Rennin hydrolysis of κ-casein during milk clotting releases caseinomacropeptide (CMP) that contain peptide sequences aiding blood clotting by inhibiting platelet aggregation and binding of γ chain of fibrinogen to fibrinogen receptors [119]. The sequences LSFMAIPPK, MAIPPKKNQDDK, MAIPPKK and NQDK in primary structure of κ-caseins were found to be involved in antithrombotic activity [119, 120]. κ-Caseinoglycopeptide from sheep casein exhibited antithrombotic property [121]. Undecapeptide (106–116) of bovine κ-casein exhibits structural similarities to dodecapeptide of human fibrinogen γ chain at C-terminal. Ile^{108}, Lys^{112} and Asp^{115} residues of undecapeptide are positionally homologous with r-chain of human fibrinogen for platelet receptors thereby inhibiting clot formation [122].

5.1.6. Antimicrobial activity

Conventional antibiotics use is increasingly turning deterrent due to antibiotic resistance by microorganisms resulting in treatment impediments and augmented healthcare expenses. Antimicrobial peptides (AMPs) resulting from milk have the advantage due to their precise infected cell targeting, broad spectrum nature, safe and economical source with vast industrial potential [17, 20, 71]. Antimicrobial properties of milk was first identified in 1930 when lactenin was identified to possess antibacterial property against scarlet fever *Streptococcus* [123]. Research has shown that milk proteins contain several motifs that can be released by proteolytic enzymes to enhance the antimicrobial potential of these proteins [124]. Disease prevention/control through BAPs primarily involves microbial interaction as AMPs or immunomodulation [125]. α_1-Casein-derived peptides, casecidins, were the first purified antibacterial peptides from casein and were effective against a wide range of Gram-positive bacteria including *Staphylococci*. Isracidin, N-terminal peptide of α_1-casein was found to be effective against Lactobacilli and Gram-positive bacteria *in vitro* and yielded a strong protection against *Staphylococcus aureus, Listeria monocytogenes* and *Streptococcus pyogenes in vivo*. Isracidin is comparable in its antibiotic action to standard commercial antibiotics in protecting cows and sheep against mastitis [126]. Caseicin A, B and C from α_1-casein share common characteristics with isracidin in their AMPs. Caseicin A and B were effective against *E. coli* and *Enterobacter sakazakii* while caseicin C only showed minor activity against *Listeria innocua* [127]. Peptides obtained from bovine α_1-casein by *Lactobacillus acidophilus* DPC6026 fermentation have been active against infectious pathogenic strains in neonate, *E. sakazakii* and *E. coli* suggesting the possible bioprotective applications of these AMPs in infant milk formula [128]. Digestion of casein with pepsin yielded two AMPs, Cp1 from α_1-casein and Cp2 from α_2-casein.

Cp1 was effective against both Gram-positive and negative bacteria whereas Cp2 exhibited inhibition generally more towards Gram-positive bacteria [129]. Casocidin-I (165–203) characterised and isolated from bovine α_2-casein was effective against *E. coli* and *S. cornosu* in infants. Human milk does not contain α_2-casein and hence it was proposed that human intestinal flora was influenced by casocidin-I ingestion [29]. Chymosin digestion of sodium caseinate resulted in the release of five antibacterial peptides (Cr1, Cr3, Cr4, Cr5 and Cr7) that were found potent against Gram-positive bacteria [130]. Peptide from human β-casein (184–210) showed broad spectrum antibiotic activity against Gram-positive and negative bacteria. Fragment from rabbit β-casein (64–77) exhibited potent action against Gram-positive bacteria [105]. Ser(P)149, a mono phosphorylated sequence of κ-casein (138–158), also displays antibacterial effects against *Porphyromonas gingivalis, E. coli* and *S. Mutans* [74]. κ-Casein-derived AMPs, kappacin, is a non-glycosylated and phosphorylated form of CMP. These peptides limit GI tract infections in neonates possibly by transmembrane cation gradient collapse and by elevating bacterial sensitivity to gastric acids [105]. CMP by itself is a powerful health booster. It is effective against cholera toxins as well as influenza virus. Prevention of bacteria adhesion to oral cavity by CMP controls dental plaque consequently promoting teeth remineralisation. CMP has also been patented in hygiene products for dental caries prevention [131]. κ-Casecidin, trypsin digest of κ-casein, exhibited both antimicrobial potential and mammalian cell cytotoxicity including cell lines of human leukaemia probably through apoptosis [127]. Whey protein Lf has been considered as a good source of AMPs. A cationic peptide, from Lf N-terminal region, lactoferricin 27, binds to microbial membrane and increases its permeability causing cell death. Potential to bind iron, a pathogenic virulence factor of many bacteria, and depriving it from cell vicinity has been another mechanism by some whey-derived AMPs especially on *E. coli* and *L. monocytogenes* [132]. Three α-LA hydrolysates by trypsin and chymotrypsin was found potent against many Gram-positive bacteria, especially *Bacillus subtilis*. Four peptide fragments from trypsin digests of β-LG exerted bactericidal property only against *B. subtilis* [132]. Targeted modifications after partial hydrolysis by pancreatic enzymes improved antimicrobial property by β-LG towards Gram-negative bacteria [132]. Whey proteins and their chemically modified analogues were effective against *Helicobacter pylori* infections and exhibited antiviral activity towards human herpes simplex virus type 1 (HSV-1) during and after infection [35]. Fungal stasis and antifungal role of Lf and its synthetic peptides are well established against *Candida albicans, Candida glabrata* and other *Candida* species [39]. Whey from goat milk on hydrolysis with gastric and duodenal juices exhibited antimicrobial properties. Nature of protein substrate and proteases affect AMP generation and their efficacy. Camel milk whey protein exhibited higher inhibition of *E. coli* in comparison with that of bovine. Camel whey hydrolysates by trypsin, thermolysin and chymotrypsin showed increased antimicrobial activity [133]. Peptic and pancreatic hydrolysates of camel colostrum proteins were inhibitory to growth of *E. coli* and *L. innocua* signifying their natural antimicrobial potential [92].

The biggest challenge faced by researchers in food-borne pathogen control is tolerance to high salt and low pH by a wide range of microorganisms (bacteria, fungi and viruses). Novel antimicrobial agents from natural sources are therefore the need of the hour to overcome this challenge [134]. Recent past has witnessed a rise in interest towards the use of BAPs as biopreservative for food or as functional food that is more prophylactic, nutritious and healthy [128]. This extended application of AMPs is advantageous over chemical food preservatives as they introduce fewer antagonistic effect, retain the organoleptic flavours of food along with their nutritional benefits and require decreased heat treatment for minimal processing [74, 105]. Milk-derived AMPs find immense potential in therapeutic healthcare and food preservation.

5.1.7. Immunomodulatory property

Body's immune system protects it from pathogens through its specialised cells, antibodies and lymphatic system. Dietary composition and their intake play a vital role in body's immune system. Immunomodulation includes stimulation or suppression of immune functions in the human body. BAPs from innumerable sources have revealed their immunomodulatory role in humans. They are identified to boost the immunological activity by regulating antibody production, cytokine regulation and reactive oxygen species induced immune functions [127]. Neonates' resistance to microbial infections is through physical transmission of passive immunity through breast feeding via a number of multifunctional factors including casein, which on digestion in GI tract releases peptides with immunomodulatory functions [72]. β-Casomorphin-11, a hexapeptide corresponding to Val-Glu-Pro-Ile-Pro-Tyr possessing immunostimulatory effect, was isolated from tryptic digest of β-casein of human milk. Several other hydrolysates from β-casein (191–193) and α_1-casein (194–199) including β-casomorphin-11 exert protection against *Klesieblla pneumoniae* in mice *in vivo* and activate phagocytic action in human and murine macrophages *in vitro* [30]. β-Casokinin-10, obtained via pepsin–chymosin hydrolysis of bovine casein, stimulated proliferation of peripheral blood lymphocytes in rats. Stimulatory effect on proliferation of human peripheral blood lymphocytes was presented by synthetic peptide sequence analogue from bovine κ-casein (Tyr-Gly) [135]. Chymosin-mediated bovine casein hydrolysis released fragments (1–23 and 193–209 of α_{1-} and β-caseins, respectively) with immunomodulatory capacities. CPP III, a commercially available CMP containing mainly α_2-casein (1–32) and β-casein (1–28), increases immunoglobulin production in spleen cell cultures of mouse. Further, it induces proliferation by lipopolysaccharide, concavalin A (Con A) and phytohaemagglutinin. *O*-Phospho-l-serine residue of CPP III was found to be responsible for this immunostimulating activity and therefore study suggested relative stability of this peptide even with GI tract proteinase action. These observations were considered vital for developing infants' formula with optimised immunomodulatory activities [72]. Enhanced mucosal immunity was observed in mice with oral administration of CMP [15]. Immunotherapy against HIV virus displayed positive results in decreasing the tendency to progress to AIDS (in patients with pre-AIDS) probably by inhibiting the infection development [136]. Eight casein hydrolysates' (titled a–h) bioactivities were assessed for their antioxidant and immunomodulatory response. Hydrolysates (d–h) exhibited increased Con A-stimulated IL-2 production; however, IL-10 was least stimulated in Jurkat cells in culture [47]. Further studies revealed cross-linking of sodium caseinate with transglutaminase before hydrolysis by Prolyve 1000 (microbial proteolytic preparation) had a positive impact on anti-inflammatory functions in Jurkat cells [137]. Peptides with immunosuppressive property are utilised during medical conditions like grafting and transplantations and also during autoimmune disorders whereas stimulative immune peptides are required for overall immune strength. Immunopeptides released after digestion in adult human are not significant to stimulate immune response for therapeutic purposes [138]. Whey peptides have exhibited its positive impact on immune system through various studies both *in vivo* and *in vitro* [15]. GMP and its peptic hydrolysates were found to stimulate proliferation and phagocytic activity of human macrophage-like cells U 937 [139]. Lymphocyte activation, antibody production, non-immune responses like NK cell and macrophage functions are all reviewed extensively by Gauthier et al. [138]. Cow and goat whey protein samples revealed *in vitro* a dose-dependent inhibition of T-lymphocyte proliferation indicating the possible role of whey peptides probably hindering some important proliferation activating signals [140]. Synthetic peptides hydrolysed by trypsin or chymotrypsin on theoretical basis from α-LA (fragments 10–16 and 104–108) and β-LG (fragments 1–8, 15–20, 55–60, 78–83, 84–91, 92–105, 139–148, 142–148 and 102–105) were evaluated for their immunomodulatory functions. Different extent of proliferation stimulation was observed by fragments 15–20, 55–60, 84–91, 92–105, 139–148 and 142–148 from β-LG and α-LA 10–16, while inhibition and/or stimulation on cytokine secretion was also observed by fragments 15–20, 55–60 and 139–148. Cytotoxic effect was predominant in β-LG 1–8, 102–105 and α-LA 104–108. Studies show whey-derived peptides to have prospective to modulate specific immune response [141]. Rat and mice models have shown that dietary intake of whey adds positive influence in combating small bowels by reducing ulcers and having high IgA in gut advocating their protective role in patients with intestinal disorders and perioperative damage to small intestine [142, 143].

5.1.8. *Anticancer property*

In spite of promising results from anticancer studies on casein peptides, casein is not regarded as chemo preventive. β-Casein derivatives, β-casomorphin-7 and 5, along with their opioid properties, have displayed arrest of cell cycle and anti-proliferative properties in colon and breast cancer cells [144, 145]. CPPs inhibit intestinal tumour proliferation through activation of calcium channels and stimulate apoptosis [146]. Calmodulin (CaM), a calcium activated protein, plays a major role in physiological maintenance of the body functions, such as cell proliferation and neurotransmission. CaM binding properties of food-derived peptides prove vital in preventing many chronic diseases including Alzheimer's and cancer [12]. Nitric oxide synthases (NOSs), protein kinase II and phosphodiesterase I (PDE) are some of the major enzymes that require CaM for their activation. Nitric oxide produced from arginine by CaM-dependent NOS is involved in neurotransmission, immune response and other vital muscle movements along with several other functions in the body. In excessive levels, it may lead to degenerative nervous disorders, such as stroke, Parkinson's, Alzheimer's, etc. NOS exists in three isoforms and depends on CaM for their activation. Inhibitors of CaM are effective against these isoforms. One isoform is involved in tumour necrosis factor production by macrophages and hence CaM inhibitory substances during diseased conditions could provide therapeutic benefits [147]. Similarly, CaM-PDE catalyses the breakdown of cAMP to linear nucleotide phosphate. Apoptotic cell death of normal and cancer cells is induced by cAMP in human body. CaM-PDE in excess causes depletion of cAMP leading to physiological imbalance in cell death. Inhibitors of CaM-PDE increases cAMP cellular levels that could offer therapeutic relief during cancer treatment. BAPs can exhibit effective CaM inhibition when they have hydrophobic AA residues and positive charge at physiological pH to interact with CaM [148]. Pepsin mediated hydrolysis of α-casein released such five peptides that were potential CaM-PDE inhibitors [12]. Whey protein and its derivatives suppress tumour growth by elevating glutathione levels and consequent free radical scavenging. The significant abundance of cysteine, an essential AA required for glutathione synthesis, in β-LG may be contributing to peptide's tumour protective role [15]. Tumour preventive role by peptides could also be through their modulatory effect on immunity and cell viability [149]. Tumour development and progression of lung, colon and oesophagus was found to be inhibited in both rat and mouse models after oral administration of bovine Lf. However, studies could not resolve its mode of action [15]. Apoptosis-inducing property of α-LA was exhibited on tumour and immature cells but not on mature cells. This property was attributed to its 3D protein folding that is unlikely of its native form [150, 151]. Case reports presented anti-tumour effect by whey proteins on urogenital cancer when administered as dietary supplements. However, further clinical trials were required to confirm its efficacy as therapeutics [152]. N-terminal peptides of Lf are investigated for their anticancer activity. Fragments of bovine Lf 17–38 induced apoptosis in HL-60 cell lines and fragments 17–41 exerted cytotoxicity in melanoma, colon carcinoma and fibrosarcoma cell lines. Cytotoxicity of Lf was found to be located within FKCRRWQWRM sequence. *In vivo* and *in vitro* cytotoxic effects on neuroblastoma cells and breast cancer cells were exhibited by lactoferricin B and lactoferricin, respectively. Lf further displayed capacity to inhibit angiogenesis required for tumour growth [74]. Human α-LA made lethal to tumours (HAMLET) and bovine α-LA made lethal to tumours (BAMLET) are complexes of calcium depleted or apoform of α-LA that exhibited anti-tumour activity in human studies. Skin papillomas treated with HAMLET (skin application) for 3 weeks exhibited reduction in lesion volume in relatively 75% of test group and further complete lesion resorption with 2-year symptom-free period [153]. Apoptosis along with decreased size and surface atrophy of tumours were observed in bladder cancer patients treated with HAMLET [154]. Daily supplementation of bovine Lf for a year significantly arrested growth of colorectal polyps in animals possibly through its iron chelating property [155]. Bovine Lf-treated colorectal cancer patients undergoing chemotherapy showed relief from chemotherapy side effects (kidney and liver toxicity, mucositis and anaemia) compared to the controls with no impact on normal cells; however the results did not achieve statistical significance [156]. WPH exhibited low tumour multiplicity in rats with chemically induced mammary tumours when fed as diet in comparison with casein hydrolysates, 20% casein and intact whey [157].

5.1.9. Miscellaneous properties

Various other health potentials of casein/whey peptides are also explored by researchers worldwide. Type 2 diabetes, a globally prevalent metabolic disorder characterised by impaired insulin secretion by pancreatic β cells is considered an epidemic in recent past. Its association with increased risk of chronic health issues including CVD and hypertension makes it a public health burden resulting in elevated healthcare cost globally. This has led to research towards finding an economic nutritional approach to improve postprandial glycaemia. Dietary modifications alone in numerous million cases could prevent this disorder [158, 159].

Epidemiological studies support positive correlation between intake of milk and dairy products with its lesser incidence. The positive impact of milk on insulin resistance is majorly attributed to its proteins and peptides that is validated through *in vitro* and *in vivo* studies [101]. Human studies on metabolic effects of milk proteins on glycaemia and insulinaemia exhibited higher insulinotropic activity by whey in comparison with casein, other protein-rich food and supplements possibly owing to faster rate of whey digestion [158]. The positive effect of WPHs as an insulinogenic agent has also been evaluated. Whey proteins are considered as an excellent source of insulin regulators due to its rich (40–50%) essential AA composition. These AAs are primarily involved in inducing and modifying insulin secretion by β cells of pancreas [160]. Experimental evidence suggested faster gastric emptying and subsequent stimulation of incretin, an insulin secretagogue produced by small intestine (SI), in the presence of hydrolysates of whey [71, 161]. Gastric emptying studies revealed influence of different protein sources and their degree of hydrolysis on insulin response. WPHs lacking carbohydrate stimulated rapid release of AAs, dipeptides (Val-Leu and Ile-Leu), along with a significant increase in insulin secretion. Insulinotropic effect of whey hydrolysate is attributed to increased postprandial concentration of AAs, such as Ile, Leu, Lys, Thr, Val, all of which stimulate insulin secretion [162]. Animal studies on diabetes and β cells of pancreas were also appraised [163]. Incretin [glucagon-like peptide (GLP 1) and glucose-dependent insulinotropic polypeptide (GIP)] hormones contribute ~70% insulin release from pancreatic β cells. Dipeptidyl peptidase (DPP IV), an abundant enzyme present in circulation and on cell surfaces, acts on incretin decreasing their plasma levels. Inhibitors of DPP IV could improve half-life of GLP-1 whose activity is preserved even during type 2 diabetes thereby stimulating its insulinotropic effect. BAPs of whey from bovine, caprine and ovine are considered insulinogenic by acting as potential inhibitors of DPP IV [94]. Brandelli et al. have extensively reviewed DPP IV inhibitors from whey proteins and their hydrolysates [40]. Yet another approach in diabetes management involves inhibition of membrane bound α-glucosidase (carbohydrate hydrolysing enzymes) from SI that catalyses release of monosaccharides from complex carbohydrates. Inhibition of α- glucosidase delays carbohydrates degradation and thereby decreasing the postprandial plasma glucose levels [164]. WPI, α-LA, β-LG and Lf hydrolysates from peptic digestion were investigated for their DPP IV and α-glucosidase inhibition potential. α-LA, β-LG and WPI hydrolysates were found to be potential inhibitors for both, whereas Lf and serum albumin could inhibit DPP IV alone suggesting WPHs potential antidiabetic property [165]. β-LG- and WPH-derived peptides by serine protease from *Cucurbita ficifolia* were found to be potent natural inhibitors of α-glucosidase, DPP-IV and ACE. The study supported antidiabetic nature of these peptides through inhibiting incretin and carbohydrate digestion [166].

Muscles play a key role in whole protein body metabolism. They serve as reservoir of AAs, which are required by vital organs and tissues for their protein synthesis when adequate AAs are not available from gut absorption. It also provides hepatic glycogenic precursors. Any impairment in muscle metabolism may thus check pathologic and chronic disease conditions. In order to maintain balance in rate of breakdown and synthesis of protein in vital organs, postprandial absorption of proteins become vital especially for elders and sportsmen [167]. Protein homoeostasis by a suitable diet strategy is critical for growth, maintenance and/or limit loss of protein after stress. The quantity and quality of protein ingested could modulate protein metabolism by its AA composition and digestibility [168]. Dietary protein from milk, meat, etc., provides AAs that intensely stimulate protein synthesis in skeletal muscles [169]. Milk proteins were investigated for their role in muscle mass maintenance and their possible role as anti-aging factor. Whey proteins have proved to possess greater postprandial retention than casein proteins. Improved anabolic properties of whey than that of casein was mainly attributed to their faster digestion and absorption leading to increased availability of AAs to stimulate protein synthesis [170]. Leucine has been recognized as important nutritional signal to stimulate postprandial muscle protein deposit [171]. During aging, impairment of protein synthesis in skeletal muscle postprandial results in a progressive decline in muscle mass leading to sarcopenia [172]. Anabolic resistance towards muscle protein synthesis is considered a major causative factor in sarcopenia [173]. Clinical trial has been registered for whey proteins' effective stimulation of protein accretion in muscle after ingestion of food in comparison with casein [174]. Recent research on myofibrillar protein synthesis after ingestion of whey in young men during rest and after exercise exhibited a 49–56% increase. The study concluded that 20g dose of whey protein is adequate for maximum stimulation of myofibrillar protein synthesis [175].

5.2. Dairy products

Starter cultures employed traditionally in dairy industry for milk fermentation release BAPs. End products of dairy industry, such as fermented milk, cheese, etc., are found to possess peptides that are biophysiologically active. Cheese is the oldest and widely used diary product. It contains abundant casein-derived peptides formed during its production due to rennet-based proteolysis. Subsequent ripening associated secondary proteolysis contributes to the formation of diverse biologically active peptides whose function depends on the nature of proteolytic enzyme used and the ripening stage of cheese [176]. Hydrolysates produced by proteases differ according to their casein specificity. CPPs have been identified in Cheddar, Comte and Herrgard cheese [21]. Various biological activities similar to casein and whey peptides have been associated with varieties of cheese. **Table 1** represents compilation of antihypertensive peptides from selected commercially available cheese. VPP and IPP sequence of several hard, semi-hard and soft cheese were also studied for their ACE inhibitory properties [15, 177–182].

Antioxidant and antimicrobial activity were presented by cheddar and artisanal Coahlo cheese peptides [182, 183]. Studies identified radical scavenging peptides from Burgos cheese made by different proteases of animal (rennet), plant (*C. cardunculus*) and microbial (*Mucor meihei*) origin. Cheese made by animal rennet could yield an antioxidant peptide from α_1-casein [184]. Antihypertensive peptides from Calpis, a Japanese soft drink, were identified as VPP and IPP from α_1-casein and β-casein, respectively [136]. Peptides from fermented milk and yogurt from bovine and sheep's milk were evaluated for their antihypertensive and antioxidant activities [97, 185]. A recent study on fermented camel milk (chal) and bovine milk using protease from LAB exhibited significant antioxidant property in camel milk in comparison with its bovine counterpart [186]. Research stays on to find novel BAPs from various dairy products and to disclose their probable values and health benefits.

Cheese type	Source and peptide information
Cheddar	α_1-casein:1–6, 1–7, 1–9, 24–32, 102–110 β-casein: 47–52, 193–209
Fresco cheese	α_1-casein:1–15, 1–22, 14–23 and 24–34 β-casein:193–205, 193–207, 193–209
Spanish cheese	Several fragments of α1 and β casein
Gouda cheese	α_1-casein: 1–9, β-casein: 60–68
Herrgard cheese	β-casein: 29–105, 29–107, 1–105, 1–107, 29–93, 30–93
Italian cheese mozzarella, crescenza, Italico and Gorgonzola	β-casein: 58–72
Parmigiano-Reggiano cheese	β-casein:8–16, 58–77, α_2-casein: 83–33

Table 1. Antihypertensive peptides from diverse commercial cheese [180–182].

6. Commercial products

Extensive evidence found in literature suggests potentiality of BAPs as therapeutic or dietary supplements/functional foods. Commercial products enriched with BAPs are available in the market with empirical knowledge from research [187]. Abundant potential is found in peptides to be used as dietary supplements or as functional foods; however, they may not always be included in regular intake as additional information regarding safety, cytotoxicity and hypersensitivity as mandatory [47]. Food and oral hygiene products along with various ingredients enriched with BAPs from milk proteins are available commercially. **Table 2** represents selected products from milk-derived peptides with their claimed potential health benefits. Various other food-derived BAPs are also available in market with substantial health benefits associated with them. A few have been approved by FDA for therapeutic uses. One of the major challenges for existing and future commercial peptide products is valid documentation of the functional and health promoting properties associated with them [26, 70].

Protein	Peptide sequence	Commercial product	Food type	Health claim	Website/ manufacturer
Milk proteins	Dipeptide and tripeptides	PeptoPro®	Ingredient	Boosts muscle protein synthesis	http://www.dsm.com/markets/foodandbeverages/en_US/products/nutraceuticals/peptopro.html DSM, Netherlands
α_1-casein	YLGYLEQLLR	Prodiet F200	Confectionary,	Stress reliever	www.ingredia.com/

Protein	Peptide sequence	Commercial product	Food type	Health claim	Website/ manufacturer
		Lactium	drink		Ingredia, France
$\alpha_{1,}$ β-Casein	IPP and VPP	Calpis	Sour milk	Hypotensive	www.calpis.net/ CalpisCo. Japan
β and κ-casein	IPP and VPP	Calpico/ Calpis AMEAL s	Fermented milk	Hypotensive	www.calpis.net/ Calpis Co., Japan
Casein-derived dodeca peptide	FFVAPFPE VFGK	C12 peptide	Peptide ingredient	Hypotensive	www.dmv.nl/en/ DMV, Netherlands
Casein	Casein hydrolysate	Excellion Calcium Caseinate A	Powder	Nutraceutical	www.dmv.nl/en/ DMV, Netherlands
	Sodium caseinate from curd	Excellion OPTIMIX 100	Dried milk protein	Nutraceutical	www.dmv.nl/en/ DMV, Netherlands
	Casein hydrolysates	PROTARMOR™ 80	Ingredient	Weight loss Sports nutrition	http://www.armor-proteines.com/en/products/ protarmor-80/ Armor Proteines, France
	CPP	Caploc	CPP ingredient	Mineral absorption	www.arlafoodsingredients.com/ Arla Foods, Denmark
	CPP	CE90CPP	ingredient	Mineral absorption	www.dmv.nl/en/ DMV, Netherland
	CPP	Kotsu calcium	Soft drink	Mineral absorption	www.asahiinryo.co.jp/ Asahi Soft Drinks Co. Ltd, Japan
	CPP	Tekkotsu Inryou	Soft drink	Mineral absorption	www.suntory.com/softdrink/ Suntory, Japan
	CPP-ACP	Recaldent	Tooth paste	Remineralisation of enamel	http://www.recaldent.com/ p_welcome.asp Cadbury enterprises
Whey protein isolate	GMP (106–109)	BiPRO WPI	Ingredient and infant formula	Antithrombic, anticariogenic Antimicrobial	www.daviscofoods.com/ specialty/bipro.html Davisco foods, Minnesota
Whey protein hydrolysate	β-LG fragments	Biozate 3 and 7	Ingredient	Hypotensive	www.daviscofoods.com/ specialty/biozate.html Davisco foods, Minnesota
α-Lactalbumin	α-LA fragments	Lacprodan® ALPHA-10 and	Health food	Improves sleep and memory	www.arlafoodsingredients.com/ products/alpha-lactalbumin/

Protein	Peptide sequence	Commercial product	Food type	Health claim	Website/ manufacturer
		Lacprodan® ALPHA-20			health-foods/, Arla foods, Denmark
Whey protein	WPC	Hiprotal® Whey Protein 35 and Hiprotal® Whey Protein 45	Ingredient for infant nutrition	Muscle building	www.domo.nl/en/markets-ingredients/list/infant-nutrition-ingredients-list/ FrieslandCampina Domo, Netherlands
Whey peptide	CMP	Lacprodan® CGMP-10	Infant food	Cognitive and neuronal development	www.arlafoodsingredients.com/products/caseinoglycomacropeptide/infant-nutrition/ Arla foods, Denmark
Whey protein	WPH	Peptigen® IF-3087/3090/3012/3080	Infant food	Mineral binding	www.arlafoodsingredients.com/products/whey-hydrolysates/infant-nutrition/Arla foods, Denmark
	WPH	Lacprodan® DI-3071/3021 Lacprodan® HYDRO.milk	Sports nutrition	Vitamins and minerals uptake; oral hygiene improvement	www.arlafoodsingredients.com/products/whey-hydrolysates/sport-nutrition/ Arla foods, Denmark

Table 2. Commercial products from milk protein/peptides [70, 187].

7. Summary and future prospects

BAPs derived from milk proteins have been a subject of growing interest as health-supporting foods due to their varied nutritional and biological properties. Several studies have shown the role of BAPs as functional food and nutraceuticals. Still, the existence of milk-derived peptides in the gastrointestinal tract has been demonstrated through only a handful of studies. In addition, very little or no information about the stability, bioavailability and efficacy of these BAPs generate a main breach in knowledge to allow an improved understanding on their role in human health. There is still a need for the development of integrated research platforms involving inter-disciplinary skills to bring more clarity on the role and mechanism of action of milk-derived BAPs in humans. Developing tools to preserve/augment the activity of BAPs and favour their optimum utilisation in food systems appear to be yet another prerequisite at this stage. Human-volunteer-based clinical studies on standard recommendations to improve the oversight and safety are also required to confirm the likely results obtained at the *in vitro* and animal model systems. To conclude, milk BAPs with their multifunctional assets appear to offer substantial upcoming prospective for the development of products to support health.

Acknowledgements

The authors thank Prof. Leela Iyengar (Adjunct Professor, Jain University) for her valuable suggestions and fruitful discussions. Authors are thankful to Jain University for the support.

Author details

R. Anusha and O.S. Bindhu*

*Address all correspondence to: bindhupradip@yahoo.co.in and os.bindhu@jainuniversity.ac.in

Department of Biochemistry, Centre for Post Graduate Studies, Jain University, Bangalore, India

References

[1] Mendis S, Davis S, Norrving B. Organizational update the world health organization global status report on noncommunicable diseases 2014; one more landmark step in the combat against stroke and vascular disease. Stroke. 2015; 46(5): e121–e122.

[2] World Health Organisation. 2008–2013 action plan for the global strategy for the prevention and control of noncommunicable diseases. 2008.

[3] Roberfroid MB. A European consensus of scientific concepts of functional foods. Nutrition. 2000; 16(7–8): 689–691.

[4] Coppens P, Da Silva MF, Pettman S. European regulations on nutraceuticals, dietary supplements and functional foods: A framework based on safety. Toxicology. 2006; 221(1): 59–74. doi: 10.1016/j.tox.2005.12.022.

[5] Henry C. Functional foods. European Journal of Clinical Nutrition. 2010; 64(7): 657–659.

[6] Defelice S. Fim, Rationale and proposed guidelines for the nutraceutical research & education Act- NREA, foundation for innovation in medicine. Presented at FIM's 10th Nutraceutical Conference, November 10-11, 2002, The Waldorf-Astoria, New York City & education Act- NREA, foundation for innovation in medicine. Presented at

FIM's 10th Nutraceutical Conference, November 10-11, 2002, The Waldorf-Astoria, New York City

[7] Zeisel SH. Regulation of "nutraceuticals." Science. 1999; 285(5435): 1853–1855.

[8] Jones PJ. Clinical nutrition: 7. Functional foods—more than just nutrition. CMAJ. 2002; 166(12): 1555–1563.

[9] Elfahri K. Release of bioactive peptides from milk proteins by *Lactobacillus* species. Melbourne:Victoria University; 2012.

[10] Arai S. Studies on functional foods in Japan—state of the art. Bioscience, Biotechnology, and Biochemistry. 1996; 60(1): 9–15.

[11] Udenigwe CC. Bioinformatics approaches, prospects and challenges of food bioactive peptide research. Trends in Food Science & Technology. 2014; 36(2): 137–143.

[12] Aluko R. Bioactive peptides. In: Functional foods and nutraceuticals. New York: Springer; 2012. pp. 37–61. doi: 10.1007/978-1-4614-3480-1_3.

[13] Udenigwe CC, Aluko RE. Food protein-derived bioactive peptides: Production, processing, and potential health benefits. Journal of Food Science. 2012; 77(1): R11–R24. doi: 10.1111/j.1750-3841.2011.02455.x.

[14] Korhonen H, Pihlanto A. Bioactive peptides: Production and functionality. International Dairy Journal. 2006; 16(9): 945–960.

[15] Pihlanto A, Korhonen H. Bioactive peptides and proteins. Advances in Food and Nutrition Research. 2003; 47: 175–276.

[16] Guzmán F.Barberis S.Illanes A. Peptide synthesis: chemical or enzymatic Electronic Journal of Biotechnology. 2007 April 15; [Cited 2016 Jan 29]; 10(2): fulltext 13 Available from: http://www.ejbiotechnology.info/index.php/ejbiotechnology/article/view/89

[17] Agyei D, Danquah MK. Industrial-scale manufacturing of pharmaceutical-grade bioactive peptides. Biotechnology Advances. 2011; 29(3): 272–277.

[18] Power O, Jakeman P, Fitzgerald R. Antioxidative peptides: Enzymatic production, in vitro and in vivo antioxidant activity and potential applications of milk-derived antioxidative peptides. Amino Acids. 2013; 44(3): 797–820.

[19] Kitts DD, Weiler K. Bioactive proteins and peptides from food sources. Applications of bioprocesses used in isolation and recovery. Current Pharmaceutical Design. 2003; 9(16): 1309–1323.

[20] Hartmann R, Meisel H. Food-derived peptides with biological activity: From research to food applications. Current Opinion in Biotechnology. 2007; 18(2): 163–169.

[21] Korhonen HJ. Bioactive components in bovine milk.In: Bioactive components in milk and dairy products; 2009. pp. 15–42.

[22] Haque, Emily and Chand, Rattan (2006). Milk protein derived bioactive peptides. [Online]. Available from:https://www.dairyscience.info/index.php/exploitation-of-anti-microbial-proteins/111-milk-protein-derived-bioactivepeptides.html?showall=1&imitstart=. Accessed: 26 October, 2015

[23] Law BA, Tamime AY. Technology of cheese making, vol. 18., Chichester, West Sussex, UK : John Wiley & Sons Ltd. 2011. Barry A. Law and Adnan Y. Tamime (Editors). Technology of Cheesemaking, 2nd Edition 2010, Wiley-Blackwell

[24] Park YW. Bioactive components in milk and dairy products. Chichester, West Sussex, UK: John Wiley & Sons Ltd. 2009. Park Y W (Editor). Bioactive components in milk and dairy products. Wiley-Blackwell, Iowa, USA

[25] Holt C. The milk salts and their interaction with casein. In: Advanced dairy chemistry, vol.3. New York, USA. Springer .1997. p. 233–256

[26] Dziuba B, Dziuba M. Milk proteins-derived bioactive peptides in dairy products: Molecular, biological and methodological aspects. Acta Scientiarum Polonorum Technologia Aliment. 2014; 13(1): 5–25.

[27] Akuzawa R, Miura T, Kawakami H. Bioactive components in caseins, caseinates, and cheeses.In: Bioactive components in milk and dairy products. Wiley-Blackwell, John Wiley & Sons Ltd, Chichester, West Sussex, UK; 2009. 8 p. 215–233. Doi: 10.1002/9780813821504.ch8

[28] Mellander O. The physiological importance of the casein phosphopeptide calcium salts. 2. Peroral calcium dosage of infants. Some aspects of the pathogenesis of rickets. Acta Societatis Botanicorum Poloniae. 1950; 55(5–6): 247–255.

[29] Meisel H, Bockelmann W. Bioactive peptides encrypted in milk proteins: Proteolytic activation and thropho-functional properties. Antonie van Leeuwenhoek. 1999; 76(1–4): 207–215. doi: 10.1023/A:1002063805780.

[30] Fiat A-M, Jollès P. Caseins of various origins and biologically active casein peptides and oligosaccharides: Structural and physiological aspects. Molecular and Cellular Biochemistry. 1989; 87(1): 5–30.

[31] Meisel H. Biochemical properties of peptides encrypted in bovine milk proteins. Current Medicinal Chemistry. 2005; 12(16): 1905–1919.

[32] Polat Z. Integrated approach to whey utilization through natural zeolite adsorption/desorption and fermentation. İzmir Institute of Technology; 2009.

[33] Kosikowski FV. Whey utilization and whey products. Journal of Dairy Science. Netherlands 1979; 23 62(7): 1149–1160. http://dx.doi.org/10.3168/jds.S0022-0302(79)83389-5

[34] Wolfe RR. Protein supplements and exercise. The American Journal of Clinical Nutrition. 2000; 72(2): 551s–557s.

[35] Madureira AR, Pereira CI, Gomes AM, Pintado ME, Malcata FX. Bovine whey proteins–overview on their main biological properties. Food Research International. 2007; 40(10): 1197–1211.

[36] Smithers GW. Whey and whey proteins—from 'gutter-to-gold'. International Dairy Journal. 2008; 18(7): 695–704. doi: 10.1016/j.idairyj.2008.03.008.

[37] Solak BB, Akin N. Health benefits of whey protein: A review. Journal of Food Science and Engineering. 2012; 2(3): 129.

[38] Spălățelu C. Biotechnological valorisation of whey. Innovative Romanian Food Biotechnology. 2012; 10: 1–8.

[39] Madureira AR, Tavares T, Gomes AMP, Pintado ME, Malcata FX. Invited review: Physiological properties of bioactive peptides obtained from whey proteins. Journal of Dairy Science. 2010; 93(2): 437–455. doi: 10.3168/jds.2009-2566.

[40] Brandelli A, Daroit DJ, Corrêa APF. Whey as a source of peptides with remarkable biological activities. Food Research International. 2015; 73: 149–161.

[41] Widyastuti Y, Febrisiantosa A. The role of lactic acid bacteria in milk fermentation. Food and Nutrition Sciences. 2014; 5(4): 435.

[42] Pihlanto A. Lactic fermentation and bioactive peptides. Rijeka, Croatia: INTECH Open Access Publisher, 2013. DOI: 10.5772/51692.ch13

[43] Choi J, Sabikhi L, Hassan A, Anand S. Bioactive peptides in dairy products. International Journal of Dairy Technology. 2012; 65(1): 1–12. doi: 10.1111/j.1471-0307.2011.00725.x.

[44] Ebner J, Arslan AA, Fedorova M, Hoffmann R, Küçükçetin A, Pischetsrieder M. Peptide profiling of bovine kefir reveals 236 unique peptides released from caseins during its production by starter culture or kefir grains. Journal of Proteomics. 2015; 117: 41–57.

[45] Gill HS, Doull F, Rutherfurd K, Cross M. Immunoregulatory peptides in bovine milk. British Journal of Nutrition. 2000; 84(S1): 111–117.

[46] Ueno K, Mizuno S, Yamamoto N. Purification and characterization of an endopeptidase that has an important role in the carboxyl terminal processing of antihypertensive peptides in *Lactobacillus helveticus* cm4. Letters in Applied Microbiology. 2004; 39(4): 313–318.10.1111/j.1472-765X.2004.01560.x.

[47] Phelan M, Aherne A, Fitzgerald RJ, O'brien NM. Casein-derived bioactive peptides: Biological effects, industrial uses, safety aspects and regulatory status. International Dairy Journal. 2009; 19(11): 643–654. doi: 10.1016/j.idairyj.2009.06.001.

[48] Silva SV, Pihlanto A, Malcata FX. Bioactive peptides in ovine and caprine cheeselike systems prepared with proteases from cynara cardunculus. Journal of Dairy Science. 2006; 89(9): 3336–3344. doi: 10.3168/jds.S0022-0302(06)72370-0.

[49] Tavares TG, Malcata FX. Whey proteins as source of bioactive peptides against hypertension. Rijeka, Croatia: INTECH Open Access Publisher. 2013. DOI: 10.5772/3318 ch4

[50] Elmazar MME, El-Sayed ST, Al-Azzouny RA. Screening some local Egyptian seeds for different proteolytic enzymes and their application to produce potent ACE-in milk hydrolysates. Journal of American Science. 2012; 8(3): 644-650.

[51] Di Pierro G, O'keeffe MB, Poyarkov A, Lomolino G, Fitzgerald RJ. Antioxidant activity of bovine casein hydrolysates produced by ficus carica l.-derived proteinase. Food Chemistry. 2014; 156: 305–311.

[52] Gonzalez-Gonzalez C, Gibson T, Jauregi P. Novel probiotic-fermented milk with angiotensin i-converting enzyme inhibitory peptides produced by *Bifidobacterium bifidum* mf 20/5. International Journal of Food Microbiology. 2013; 167(2): 131–137.

[53] Guinane CM, Kent RM, Norberg S, O'connor PM, Cotter PD, Hill C, Fitzgerald GF, Stanton C, Ross RP. Generation of the antimicrobial peptide caseicin a from casein by hydrolysis with thermolysin enzymes. International Dairy Journal. 2015; 49: 1–7.

[54] De Gobba C, Tompa G, Otte J. Bioactive peptides from caseins released by cold active proteolytic enzymes from *Arsukibacterium ikkense*. Food Chemistry. 2014; 165: 205–215.

[55] Ahtesh F, Stojanovska L, Shah N, Mishra VK. Effect of flavourzyme((r)) on angiotensin-converting enzyme inhibitory peptides formed in skim milk and whey protein concentrate during fermentation by lactobacillus helveticus. Journal of Food Science. 2016; 81(1): M135–M143. doi: 10.1111/1750-3841.13177.

[56] Kimmerlin T, Seebach D, Hilvert D. Synthesis of β3-peptides and mixed α/β3-peptides by thioligation. Helvetica Chimica Acta. 2002; 85(6): 1812–1826. doi: 10.1002/1522-2675(200206)85:6<1812::AID-HLCA1812>3.0.CO;2-E.

[57] Gracia SR, Gaus K, Sewald N. Synthesis of chemically modified bioactive peptides: Recent advances, challenges and developments for medicinal chemistry. Future Medicinal Chemistry. 2009; 1(7): 1289–1310.

[58] Perez Espitia PJ, De Fátima Ferreira Soares N, Dos Reis Coimbra JS, De Andrade NJ, Souza Cruz R, Alves Medeiros EA. Bioactive peptides: Synthesis, properties, and applications in the packaging and preservation of food. Comprehensive Reviews in Food Science and Food Safety. 2012; 11(2): 187–204. doi: 10.1111/j. 1541-4337.2011.00179.x.

[59] Saito H, Inui K-L. Dipeptide transporters in apical and basolateral membranes of the human intestinal cell line caco-2. American Journal of Physiology-Gastrointestinal and Liver Physiology. 1993; 265(2): G289–G294.

[60] Rubio-Aliaga I, Daniel H. Peptide transporters and their roles in physiological processes and drug disposition. Xenobiotica. 2008; 38(7–8): 1022–1042.

[61] Daniel H. Molecular and integrative physiology of intestinal peptide transport. Annual Review of Physiology. 2004; 66: 361–384.

[62] Adibi SA. Regulation of expression of the intestinal oligopeptide transporter (pept-1) in health and disease. American Journal of Physiology-Gastrointestinal and Liver Physiology. 2003; 285(5): G779–G788.

[63] Wada Y, Lönnerdal B. Bioactive peptides derived from human milk proteins—mechanisms of action. The Journal of Nutritional Biochemistry. 2014; 25(5): 503–514.

[64] Shen W-C, Wan J, Ekrami H. Penetration enhancement for polypeptides through epithelia (c) means to enhance penetration. Advanced Drug Delivery Reviews. 1992; 8(1): 93–113. doi: 10.1016/0169-409X(92)90016-J.

[65] Shimizu M, Tsunogai M, Arai S. Transepithelial transport of oligopeptides in the human intestinal cell, caco-2. Peptides. 1997; 18(5): 681–687.

[66] Satake M, Enjoh M, Nakamura Y, Takano T, Kawamura Y, Arai S, Shimizu M. Transepithelial transport of the bioactive tripeptide, val-pro-pro, in human intestinal caco-2 cell monolayers. Bioscience, Biotechnology, and Biochemistry. 2002; 66(2): 378–384.

[67] Miner-Williams WM, Stevens BR, Moughan PJ. Are intact peptides absorbed from the healthy gut in the adult human? Nutrition Research Reviews. 2014; 27(02): 308–329.

[68] Baker J, Hindalgo IJ, Borchardt RT. Intestinal epithelial barrier and vascular endothelial barrier to peptide and protein delivery. In: Lee V, editor. Peptide and protein drug delivery, vol. 4. NewYork: CRC Press; 1990.

[69] Jao C-L, Huang S-L, Hsu K-C. Angiotensin i-converting enzyme inhibitory peptides: Inhibition mode, bioavailability, and antihypertensive effects. BioMedicine. 2012; 2(4): 130–136.

[70] Nongonierma AB, O'keeffe MB, Fitzgerald RJ. Milk protein hydrolysates and bioactive peptides. In: McSweeney PLH, O'Mahony JA, editors. Advanced dairy chemistry, New York: Springer; 2016. pp. 417–482.10.1007/978-1-4939-2800-2_15.

[71] Nongonierma AB, Fitzgerald RJ. The scientific evidence for the role of milk protein-derived bioactive peptides in humans: A review. Journal of Functional Foods. 2015; 17: 640–656.

[72] Silva SV, Malcata FX. Caseins as source of bioactive peptides. International Dairy Journal. 2005; 15(1): 1–15.

[73] Mohanty D, Mohapatra S, Misra S, Sahu P. Milk derived bioactive peptides and their impact on human health—A review. Saudi Journal of Biological Sciences. 2015. 1-6. (Article in Press) Available online 17 June 2015. http://dx.doi.org/10.1016/j.sjbs.2015.06.005

[74] Lopez-Exposito I, Recio I. Protective effect of milk peptides: Antibacterial and antitumor properties. In: Bioactive components of milk. New York, USA: Springer. 2008. p. 271–294.ch11

[75] Fitzgerald R. Potential uses of caseinophosphopeptides. International Dairy Journal. 1998; 8(5): 451–457.

[76] Bouhallab S, Bouglé D. Biopeptides of milk: Caseinophosphopeptides and mineral bioavailability. Reproduction Nutrition Development. 2004; 44(5): 493–498.

[77] Iijima Y, Cai F, Shen P, Walker G, Reynolds C, Reynolds E. Acid resistance of enamel subsurface lesions remineralized by a sugar-free chewing gum containing casein phosphopeptide-amorphous calcium phosphate. Caries Research. 2004; 38(6): 551–556.

[78] García-Nebot MJ, Alegría A, Barberá R, Clemente G, Romero F. Addition of milk or caseinophosphopeptides to fruit beverages to improve iron bioavailability? Food Chemistry. 2010; 119(1): 141–148.

[79] Vegarud GE, Langsrud T, Svenning C. Mineral-binding milk proteins and peptides; occurrence, biochemical and technological characteristics. British Journal of Nutrition. 2000; 84(S1): 91–98.

[80] Rui XU. Calcium binding of peptides derived from enzymatic hydrolysates of whey protein concentrate. International Journal of Dairy Technology. 2009; 62(2): 170–173.10.1111/j.1471-0307.2009.00477.x.

[81] Pan D, Lu H, Zeng X. A newly isolated ca binding peptide from whey protein. International Journal of Food Properties. 2013; 16(5): 1127–1134.

[82] Stadtman ER. Protein oxidation and aging. Free Radical Research. 2006; 40(12): 1250–1258.

[83] Pan D, Guo Y, Jiang X. Anti-fatigue and antioxidative activities of peptides isolated from milk proteins. Journal of Food Biochemistry. 2011; 35(4): 1130–1144.

[84] Suetsuna K, Ukeda H, Ochi H. Isolation and characterization of free radical scavenging activities peptides derived from casein. The Journal of Nutritional Biochemistry. 2000; 11(3): 128–131. doi: 10.1016/S0955-2863(99)00083-2.

[85] Rival SG, Boeriu CG, Wichers HJ. Caseins and casein hydrolysates. 2. Antioxidative properties and relevance to lipoxygenase inhibition. Journal of Agricultural and Food Chemistry. 2001; 49(1): 295–302.

[86] Kim G-N, Jang H-D, Kim C-I. Antioxidant capacity of caseinophosphopeptides prepared from sodium caseinate using alcalase. Food Chemistry. 2007; 104(4): 1359–1365.

[87] Luo Y, Pan K, Zhong Q. Physical, chemical and biochemical properties of casein hydrolyzed by three proteases: Partial characterizations. Food Chemistry. 2014; 155: 146–154.

[88] Peña-Ramos EA, Xiong YL, Arteaga GE. Fractionation and characterisation for antioxidant activity of hydrolysed whey protein. Journal of the Science of Food and Agriculture. 2004; 84(14): 1908–1918.

[89] Pena-Ramos E, Xiong Y. Antioxidative activity of whey protein hydrolysates in a liposomal system. Journal of Dairy Science. 2001; 84(12): 2577–2583.

[90] Sadat L, Cakir-Kiefer C, N'negue M-A, Gaillard J-L, Girardet J-M, Miclo L. Isolation and identification of antioxidative peptides from bovine α-lactalbumin. International Dairy Journal. 2011; 21(4): 214–221.

[91] Zhang Q-X, Wu H, Ling Y-F, Lu R-R. Isolation and identification of antioxidant peptides derived from whey protein enzymatic hydrolysate by consecutive chromatography and q-tof ms. Journal of Dairy Research. 2013; 80(03): 367–373. doi:10.1017/S0022029913000320.

[92] Jrad Z, El Hatmi H, Adt I, Girardet J-M, Cakir-Kiefer C, Jardin J, Degraeve P, Khorchani T, Oulahal N. Effect of digestive enzymes on antimicrobial, radical scavenging and angiotensin i-converting enzyme inhibitory activities of camel colostrum and milk proteins. Dairy Science & Technology. 2014; 94(3): 205–224. doi: 10.1007/s13594-013-0154-1.

[93] Chatterjee A, Kanawjia SK, Khetra Y, Saini P, Mann B. Response surface analyses for administering production of whey protein hydrolysate with hypotensive and antioxidant bioactivity. Indian Journal of Dairy Science. 2015; 68(2): 111-119.

[94] Power O, Fernández A, Norris R, Riera F, Fitzgerald R. Selective enrichment of bioactive properties during ultrafiltration of a tryptic digest of β-lactoglobulin. Journal of Functional Foods. 2014; 9: 38–47.

[95] Corrêa APF, Daroit DJ, Fontoura R, Meira SMM, Segalin J, Brandelli A. Hydrolysates of sheep cheese whey as a source of bioactive peptides with antioxidant and angiotensin-converting enzyme inhibitory activities. Peptides. 2014; 61: 48–55.

[96] Taylor S, Brock J, Kruger C, Berner T, Murphy M. Safety determination for the use of bovine milk-derived lactoferrin as a component of an antimicrobial beef carcass spray. Regulatory Toxicology and Pharmacology. 2004; 39(1): 12–24.

[97] Hernández-Ledesma B, Miralles B, Amigo L, Ramos M, Recio I. Identification of antioxidant and ace-inhibitory peptides in fermented milk. Journal of the Science of Food and Agriculture. 2005; 85(6): 1041–1048.

[98] Peng X, Kong B, Xia X, Liu Q. Reducing and radical-scavenging activities of whey protein hydrolysates prepared with alcalase. International Dairy Journal. 2010; 20(5): 360–365.

[99] Contreras MDM, Sevilla MA, Monroy-Ruiz J, Amigo L, Gómez-Sala B, Molina E, Ramos M, Recio I. Food-grade production of an antihypertensive casein hydrolysate and

resistance of active peptides to drying and storage. International Dairy Journal. 2011; 21(7): 470–476. doi: 10.1016/j.idairyj.2011.02.004.

[100] Nguyen DN, Sangild PT, Li Y, Bering SB, Chatterton DE. Processing of whey modulates proliferative and immune functions in intestinal epithelial cells. Journal of Dairy Science. 2015. 99(2): 959-969

[101] Hsieh C-C, Hernández-Ledesma B, Fernández-Tomé S, Weinborn V, Barile D ,De Moura Bell JMLN. Milk proteins, peptides, and oligosaccharides: Effects against the 21st century disorders. BioMed research international. 2015. Article ID 146840, 16 pages, 2015. doi:10.1155/2015/146840

[102] Saito T. Antihypertensive peptides derived from bovine casein and whey proteins. In: Bioactive components of milk., New York, USA: Springer 2008. p. 295–317.doi: : 10.1007/978-0-387-74087-4ch12

[103] Aihara K, Ishii H, Yoshida M. Casein-derived tripeptide, val-pro-pro (vpp), modulates monocyte adhesion to vascular endothelium. Journal of Atherosclerosis and Thrombosis. 2009; 16(5): 594–603. doi: 10.5551/jat.729.

[104] Recio I, Contreras MDM, Gómez-Sala B, Vázquez M, Fernández-Escribano M, Campo RD. Effect of a casein hydrolysate containing novel peptides in hypertensive subjects. *Annals of Nutrition and Metabolism* 2011;58(3): 16-17

[105] López-Expósito I, Quirós A, Amigo L, Recio I. Casein hydrolysates as a source of antimicrobial, antioxidant and antihypertensive peptides. Le Lait. 2007; 87(4–5): 241–249.

[106] He R, Ma H, Zhao W, Qu W, Zhao J, Luo L, Zhu W. Modelling the QSAR of ACE-inhibitory peptides with ANN and its applied illustration. International Journal of Peptides.2012. Article ID 620609, 9 pages. 2012:1-9 doi: 10.1155/2012/620609

[107] Li Y, Yu J. Research progress in structure-activity relationship of bioactive peptides. Journal of Medicinal Food. 2015; 18(2): 147–156.

[108] Norris R, O'keeffe MB, Poyarkov A, Fitzgerald RJ. Peptide identification and angiotensin converting enzyme (ace) inhibitory activity in prolyl endoproteinase digests of bovine αs-casein. Food Chemistry. 2015; 188: 210–217. doi: 10.1016/j.foodchem.2015.04.130.

[109] Hernández-Ledesma B, Recio I, Ramos M, Amigo L. Preparation of ovine and caprine β-lactoglobulin hydrolysates with ace-inhibitory activity. Identification of active peptides from caprine β-lactoglobulin hydrolysed with thermolysin. International Dairy Journal. 2002; 12(10): 805–812. doi: 10.1016/S0958-6946(02)00080-8.

[110] Hernández-Ledesma B, Miguel M, Amigo L, Aleixandre MA, Recio I. Effect of simulated gastrointestinal digestion on the antihypertensive properties of synthetic β-lactoglobulin peptide sequences. Journal of Dairy Research. 2007; 74(03): 336–339. doi: 10.1017/S0022029907002609.

[111] Fluegel SM, Shultz TD, Powers JR, Clark S, Barbosa-Leiker C, Wright BR, Freson TS, Fluegel HA, Minch JD, Schwarzkopf LK, Miller AJ, Di Filippo MM. Whey beverages decrease blood pressure in prehypertensive and hypertensive young men and women. International Dairy Journal. 2010; 20(11): 753–760. doi: 10.1016/j.idairyj.2010.06.005.

[112] Guesdon, B., Pichon, L, & Tomé, D. Opioid peptides. In: Y. Mine, F. Shahidi, editors,Nutraceutical proteins and peptides in health and disease Boca Raton, Florida, USA: CRC Press(2005). p. 367-376. doi: 10.1201/9781420028836 ch17

[113] Teschemacher H. Atypical opioid peptides. In: Herz A, Akil H, Simon E, editors. Opioids. Springer Berlin Heidelberg; 1993. pp. 499–528. doi: 10.1007/978-3-642-77460-7_21.

[114] Sharma S, Singh R, Rana S. Bioactive peptides: A review. International Journal Bio Automation. 2011; 15(4): 223–250.

[115] Trivedi M, Li Y, Lopez-Toledano M, Clarke A, Deth R. Differential neurogenic effects of casein derived opioid peptides on neuronal stem cells: Implications for redox based epigenetic changes. The Journal of Nutritional Biochemistry. 2015. (Article in Press) published online November 06, 2015. doi: 10.1016/j.jnutbio.2015.10.012.

[116] Teschemacher H. Opioid receptor ligands derived from food proteins. Current Pharmaceutical Design. 2003; 9(16): 1331–1344.

[117] Martínez-Maqueda D, Miralles B, De Pascual-Teresa S, ReverÓN IS, MuñOz R, Recio I. Food-derived peptides stimulate mucin secretion and gene expression in intestinal cells. Journal of Agricultural and Food Chemistry. 2012; 60(35): 8600–8605.

[118] Martínez-Maqueda D, Miralles B, Ramos M, Recio I. Effect of β-lactoglobulin hydrolysate and β-lactorphin on intestinal mucin secretion and gene expression in human goblet cells. Food Research International. 2013; 54(1): 1287–1291.

[119] Fiat A-M, Migliore-Samour D, Jollès P, Drouet L, Sollier CBD, Caen J. Biologically active peptides from milk proteins with emphasis on two examples concerning antithrombotic and immunomodulating activities. Journal of Dairy Science. 1993; 76(1): 301–310.

[120] Fiat A, Migliore-Samour D, Jollès P. Biologically active peptides from milk proteins with emphasis on antithrombosis and immunomodulation. Journal of Dairy Science. 1990;73 Supplement 1: p.112.

[121] Qian Z-Y, Jollès P, Migliore-Samour D, Schoentgen F, Fiat A-M. Sheep κ-casein peptides inhibit platelet aggregation. Biochimica et Biophysica Acta (BBA)-General Subjects. 1995; 1244(2): 411–417.

[122] Nagpal R, Behare P, Rana R, Kumar A, Kumar M, Arora S, Morotta F, Jain S, Yadav H. Bioactive peptides derived from milk proteins and their health beneficial potentials: An update. Food & Function. 2011; 2(1): 18–27. doi: 10.1039/C0FO00016G.

[123] Jones FS, Simms HS. The bacterial growth inhibitor (lactenin) of milk: I. The preparation in concentrated form. The Journal of Experimental Medicine. 1930; 51(2): 327–339.

[124] Mills S, Ross R, Hill C, Fitzgerald G, Stanton C. Milk intelligence: Mining milk for bioactive substances associated with human health. International Dairy Journal. 2011; 21(6): 377–401.

[125] Hancock RE, Sahl H-G. Antimicrobial and host-defense peptides as new anti-infective therapeutic strategies. Nature Biotechnology. 2006; 24(12): 1551–1557.

[126] Lahov E, Regelson W. Antibacterial and immunostimulating casein-derived substances from milk: Casecidin, isracidin peptides. Food and Chemical Toxicology. 1996; 34(1): 131–145.

[127] Tidona F, Criscione A, Guastella AM, Zuccaro A, Bordonaro S, Marletta D. Bioactive peptides in dairy products. 2009. 2009; 8(3): 26. doi: 10.4081/ijas.2009.315.

[128] Benkerroum N. Antimicrobial peptides generated from milk proteins: A survey and prospects for application in the food industry. A review. International Journal of Dairy Technology. 2010; 63(3): 320–338.

[129] Mccann KB, Shiell BJ, Michalski WP, Lee A, Wan J, Roginski H, Coventry MJ. Isolation and characterisation of a novel antibacterial peptide from bovine αs1-casein. International Dairy Journal. 2006; 16(4): 316–323. doi: 10.1016/j.idairyj.2005.05.005.

[130] Mccann KB, Shiell BJ, Michalski WP, Lee A, Wan J, Roginski H, Coventry MJ. Isolation and characterisation of antibacterial peptides derived from the f(164–207) region of bovine αs2-casein. International Dairy Journal. 2005; 15(2): 133–143. doi: 10.1016/j.idairyj.2004.06.008.

[131] Aimutis WR. Bioactive properties of milk proteins with particular focus on anticariogenesis. The Journal of Nutrition. 2004; 134(4): 989S–995S.

[132] Gobbetti M, Minervini F, Rizzello CG. Angiotensin i-converting-enzyme-inhibitory and antimicrobial bioactive peptides. International Journal of Dairy Technology. 2004; 57(2–3): 173–188.

[133] Salami M, Moosavi-Movahedi AA, Ehsani MR, Yousefi R, Haertlé T, Chobert J-M, Razavi SH, Henrich R, Balalaie S, Ebadi SA. Improvement of the antimicrobial and antioxidant activities of camel and bovine whey proteins by limited proteolysis. Journal of Agricultural and Food Chemistry. 2010; 58(6): 3297–3302.

[134] Shahidi F, Zhong Y. Bioactive peptides. Journal of AOAC International. 2008; 91(4): 914–931.

[135] Meisel H. Biochemical properties of regulatory peptides derived from mil proteins. Peptide Science. 1997; 43(2): 119–128.

[136] Gobbetti M, Stepaniak L, De Angelis M, Corsetti A, Di Cagno R. Latent bioactive peptides in milk proteins: Proteolytic activation and significance in dairy processing. Critical Reviews in Food Science and Nutrition. 2002; 42(3): 223–239.

[137] O'sullivan D, Lahart N, O'callaghan Y, O'brien NM, Fitzgerald RJ. Characterisation of

the physicochemical, residual antigenicity and cell activity properties of transglutaminase cross-linked sodium caseinate hydrolysates. International Dairy Journal. 2013; 33(1): 49–54.

[138] Gauthier SF, Pouliot Y, Saint-Sauveur D. Immunomodulatory peptides obtained by the enzymatic hydrolysis of whey proteins. International Dairy Journal. 2006; 16(11): 1315–1323. doi: 10.1016/j.idairyj.2006.06.014.

[139] Li EW, Mine Y. Immunoenhancing effects of bovine glycomacropeptide and its derivatives on the proliferative response and phagocytic activities of human macrophagelike cells, u937. Journal of Agricultural and Food Chemistry. 2004; 52(9): 2704–2708.

[140] Eriksen EK, Vegarud GE, Langsrud T, Almaas H, Lea T. Effect of milk proteins and their hydrolysates on in vitro immune responses. Small Ruminant Research. 2008; 79(1): 29–37. doi: 10.1016/j.smallrumres.2008.07.003.

[141] Jacquot A, Gauthier SF, Drouin R, Boutin Y. Proliferative effects of synthetic peptides from β-lactoglobulin and α-lactalbumin on murine splenocytes. International Dairy Journal. 2010; 20(8): 514–521. doi: 10.1016/j.idairyj.2010.02.013.

[142] Yanagawa M, Fukatsu K, Mitsui T, Murakoshi S, Yasuhara H, Nishimura R. Effects of a new immune-modulating diet enriched with whey-hydrolyzed peptide, fermented milk, and isomaltulose on gut associated lymphoid tissue in mice. e-SPEN Journal. 2013; 8(6): e241–e245.

[143] Kume H, Okazaki K, Takahashi T, Yamaji T. Protective effect of an immune-modulating diet comprising whey peptides and fermented milk products on indomethacin-induced small-bowel disorders in rats. Clinical Nutrition. 2014; 33(6): 1140–1146.

[144] Hatzoglou A, Bakogeorgou E, Hatzoglou C, Martin P-M, Castanas E. Antiproliferative and receptor binding properties of α-and β-casomorphins in the t47d human breast cancer cell line. European Journal of Pharmacology. 1996; 310(2): 217–223.

[145] Pepe G, Tenore GC, Mastrocinque R, Stusio P, Campiglia P. Potential anticarcinogenic peptides from bovine milk. Journal of Amino Acids. 2013. Article ID 939804. 2013: 1-13. doi: 10.1155/2013/939804.

[146] Perego S, Cosentino S, Fiorilli A, Tettamanti G, Ferraretto A. Casein phosphopeptides modulate proliferation and apoptosis in ht-29 cell line through their interaction with voltage-operated l-type calcium channels. The Journal of Nutritional Biochemistry. 2012; 23(7): 808–816.

[147] Omoni AO, Aluko RE. Effect of cationic flaxseed protein hydrolysate fractions on the in vitro structure and activity of calmodulin-dependent endothelial nitric oxide synthase. Molecular Nutrition &Food Research. 2006; 50(10): 958–966.

[148] Aluko RE, Food protein-derived peptides as calmodulin inhibitors. In: Bioactive proteins and peptides as functional foods and nutraceuticals. E.C.Y Li-Chan, Y. Mine

and B. Jiang, editors. Singapore: Blackwell. 2010. p. 55–65. doi: 10.1002/9780813811048.ch5

[149] Wright S, Zhong J, Larrick J. Inhibition of apoptosis as a mechanism of tumor promotion. The FASEB Journal. 1994; 8(9): 654–660.

[150] Svensson M, Sabharwal H, Håkansson A, Mossberg A-K, Lipniunas P, Leffler H, Svanborg C, Linse S. Molecular characterization of α–lactalbumin folding variants that induce apoptosis in tumor cells. Journal of Biological Chemistry. 1999; 274(10): 6388–6396.

[151] Håkansson A, Andréasson J, Zhivotovsky B, Karpman D, Orrenius S, Svanborg C. Multimeric α-lactalbumin from human milk induces apoptosis through a direct effect on cell nuclei. Experimental Cell Research. 1999; 246(2): 451–460.

[152] Bounous G. Whey protein concentrate (wpc) and glutathione modulation in cancer treatment. Anticancer Research. 2000; 20(6): 4785–4792.

[153] Gustafsson L, Leijonhufvud I, Aronsson A, Mossberg A-K, Svanborg C. Treatment of skin papillomas with topical α-lactalbumin–oleic acid. New England Journal of Medicine. 2004; 350(26): 2663–2672.

[154] Mossberg AK, Wullt B, Gustafsson L, Månsson W, Ljunggren E, Svanborg C. Bladder cancers respond to intravesical instillation of (hamlet human α-lactalbumin made lethal to tumor cells). International Journal of Cancer. 2007; 121(6): 1352–1359.

[155] Kozu T, Iinuma G, Ohashi Y, Saito Y, Akasu T, Saito D, Alexander DB, Iigo M, Kakizoe T, Tsuda H. Effect of orally administered bovine lactoferrin on the growth of adenomatous colorectal polyps in a randomized, placebo-controlled clinical trial. Cancer Prevention Research. 2009; 2(11): 975–983.

[156] Moastafa TM, El-Sissy AE-DE, El-Saeed GK, Koura MSE-D. Study on the therapeutic benefit on lactoferrin in patients with colorectal cancer receiving chemotherapy. International Scholarly Research Notices. 2014; Article ID 184278, 10 pages. 2014:1-10. doi:10.1155/2014/184278

[157] Ronis MJ, Hakkak R, Korourian S, Badger TM. Whey protein hydrolysate but not whole whey protein protects against 7, 12-dimethylbenz (a) anthracene-induced mammary tumors in rats. Nutrition and Cancer. 2015; 67(6): 949–953.

[158] Pasin G, Comerford KB. Dairy foods and dairy proteins in the management of type 2 diabetes: A systematic review of the clinical evidence. Advances in Nutrition: An International Review Journal. 2015; 6(3): 245–259. doi: 10.3945/an.114.007690.

[159] Mignone LE, Wu T, Horowitz M, Rayner CK. Whey protein: The "whey" forward for treatment of type 2 diabetes? World Journal of Diabetes. 2015; 6(14): 1274–1284. doi: 10.4239/wjd.v6.i14.1274.

[160] Power O, Hallihan A, Jakeman P. Human insulinotropic response to oral ingestion of

native and hydrolysed whey protein. Amino Acids. 2009; 37(2): 333–339. doi: 10.1007/s00726-008-0156-0.

[161] Jakubowicz D, Froy O, Ahrén B, Boaz M, Landau Z, Bar-Dayan Y, Ganz T, Barnea M, Wainstein J. Incretin, insulinotropic and glucose-lowering effects of whey protein pre-load in type 2 diabetes: A randomised clinical trial. Diabetologia. 2014; 57(9): 1807–1811.

[162] Morifuji M, Ishizaka M, Baba S, Fukuda K, Matsumoto H, Koga J, Kanegae M, Higuchi M. Comparison of different sources and degrees of hydrolysis of dietary protein: Effect on plasma amino acids, dipeptides, and insulin responses in human subjects. Journal of Agricultural and Food Chemistry. 2010; 58(15): 8788–8797. doi: 10.1021/jf101912n.

[163] Gaudel C, Nongonierma AB, Maher S, Flynn S, Krause M, Murray BA, Kelly PM, Baird AW, Fitzgerald RJ, Newsholme P. A whey protein hydrolysate promotes insulinotropic activity in a clonal pancreatic beta-cell line and enhances glycemic function in ob/ob mice. The Journal of Nutrition 2013; 143(7): 1109–1114. doi: 10.3945/jn.113.174912.

[164] Kahn SE, Cooper ME, Del Prato S. Pathophysiology and treatment of type 2 diabetes: Perspectives on the past, present, and future. Lancet. 2014; 383(9922): 1068–1083. doi: 10.1016/s0140-6736(13)62154-6.

[165] Lacroix IME, Li-Chan ECY. Inhibition of dipeptidyl peptidase (dpp)-iv and α-glucosidase activities by pepsin-treated whey proteins. Journal of Agricultural and Food Chemistry. 2013; 61(31): 7500–7506. doi: 10.1021/jf401000s.

[166] Konrad B, Anna D, Marek S, Marta P, Aleksandra Z, Jozefa C. The evaluation of dipeptidyl peptidase (dpp)-iv, alpha-glucosidase and angiotensin converting enzyme (ace) inhibitory activities of whey proteins hydrolyzed with serine protease isolated from Asian pumpkin. International Journal of Peptide Research and Therapeutics. 2014; 20(4): 483–491. doi: 10.1007/s10989-014-9413-0.

[167] Wolfe RR. The underappreciated role of muscle in health and disease. The American Journal of Clinical Nutrition. 2006; 84(3): 475–482.

[168] Dangin M, Boirie Y, Guillet C, Beaufrère B. Influence of the protein digestion rate on protein turnover in young and elderly subjects. The Journal of Nutrition. 2002; 132(10): 3228S–3233S.

[169] Koopman R, Crombach N, Gijsen AP, Walrand S, Fauquant J, Kies AK, Lemosquet S, Saris WH, Boirie Y, Van Loon LJ. Ingestion of a protein hydrolysate is accompanied by an accelerated in vivo digestion and absorption rate when compared with its intact protein. American Journal of Clinical Nutrition. 2009; 90(1): 106–115. doi: 10.3945/ajcn.2009.27474.

[170] Koopman R, Walrand S, Beelen M, Gijsen AP, Kies AK, Boirie Y, Saris WH, Van Loon LJ. Dietary protein digestion and absorption rates and the subsequent postprandial muscle protein synthetic response do not differ between young and elderly men. The Journal of Nutrition. 2009; 139(9): 1707–1713. doi: 10.3945/jn.109.109173.

[171] Katsanos CS, Kobayashi H, Sheffield-Moore M, Aarsland A, Wolfe RR. A high proportion of leucine is required for optimal stimulation of the rate of muscle protein synthesis by essential amino acids in the elderly. American Journal of Physiology-Endocrinology and Metabolism. 2006; 291(2): E381–E387.

[172] Volpi E, Mittendorfer B, Rasmussen BB, Wolfe RR. The response of muscle protein anabolism to combined hyperaminoacidemia and glucose-induced hyperinsulinemia is impaired in the elderly 1. The Journal of Clinical Endocrinology & Metabolism. 2000; 85(12): 4481–4490.

[173] Cuthbertson D, Smith K, Babraj J, Leese G, Waddell T, Atherton P, Wackerhage H, Taylor PM, Rennie MJ. Anabolic signaling deficits underlie amino acid resistance of wasting, aging muscle. The FASEB Journal. 2005; 19(3): 422–424.

[174] Pennings B, Boirie Y, Senden JM, Gijsen AP, Kuipers H, Van Loon LJ. Whey protein stimulates postprandial muscle protein accretion more effectively than do casein and casein hydrolysate in older men. The American Journal of Clinical Nutrition. 2011; 93(5): 997–1005.

[175] Witard OC, Jackman SR, Breen L, Smith K, Selby A, Tipton KD. Myofibrillar muscle protein synthesis rates subsequent to a meal in response to increasing doses of whey protein at rest and after resistance exercise. The American Journal of Clinical Nutrition. 2014; 99(1): 86–95.

[176] Upadhyay V, Mcsweeney P, Magboul A, Fox P. Proteolysis in cheese during ripening. Cheese: Chemistry, Physics and Microbiology. 2004; 1: 391–433.

[177] Gómez-Ruiz JÁ, Taborda G, Amigo L, Recio I, Ramos M. Identification of ace-inhibitory peptides in different spanish cheeses by tandem mass spectrometry. European Food Research and Technology. 2006; 223(5): 595–601.

[178] Ong L, Henriksson A, Shah NP. Angiotensin converting enzyme-inhibitory activity in cheddar cheeses made with the addition of probiotic lactobacillus casei sp. Le Lait. 2007; 87(2): 149–165.

[179] Bütikofer U, Meyer J, Sieber R, Walther B, Wechsler D. Occurrence of the angiotensin-converting enzyme-inhibiting tripeptides val-pro-pro and ile-pro-pro in different cheese varieties of swiss origin. Journal of Dairy Science. 2008; 91(1): 29–38.

[180] Torres-Llanez M, González-Córdova A, Hernandez-Mendoza A, Garcia H, Vallejo-Cordoba B. Angiotensin-converting enzyme inhibitory activity in mexican fresco cheese. Journal of Dairy Science. 2011; 94(8): 3794–3800.

[181] Ardö Y, Lilbaek H, Kristiansen K, Zakora M, Otte J. Identification of large phosphopeptides from β-casein that characteristically accumulate during ripening of the semi-hard cheese herrgård. International Dairy Journal. 2007; 17(5): 513–524.

[182] Pritchard SR, Phillips M, Kailasapathy K. Identification of bioactive peptides in commercial cheddar cheese. Food Research International. 2010; 43(5): 1545–1548.

[183] Silva RA, Lima MS, Viana JB, Bezerra VS, Pimentel MC, Porto AL, Cavalcanti MT, Lima Filho JL. Can artisanal "coalho" cheese from northeastern brazil be used as a functional food? Food Chemistry. 2012; 135(3): 1533–1538.doi: 10.1016/j.foodchem.2012.06.058.

[184] Timón ML, Parra V, Otte J, Broncano JM, Petrón MJ. Identification of radical scavenging peptides (<3 kda) from burgos-type cheese. LWT-Food Science and Technology. 2014; 57(1): 359–365.

[185] Chobert J-M, El-Zahar K, Sitohy M, Dalgalarrondo M, Métro F, Choiset Y, Haertlé T. Angiotensin i-converting-enzyme (ace)-inhibitory activity of tryptic peptides of ovine β-lactoglobulin and of milk yoghurts obtained by using different starters. Le Lait. 2005; 85(3): 141–152.

[186] Soleymanzadeh N, Mirdamadi S, Kianirad M. Antioxidant activity of camel and bovine milk fermented by lactic acid bacteria isolated from traditional fermented camel milk (chal). Dairy Science & Technology. 2016: 1–15. doi: 10.1007/s13594-016-0278-1.

[187] Carrasco-Castilla J, Hernández-Álvarez AJ, Jiménez-Martínez C, Gutiérrez-López GF, Dávila-Ortiz G. Use of proteomics and peptidomics methods in food bioactive peptide science and engineering. Food Engineering Reviews. 2012; 4(4): 224–243.

6

Usefulness of Faecal Markers in Cow's Milk Protein Immunomediated Reactions

Maria Elisabetta Baldassarre, Raffaella Panza and Nicola Laforgia

Abstract

Cow's milk protein allergy (CMPA) affects children most commonly than adults, with symptoms usually developing before 1 year of age and within 1 week after the intake of cow's milk. Food allergies can be divided into: IgE mediated and non-IgE mediated. Some reactions may include both mechanisms (mixed type). The most studied faecal Mar-kers, so far, are calprotectin, Tumor necrosis factor-alpha (TNF-α), beta-defensin and eosinophil cationic protein (ECP). Calprotectin belongs to the S-100 family of calcium-binding proteins and seems to be involved in the regulation of inflammation. Faecal calprotectin (FC) values are significantly higher in infants suspected of having CMPA than in a comparison group of healthy infants. Moreover, there is a significant decrease in FC in infants with CMPA after a period of dietary antigen elimination, although levels use to remain higher than in age- and diet-matched comparisons. TNF-α is a cytokine involved in systemic inflammation and is one of the cytokines that make up the acute phase reaction. TNF-α expression in the epithelial cells and mononuclear cells in the lamina propria is markedly increased in FPIES patients. TNF-α is also increased in the stools of patients with gastrointestinal milk allergy after milk challenge. Defensins are small (~29 to 42 amino acid) cationic arginine and cysteine rich, amphipathic peptides with a molecular weight of 3–5 kDa. They can be classified into three groups: α-, β- and θ-defensins. Among them, only α- and β-defensins are expressed in humans. Defensins display various functions, including antimicrobial activity and also act as chemoattractant. They contribute to host immunity and to maintain the balance between pathogens and normal flora. Beta-defensins values detected in infants with a previous diagnosis of CMPA prior to the oral food challenge, and during each provocation do not seem to show significant changes. ECP is a single-chain, zinc-containing protein with a molecular weight ranging from 16 to 22 kDa and is one of the most important proteins in the granules of eosinophil granulocytes. Infants with atopic eczema exhibit a specific faecal protein pattern characterized by an increase in both ECP and TNF-α. The faecal concentration of ECP enhances particularly in patients with immediate-type reactions to the cow's milk challenge whereas faecal TNF-α enhances in those with delayed-type reactions, confirming the different pathogenesis (IgE mediated and non-IgE mediated) of these two types of reactions.

Keywords: Cow's milk protein allergy, faecal markers, faecal calprotectin, human β-defensins, TNF-α, eosinophilic cationic protein

1. Introduction

Food allergy is an abnormal immune response to components of the diet, in particular to proteins. Food allergy can manifest itself by configuring different clinical entities, including atopic dermatitis, gastrointestinal or respiratory symptoms, and anaphylaxis.

The clinical history and sometimes allergen-specific serum IgE tests, skin tests and/or elimination diets can help you achieve the diagnosis.

The therapy is mainly based on the elimination of the food that triggers the reaction.

The prevalence of true food allergy ranges from <1 to 3% and varies by geography and method of assessment.

It is important not to confuse food allergy with non-immune reactions to food (e.g. lactose intolerance, irritable bowel syndrome, infectious gastroenteritis) and reactions to food contaminants (e.g. latex dust in food handled by workers wearing latex gloves) or additives (e.g. monosodium glutamate, metabisulphite, tartrazine), which cause most food reactions [1].

Milk allergy is one of the most common food allergies in infants and children. It consists in an abnormal immune response towards milk and its products.

There are two main groups of proteins in cow's milk that can cause an allergic reaction:

- Casein, found in the solid part (curd) of milk that curdles
- Whey, found in the liquid part of milk that remains after milk curdles

Cow's milk is the most usual cause—as it is the most consumed milk worldwide—but milk from sheep, goats, buffalo and other mammals can cause a reaction as well. Less commonly, people allergic to cow's milk are also allergic to soya milk.

Food allergy can be mediated by IgE, T cells or both. IgE-mediated allergy (e.g. asthma, urticaria or anaphylaxis) presents acute onset, usually starts during infancy, and generally occurs in people with a strong family history of atopy. On the other hand, T-cell-mediated allergy (e.g. celiac disease or dietary protein gastroenteropathies) develops gradually and is chronic; it is common among infants and children [1]. Allergies mediated by both IgE and T cells (e.g. atopic dermatitis, eosinophilic gastroenteropathy) tend to be delayed in onset or chronic.

CMPA produces a range of symptoms and clinical entities (**Figure 1**). The principal symptoms are gastrointestinal (GI-CMPA) and dermatological and are represented by atopic dermatitis, vomiting and gastrointestinal distress such as infantile colic, gastroesophageal reflux, esophagitis, diarrhoea (typically in the very young), constipation (typically in older children), stomach pain or flatulence. In some cases, particularly in infants, it may also cause proctocolitis, with the presence of blood and/or mucus in the stools.

According to the main clinical manifestations, organic GI-CMPA in infants can be usefully classified [2] into:

- food protein-induced enterocolitis syndrome (FPIES),
- food protein-induced proctocolitis (FPIP),
- food protein-induced enteropathy.

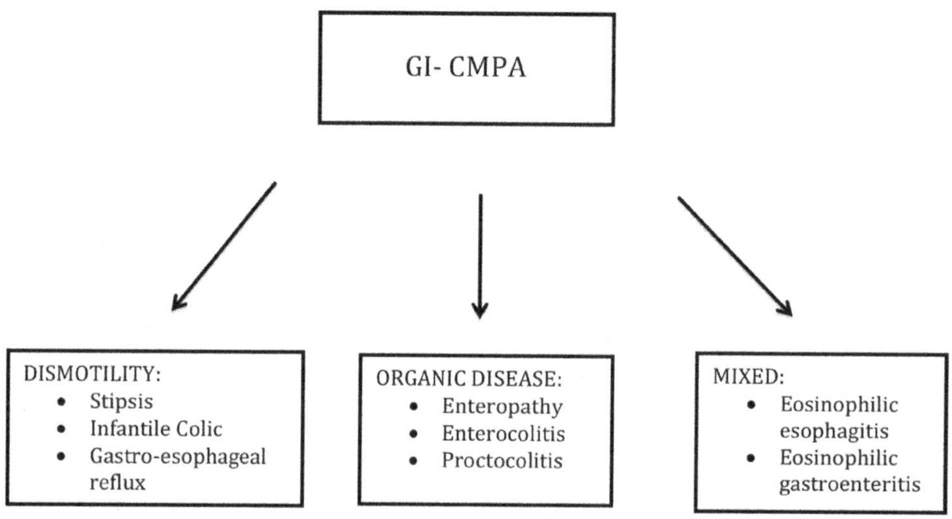

Figure 1. GI-CMPA clinical manifestations in infants can be classified into three main groups: gastrointestinal dysmotility, organic disease and mixed forms characterized by eosinophilic mucosal infiltration.

Patients with FPIES experience repetitive vomiting, starting 1 or 2 hours after the ingestion of offending foods, followed by diarrhoea. However, these patients do not develop acute cutaneous or respiratory symptoms, which commonly accompany IgE-mediated food allergy. FPIES can manifest with systemic symptoms such as lethargy, hypotension, hypothermia, pallor, ileus, bloody stools, methemoglobinemia, thrombocytosis and, sometimes, high temperature with neutrophilia. Therefore, the first diagnoses to be excluded for these patients are sepsis or surgical abdominal emergency.

Patients with FPIP typically develop grossly blood-streaked stools with mucus in the first few months of life. In contrast to FPIES, almost all patients with FPIP develop no systemic symptoms and seem to be well except for the bloody stools. They have no growth delay or poor weight gain. Mild anaemia is seen in rare cases. Many patients with FPIP are breast fed, and the cause is thought to be mainly cow's milk proteins passed through the breast milk (**Table 1**).

It is still unclear when the sensitization phase of allergic proctocolitis occurs. It is thought that dietary antigens can cross the placental barrier or enter the amniotic fluid, which is swallowed by the foetus, causing in utero sensitization. Another possibility is that variations in the concentration of immunomodulatory substances in human milk can alter the protective effect of breastfeeding against allergy. In particular, maternal leukocytes contained in human milk may play a role in antigen processing and presentation to neonatal lymphocytes in the intestine. Thus, it is possible that the ingestion of dietary food proteins excreted in the mother's milk, in case of particular physiologic conditions favouring immunogenic responses (in the neonate or maternal milk), may result in allergic sensitization. However, there are insufficient data to recommend dietary restriction during pregnancy and/or lactation in order to prevent allergy onset [3].

	IgE mediated	Non–IgE mediated
Calprotectin	++	+++
Beta-defensins	–	–
TNF-α	+	+++
ECP	+++	+/–

Table 1. Faecal markers values in IgE-mediated and non–IgE-mediated CMPA.

Cow's milk and soya-based formulas are the major causative foods in the remaining cases.

Patients with enteropathy typically develop chronic diarrhoea and show poor weight gain in the first several months of life. Mild-to-moderate anaemia and hypoproteinemia were seen in some patients with enteropathy. Enteropathy has to be distinguished from celiac disease (CD), that is associated with sensitivity to wheat protein, and is characterized by similar symptoms (diarrhoea, poor weight gain, sideropenic anaemia) usually occurring when the infant is 7–8 months old, as this is the common period of gluten introduction in the diet.

2. Diagnosis of CMPA and faecal markers

Cow's milk protein allergy (CMPA) affects children most commonly than adults [4], with symptoms usually developing before 1 year of age and within 1 week after the intake of cow's milk [5, 6]. During infancy, symptoms suggesting CMPA are observed in 5–15% of the population; however, when specific diagnostic criteria are used, the incidence of CMPA is approximately 2–5% [7].

Skin prick test and serum-specific IgE tests are helpful for diagnosis, mainly in IgE-mediated forms. On the other hand, in cases of CMPA with gastrointestinal chronic signs and symptoms (GI-CMPA), skin prick test and specific IgE test are usually negative, because these forms are mostly non-IgE mediated. This is the reason why the only reliable method of diagnosis in such cases is a food challenge, double blind and placebo controlled [7]. The food challenge needs to be done in hospital, and it could be also dangerous. This easily explains why there is a growing interest to find any faecal biomarker for diagnosis and follow-up (**Table 2**).

A National Institutes of Health working group defined a *biomarker* as 'a characteristic that is objectively measured and evaluated as an indicator of normal biological processes, pathogenic processes or pharmacological responses to a therapeutic intervention'. A biomarker has to be reproducible, accurate, easy to interpret by the clinician, acceptable for the patient, sensitive and specific for the outcome it is expected to identify [8].

Faecal inflammatory markers are represented by different molecules that leak from or are generated by the inflamed mucosa of the bowel. Therefore, these markers could represent a non-invasive means of evaluating objectively mucosal inflammation [9].

The most studied faecal markers, so far, are calprotectin, tumor necrosis factor-alpha (TNF-α), β-defensin and eosinophil cationic protein (ECP).

	GA	Birth weight	Delivery	Antibiotic	Diet	Microbiota	NEC	Gender
FC	None	– (only in term)	+ (only for C-section in preterm)	– (in very low birth weight (VLBW))	Not agree on	–		
HBD2	+		–		–		+	
TNF-α	–		–		–			
ECP								–

Table 2. Relations between faecal markers values in term and preterm newborns and different variables.

3. Calprotectin

Calprotectin was first discovered as a protein with antibacterial activity [9]. It can be found in the cytoplasm of neutrophil granulocytes where it forms about 60% of cytosolic proteins [10], but it is also expressed on the cell membranes of monocytes and in some mucosal epithelial cells [11, 12]. It is a 36.5 kDa heterodimer composed of one light (MRP8) and two heavy (MRP14) chains (8 and 14 kDa) and belongs to the S-100 family of calcium-binding proteins [13, 14]. Calprotectin also contains histidine-based zinc-binding sequences (His-X-X-X-His motif) involved in its antibacterial activity [15]. Although its exact biological function is not known, calprotectin was shown to have bactericidal and fungicidal properties [16]. It is thought that, binding calcium and zinc, calprotectin deprives microorganisms of zinc and additionally inhibits many zinc-dependent enzymes [17–21].

Various data also suggest that it may be involved in the regulation of inflammation. Calprotectin is secreted extracellularly from stimulated neutrophils [22] and monocytes [23], or is released by cell disruption or death. Once released, calprotectin may be detected in serum, body fluids and faeces [24]. A high level of calprotectin was found in extracellular fluid during several inflammatory diseases, such as rheumatoid arthritis [17], cystic fibrosis [25] and active multiple sclerosis. Once released in extracellular fluids, soluble calprotectin provides both bacteriostatic and cytokine-like effects in the local environment. When calprotectin metabolism is affected on a systemic level, the zinc-binding properties of the protein may induce severe dysregulation of zinc homeostasis causing severe clinical symptoms. Only monocytes and immature macrophages present the membrane form of calprotectin; therefore, the presence of calprotectin-positive infiltrating cells is related to the influx of mononuclear phagocytes to the site of inflammation. On the other hand, the intracellular distribution of calprotectin is influenced by the activation state of macrophages. In non-activated macrophages, the protein complex is in the cytosolic fraction; once stimulated, the complex moves towards the cell membrane, thus localizing with proteins of the cytoskeleton. Therefore, calprotectin may be related to phagocytosis, cell movement or signal transduction. Although calprotectin is classified as a specific marker for neutrophils and macrophages, it was also found in other cell types, such as in keratinocytes in inflammatory dermatoses [26] and squamous cell carcinoma. Moreover, the 14 kDa subunit of calprotectin is expressed in a subset of microglia in brain tissue of patients affected by Alzheimer's disease [27]. Since the expression of calprotectin in these cell types seemed to be up-regulated by the inflamed state of the tissue, the functional relevance of the factor to each inflammatory process was suggested [25–27]. In synthesis, calprotectin expression and release seem to be of particular importance in immune and immunopathological reactions. However, the exact biological role(s) of the factor is now under investigation.

Moreover, calprotectin presents growth-inhibitory and apoptosis-inducing activities against various cell types, even including tumour cells and normal fibroblasts. Calprotectin seems to induce apoptosis through a dual mechanism. One is the zinc exclusion from the target cells, and the other is the binding of the factor to target cell surface, possibly in a ligand-receptor fashion [28].

Several gastrointestinal diseases, inflammatory bowel diseases (IBD) among them, can cause a higher release of leukocytes in stools [29, 30].

In adults, a strong correlation between 4-day faecal excretion of 111-indium leukocytes, considered the standard criterion faecal marker of inflammation [31] and faecal calprotectin levels, raised the interest in calprotectin as a marker for intestinal mucosa inflammation [9].

Calprotectin can be easily measured in stools. In fact, the calcium saturated form of calprotectin is highly resistant to proteolysis and colonic bacterial degradation, allowing faeces sample to be kept for up to 1 week at room temperature without any significant degradation [9, 20, 32–34]. Moreover, calprotectin is stable at −20°C for at least 6 months. Several enzyme-linked immunoabsorbent assays (ELISA) using small stool samples (0.1 g) are commercially available. The reference value is 50 μg/g faeces for healthy adults and children aged from 4 to 17 years, regardless of sex [35].

Less agreement has been shown in regard of cut-off values in neonatal age, for both term and preterm newborns. This is mostly due to the high interindividual variations of calprotectin values in this population. Calprotectin is already present at high levels in the first passed meconium, indicating the capability of the foetal GI tract to produce and secrete this protein [9].

Several factors can influence faecal values of calprotectin in newborns, such as gestational age, postnatal age, delivery mode, antibiotic treatment, diet and gut microbiota.

In term infants, there is a negative correlation between calprotectin levels in meconium and gestational age or birthweight [9]. Such correlation was not confirmed in preterm infants, without any obvious reason for this discrepancy [36, 37]. Healthy full-term and preterm infants, especially younger than 3 months old, present high faecal calprotectin levels, comparable to those seen in children or adults affected by IBDs. Full-term and preterm neonates at the same postnatal age do not show significant differences in calprotectin values [38].

Calprotectin values in full-term newborns seem to slightly but significantly increase at day of life 7 compared with values at day of life 3, the levels then remaining similar for the first month of life [24]. Then, a decrease occurs between 6 weeks and 6 months of life [39].

The mode of delivery was not found to influence faecal calprotectin in full-term neonates [24, 40], even if a positive correlation was found with caesarean delivery in preterm infants [37]. No relationships were found with gestational age [37, 41, 42].

Faecal calprotectin was also found to correlate negatively with antibiotic treatments in this population of very low birthweight infants [37].

The influence of diet on calprotectin levels is not agreed on. Some researchers [24, 40] found no differences between breast fed and formula-fed infants during the first month of life in full-term infants, while in another study [43], faecal calprotectin concentrations were found to be significantly lower in breast fed than in formula-fed infants during the 'preweaning' period. Finally, some other researchers [44, 45] described opposite results with higher calprotectin levels in exclusively breast fed infants compared with mixed-fed or formula-fed ones. However, the comparison between the different studies is difficult because infants were recruited at various ages. As calprotectin levels change during the first year of age, the different inclusion criteria could explain the obtained results. Moreover, most of these studies gave poor information about the infants' background and the composition of the formulas, which could affect calprotectin levels [9].

Gut microbiota has been shown to affect faecal calprotectin values. Oral supplementation by Bifidobacterium lactis Bb12, which modified the equilibrium of the gut microbiota, led to a significant decrease in calprotectin levels in the probiotic group compared with the placebo one [46].

As high faecal calprotectin values in newborns show an increment of granulocytes in the intestinal lumen due to enhanced intestinal permeability and/or development of the bowel-associated lymphoid tissue, interindividual variations should be related to environmental factors (e.g. mode of feeding, intestinal colonization or response to dietary antigens) which could individually alter this process [9].

In the light of this, we can now analyze calprotectin usefulness in the management of CMPA.

4. Calprotectin as a predictive factor in CMPA

It was hypothesized that alterations in newborns faecal calprotectin (FC) could be associated with specific disorders in infancy such as atopic dermatitis, cow's milk intolerance, severe infantile colic and gastroesophageal reflux [47], so that calprotectin was proposed as a predictive marker in these pathological conditions.

The predictive value of calprotectin measured at birth as a possible marker of allergic predisposition in the first 2 years of age was tested, but the comparison of calprotectin concentration at birth did not lead to any statistically significant result in allergic vs. non-allergic children at 2 years of age. The levels of FC in the first month of life are not influenced by a possible individual predisposition to atopy [48].

5. Calprotectin, diagnosis and management of CMPA

No single laboratory test is either sensitive or specific enough to be diagnostic of allergic colitis, but the finding of either peripheral eosinophilia or eosinophils in stool samples is often considered suggestive of this condition. It was widely demonstrated that faecal calprotectin (FC) values are significantly higher in infants suspected of having CMPA than in a comparison group of healthy infants. Moreover, there is a significant decrease in faecal calprotectin in infants with CMPA after a period of dietary antigen elimination, although levels use to remain higher than in age- and diet-matched comparisons [7, 49].

FC levels before the CMP elimination diet seem to be higher both in the IgE-mediated CMPA group and in the non-IgE-mediated CMPA group compared with the control group. However, the difference seems to be statistically significant only in the non-IgE-mediated CMPA group (**Table 3**).

Calprotectin	Useful for diagnosis and treatment follow up
	Not susceptible to degradation
HBD2	Not useful
TNF-α	Useful for diagnosis and treatment follow up
	Highly susceptible to degradation
ECP	Usefulness not agreed on

Table 3. Usefulness of faecal markers in CMPA.

On the other hand, a statistically significant difference was found between FC levels before and those after the cow milk's proteins elimination diet both in the IgE-mediated CMPA group and in the non-IgE-mediated CMPA group. According to these findings, FC levels may be useful only in treatment follow-up for the IgE-mediated group, while, in the non-IgE-mediated group, FC may be useful both for the follow-up of treatment and recurrence determination [49].

Moreover, a comparison of the IgE-mediated and non-IgE-mediated groups revealed significantly higher FC levels in the non-IgE-mediated group.

It was supposed that FC levels within the non-IgE-mediated group are higher because gastrointestinal symptoms and colitis are predominant in these patients. Therefore, faecal calprotectin may be more useful to detect relapses during the follow-up of patients in the non-IgE-mediated group, as gastrointestinal involvement is more common in this patients' population.

Given that FC can increase in case of several inflammatory bowel conditions, FC can be useful only to determine relapses and follow ups after diagnosing patients as CMPA particularly with gastrointestinal involvement [7].

It is to say that the addition of Lactobacillus GG (LGG) to an extensively hydrolysed casein formula significantly improves the recovery of the inflamed colonic mucosa as indicated indirectly by greater decreases in faecal calprotectin and in the number of infants with the persistence of occult blood in stools after 1 month. The mechanisms of this beneficial effect are not well known but may be linked to the effects that LGG has on enhancing the intestinal mucosa's barrier function, cooperating in breakdown of protein antigens, competing with bacterial pathogens and fostering early immune system development towards nonallergy, as well as easing symptoms of eczema attributed to CMPA [49–53].

6. Beta-defensins

The sterile amniotic fluid fills the foetal gut and delivery triggers a rapid transition to bacterial colonization, a crucial challenge for the immune system of the newborn. Despite a naive adaptive immune system, infants rarely become infected, suggesting strong innate defence mechanisms [54]. Several peptides have been identified in meconium and faeces from neonates during the first weeks of life, suggesting their participation in the gut barrier against infection [55].

Defensins are small (~29 to 42 amino acid) cationic arginine and cysteine rich, amphipathic peptides with a predominantly β-sheet structure stabilized by 3 disulfide bonds and a molecular weight of 3–5 kDa [56]. They can be classified on the basis of structure and disulfide bond organization into three groups: α-, β- and θ-defensins. Among them, only α- and β-defensins are expressed in humans. In particular, humans express six α-defensins and up to 31 β-defensins [57]. The α-defensins can be further subdivided into myeloid (HNP1-4) and enteric [human defensin (HD) 5 and 6] peptides on the basis of both expression patterns and genetic organization [58]. HNP (from 1 to 4) are mainly expressed by neutrophils but can also be expressed by B cells some T cells, natural killer (NK) cells, monocyte/macrophages and immature dendritic cells (DCs) [58, 59]. HD5 and HD6 can be found in epithelial Paneth cells belonging to the small bowel [60, 61]. HD5 is also expressed by epithelial cells in the genitourinary tract [62–65].

Human β-defensins (HBDs) are largely expressed by skin epithelial cells and at mucosal surfaces in contact with the environment [66, 67]. They are also expressed by monocytes, macrophages, and certain DCs, and a subset of β-defensins are only expressed in the male reproductive tract [68, 69].

Defensins display various functions, including antimicrobial activity towards Gram-positive and Gram-negative bacteria as well as towards enveloped viruses and fungi [70], and also act as chemoattractant. They also trigger histamine release, wound repair and apoptosis. Defensins contribute to host immunity and to maintain the balance between pathogens and normal flora [71], creating small micropores in the bacterial membranes: this causes a damage to the cell structural integrity and the consequent breakdown of the bacterial cell. Therefore, defensins, thanks to this antimicrobial quality, protect the host epithelium and stem cells from virulent pathogens and also contribute to regulate the number and composition of commensal microbiota [72]. It seems that α-defensins influence the composition of the small intestinal commensal microbiota and the presence of interleukin-17-producing T cells in the lamina propria [73]. In the intestine, α-defensins are highly expressed by Paneth cells and largely confined to the small intestine, whereas β-defensins are expressed by epithelial cells at multiple sites. In general, α-defensins and β-defensin 1 are constitutively expressed, whereas β-defensin 2 (HBD2) to 4 are inducible at sites of infection or inflammation [74, 75]. Recent findings have suggested that, in addition to genetic factors, IBD pathogenesis may result from a breach in the effective mucosal barrier to constituents of the commensal microbiota, thus eliciting pathologic responses from the normal mucosal immune system. In particular, Crohn's disease may, at least in part, be due to a relative defensin deficiency related to reduced expression of Paneth cell α-defensin in disease of the ileum and reduced secretion of inducible β-defensin, namely due to lower HBD2 gene copy number in colonic Crohn's disease [76], allowing intestinal microbes to invade the mucosa and stimulate uncontrolled pro-inflammatory immune responses [77].

Very few data are yet available on innate defence in neonates. In this population, defensins should provide a first-line defence against infection, promoting interactions between the innate and adaptive immunity in newborn infants [78]. Low levels of two enteric α-defensins, that is HD-5 and HD-6, were found in foetus at a gestational age corresponding to preterm infants suggesting that an immaturity of local defence could predispose infants born prematurely to infection from intestinal microorganisms [79]. Besides, these α-defensins were up-regulated during necrotizing enterocolitis (NEC) [80]. Moreover, the expression of mucosal HBD2 mRNA seems to be increased in colonic inflamed mucosa in adults [81].

HBD2 can be easily measured in faeces of adults and children using a commercially available ELISA [82–84].

HBD2 is always detectable in the faeces of full-term and preterm neonates and provides a first kinetic analysis throughout the first weeks of life. The levels are higher than those observed in healthy children and adults and are positively correlated with gestational age at birth [85]. Mode of delivery and mode of feeding do not seem to influence HBD2 values in healthy infants. After 2 weeks of post-natal age, the levels are identical in full-term and preterm infants. These time course mimic partially the one previously described with faecal calprotectin [9, 24], suggesting that birth is associated with a 'physiological' inflammation, which might represent a trait of the gut neonatal adaptation to the various encountered antigens.

A reduction in β-defensin production, as seen in preterm newborns, is associated with an alteration of the colonic microbiota that can determine a colonic inflammation [77] and may be related with a high risk of NEC [80, 86].

Moreover, HBD2 is up-regulated in infants suffering from severe intestinal distress due to NEC, indicating an activation of the mucosal innate defence. As HBD2 has a chemoattractant activity for cells expressing the chemokine receptor CCR-6, such as DCs, defensin could serve as a bridge between the innate immunity at the intestinal mucosa and subsequent adaptive immune responses during NEC [87].

7. Beta-defensins and CMPA

Few studies have been performed to evaluate the correlation between faecal β-defensins values in infants affected by CMPA. In particular, β-defensins values detected in infants with a previous diagnosis of CMPA prior to the oral food challenge, and during each provocation period (3–5 days after the start of either the active or placebo provocation) do not seem to show significant changes [88].

Therefore, nowadays β-defensins are not considered useful in the management of CMPA.

8. Faecal Tumor necrosis factor-alpha

TNF-α is a cytokine involved in systemic inflammation and is one of the cytokines that make up the acute phase reaction.

TNF is an endogenous pyrogen, as it can provoke fever, but it also shows other properties such as apoptosis, cachexia and inflammation induction. It inhibits tumorigenesis and viral replication and responds to sepsis through IL1 and IL6 producing cells. Several human diseases including Alzheimer's disease [89], cancer [90], major depression [91] and IBD [92] seem to be linked to an impaired TNF production. Increased faecal TNF-α levels have been found in Shigella enteritis [93] and in children with IBD and correlate with the severity of colitis in the latter [94].

TNF is primarily produced as a 212-amino acid-long type II transmembrane protein arranged in stable homotrimers [95, 96]. Then, metalloprotease TNF-alpha converting enzyme (TACE, also called ADAM17) generates the soluble homotrimeric cytokine (sTNF) by cleaving the membrane-integrated form [97]. Both the secreted and the membrane bound forms are biologically active, although the specific functions of each are controversial.

TNF was thought to be produced primarily by macrophages [98], but it is produced also by a broad variety of cell types including lymphoid cells, mast cells, endothelial cells, cardiac myocytes, adipose tissue, fibroblasts and neurons.

It has a number of actions on various organ systems, generally together with interleukin (IL)-1 and IL-6.

Among these actions, one of the most important is stimulating the acute phase response, leading to an increase in C-reactive protein and a number of other mediators.

It is a potent chemoattractant for neutrophils and promotes the expression of adhesion molecules on endothelial cells, helping neutrophils migrate.

A local increase in concentration of TNF causes the cardinal signs of inflammation to occur: heat, swelling, redness, pain and loss of function.

The overexpression of TNF causes many of the clinical problems associated with autoimmune disorders such as rheumatoid arthritis, IBD, psoriasis. Thus, these disorders are sometimes treated by using a TNF inhibitor, such as infliximab (Remicade), adalimumab (Humira), certolizumab pegol (Cimzia) or etanercept (Enbrel).

TNF-α time course was evaluated in preterm and term neonates. Mean values at day of life 15 were negatively related with GA with significantly higher values in preterm than in term newborns. The difference was more evident at day of life 30. Neither type of delivery nor type of feeding seem to influence TNF-α values, both in term and preterm neonates.

TNF-α is released after the contact of the macrophages with Gram-positive and Gram-negative bacteria. Gastrointestinal tract of preterm newborns is poorly colonized, harbouring no more than seven bacterial species [99], with high level of staphylococci and clostridia and low level of probiotic strains Lactobacilli and Bifidobacteria that demonstrated high capacities to reduce TNF-α concentrations in the gut [100]. However, the higher levels of TNF-α in preterm newborns may promote the increase of inducible HBD2 production: this agrees with the lower HBD2 levels demonstrated in preterm neonates, increasing over time. Defensins' expression, in fact, increases in response to TRL ligand, TNF-α, IL-1ß, INF-γ [101].

Faecal TNF-α can be determined using an ELISA kit adapted for faecal samples. In normal control cohorts of children, TNF-α levels are considered normal if they are inferior to 90 pg/g [102].

9. TNF-α and CMPA

TNF-α seems to be involved in the pathogenesis of FPIES through an alteration of intestinal permeability [2, 103, 104] that can lead to an aberrant increased absorption of luminal antigens.

TNF-α expression in the epithelial cells and mononuclear cells in the lamina propria is markedly increased in FPIES patients. In addition, TNF-α is highly secreted, antigen specifically, by peripheral blood mononuclear cells from patients with FPIES [103, 105, 106] and is also increased in the stools after milk challenge of patients with gastrointestinal milk allergy [107, 108].

Infants with atopic eczema exhibit a specific faecal protein pattern characterized by an increase in both ECP and TNF-α [109]. The faecal concentration of ECP was enhanced particularly in patients with immediate-type reactions to the cow's milk challenge, whereas faecal TNF-α was enhanced in those with delayed-type reaction, confirming the different pathogenesis (IgE mediated and non-IgE mediated) of these two types of reactions.

TNF-α is a helpful faecal marker to discriminate two main types of persistent diarrhoea with onset within the first weeks or months of life: constitutive intestinal epithelial disorders, such as epithelial dysplasia (ED) or microvillus atrophy (MVA), and immune-inflammatory disorders, such as inflammatory colitis (IC) or autoimmune enteropathy (AIE) [102, 110–112].

Only in inflammatory disorders, an increase in TNF-α levels appears, whereas it remains undetectable/normal in constitutive epithelial disorders [102]. Moreover, TNF-α levels closely correlate with the inflammatory activity of the intestinal mucosa [94].

TNF-α increase reflects that previously seen for calprotectin. Calprotectin and TNF values are dramatically increased in neonates and small infants with immune-inflammatory disorders. The main difference between TNF-α and calprotectin is that the first is highly susceptible to degradation [113]. Therefore, calprotectin measurement might be preferable in the diagnostic work-up of diarrhoea and more appropriate to the clinical setting [102].

10. Eosinophil cationic protein (ECP)

ECP is one of the most important proteins in the granules of eosinophil granulocytes together with the major basic protein (MBP), eosinophil peroxidase (EPO) and eosinophil-derived neurotoxin/eosinophil protein X (EDN/EPX). Very small amounts of ECP can also be found in neutrophil granulocytes and monocytes [114, 115].

ECP is a single-chain, zinc-containing protein with a molecular weight ranging from 16 to 22 kDa. The heterogeneity of the molecule is partially due to differences in glycosylation of three potential sites in its amino acid chain [116].

The gene that codes for ECP has been located on chromosome 14q11.2 and three polymorphisms have been identified [117].

ECP synthesis in eosinophil granulocytes begins already at the stage of promyelocytes in primary granules; from the myelocyte stage, ECP is present only in the matrix of specific granules of eosinophil granulocytes [118, 119]. Mature eosinophil granulocytes contain 13.5 mcg ECP/10^6 cells [120]. Unstimulated neutrophil granulocytes cannot produce ECP themselves, but these granulocytes can take ECP from the local environment [121]. After cellular stimulation, ECP mRNA can be found in neutrophil granulocytes [122]. On the other hand, monocytes are able to synthetize ECP, except for when macrophage differentiation is ongoing [115].

Activated eosinophil tissue granulocytes can excrete ECP, in response to two kinds of stimuli: antibody dependent (IgG, IgA) and antibody independent (C3 and C5 complement components) [123, 124]. IL like IL-5 and IL-3, and the granulocyte–monocyte colony-stimulating factor (GM-CSF) have positive impact on this secretion. During the excretion of the ECP molecule from eosinophil granules, enzymatic deglycosylation occurs and converts the inactive ECP form with high molecular mass into a cytotoxic variant with low molecular mass [125].

Inhibitory effect on ECP secretion has been confirmed for cyclosporine A, dexamethasone, rapamycin, formoterol and pemirolast [126–128].

ECP exhibits numerous biological activities that may be classified into cytotoxic and non-toxic reactions.

Cytotoxic activity of ECP is effective against a wide range of microorganisms: parasites, Gram-negative and Gram-positive bacteria, viruses [129–131].

- *Antibacterial activity*: for antibacterial ECP activity, electrostatic interactions are important between negatively charged cellular membrane or cellular bacterial wall and positively charged ECP, followed by destabilization of bacterial membrane. Other mechanisms involve the formation of transmembrane pores that allow the transition of water and osmotic cell lysis. Specific interaction of ECP with lipopolysaccharides and peptidoglycans on the bacterial cell wall may result in bacterial cell aggregation and cell death [132–136].

- *Antihelminthic activity*: eosinophilia is present in the peripheral blood of patients with parasitary diseases and granulocyte and eosinophil infiltrates have been detected in tissue biopsy. However, the actual role of ECP and eosinophil granulocytes in parasitary infection has not been clarified. ECP seems to play a role in the isolation of pathogenic agent and the entire infected region in the form of granuloma. Blood cells, particularly eosinophil granulocytes, are part of the structure of such granulomas. The chemotactic action of ECP on fibroblasts is the first step towards the remodelling of the extracellular matrix. Moreover, ECP promotes the secretion of tumor growth factor-β (TGF-β), that is a pro-fibrotic mediator synthetized by fibroblasts [121, 137].

- *Antiviral activity*: antiviral action of ECP is mediated by ribonuclease activity [121]. ECP belongs to the family of ribonucleases A (RNase A) which cleave the single-strand RNA molecules [131, 138]. Ribonuclease activity is, however, the same in all genetic and post-translational ECP variants and is entirely independent of cytotoxic activity. Most of the studies have been performed on respiratory syncytial virus (RSV) [139, 140], showing a decrease of viral infectivity due to ECP, yet ribonuclease activity does not seem sufficient to explain the whole antiviral effect attributed to eosinophil granulocytes [121, 138].

- *Antihost activity*: in addition to its important role in host immune defence, ECP may also cause undesired side-effects on the host's own tissues via its cytotoxic activity. Neuronal damage has been described, as well as the damage of muscular cells and respiratory tract epithelial cells. One of the possible mechanisms of eosinophil-induced tissue destruction is based on the apoptotic action of ECP via activation of the caspase cycle [132, 141–143]. In contrast, in several dermatoses, intracellular accumulation of ECP is crucial in damaging dermal cell, whilst RNase activity and cation-dependent cytotoxicity of the ECP molecule seem to cause skin lesions [121, 144].

- *Non-toxic activity*: several immunomodulatory properties (e.g. inhibition of T-cell proliferation, up-regulation of receptors and adhesion molecules on epithelial cells, or basophil histamine release) are involved in ECP non-toxic action [132, 137, 145]. Chemotactic action on fibroblasts is the first ECP action in tissue repair processes. Moreover, ECP stimulates the secretion of TGF-β, whose pro-fibrotic action alters the intracellular metabolism of fibroblasts. ECP mediates both the increase of proteoglycan synthesis and the inhibition of proteoglycan degradation, with consequent intracellular proteoglycan accumulation [121, 137, 146]. Moreover, ECP seems to play a role in atherogenesis as it enables adhesion of monocytes on endothelial cells, is involved in coagulation cascade, and has a stabilizing impact on the plaque [137, 145, 146].

Only activated eosinophil granulocytes release the granule content, and therefore, the determination of ECP concentration is a considerably more specific indicator of eosinophil inflammation than eosinophil granulocyte count in peripheral blood. ECP has been associated with several pathologic conditions, especially atopic diseases: allergic asthma, allergic rhinitis and perennial rhinitis, atopic eczema/dermatitis syndrome (AEDS) [116, 147–149]. Elevated serum values are proportional to the intensity of allergic inflammation and indicate acute allergen exposure. ECP levels are also augmented in several gastrointestinal disorders, some of which are IgE-associated: eosinophil diseases (esophagitis, gastro-enteritis and colitis), gastrointestinal food allergy and intestinal parasitoses. Also, ECP values are enhanced in non-IgE-dependent disorders such as non allergic asthma with aspirin intolerance, respiratory infections, sinonasal polyposis, Churg-Strauss disease and idiopathic hypereosinophilia (HES) syndrome [150].

Differential diagnostics of hypereosinophilia could be easier thanks to the evaluation of plasma ECP concentration and the ECP/eosinophil count ratio, as ECP levels and the above ratio are increased in patients with HES. Moreover, these values are higher in patients with reactive eosinophilia associated with malignancy than in those patients with reactive eosinophilia associated with inflammation. However, on the basis of ECP concentration and ECP/eosinophil count ratio, discriminating clonal from reactive eosinophilia is not possible [121, 151].

ECP is present in numerous body fluids such as plasma, serum, sputum, bronchoalveolar lavage (BAL), saliva, nasal lavage, tears, jejunal fluid, faeces, synovial fluid [145].

Faecal ECP is intensively investigated as a novel potential marker of IBD and eosinophil gastroenteritis [152, 153]. Faecal ECP levels are not dependent on the number of eosinophil granulocytes or serum ECP values, which makes this type of measurement a potential intestinal marker. Moreover, faecal ECP levels do not differ in dependence to age or gender. Serum and faecal ECP mean and median values have been evaluated in a population of healthy children. Serum values are, respectively, 13.50 and 9.54 mcg/L, whereas faecal values are 1.93 and 1.20 mcg/g [154].

11. ECP and CMPA

ECP values are a valid index to evaluate eosinophil granulocytes activity, even more than eosinophil peripheral blood count. In particular, serum ECP values have been demonstrated to strictly correlate with disease activity in CMPA: ECP levels were determined during a diet with and without cow's milk in a patient with eosinophilic enteritis. ECP levels were considerably elevated during the diet with milk, although they returned to normal values several months after milk was withdrawn [155].

To demonstrate inflammation and increased protein leakage from the gut during a cow's milk elimination-challenge test in faecal samples of infants presenting with different symptoms suggestive of cow's milk allergy, ECP levels were measured in faecal samples of 208 infants with a mean age of 7 months. Pre-challenge samples were obtained after a mean 3-weeks elimination period, while post-challenge samples were collected 4 days after starting the challenge. Among these infants, pre- and post-challenge ECP levels were increased in those reacting after 24 h than in those reacting within 1 h. Moreover, pre-challenge levels of ECP were higher in those showing intestinal symptoms. Therefore, in infants with slowly Developing gastrointestinal symptoms, enhanced faecal ECP values could be helpful in discriminating patients from those who tolerate cow's milk. Serial follow-up of faecal ECP can be useful to evaluate the degree of intestinal inflammation and to determine an appropriate time for a challenge test. Nevertheless, faecal ECP values are not diagnostic tools for cow's milk allergy [156].

Kapel et al. [107] also evaluated faecal ECP values before and after the oral challenge in 13 patients with GI-CMPA, not showing any significant change. No data are available about timing of measurement.

Kristjánsson et al., using rectal protein challenge, investigated the local inflammatory reaction to gluten and CM protein in adult patients with CD in remission, but still complaining gastrointestinal symptoms. In 20 celiac patients and 15 healthy controls, rectal challenges with wheat gluten and dried cow's milk powder were conducted. Fifteen hours after challenge, the reaction of intestinal mucosa was recorded evaluating local secretion of neutrophil and eosinophil granule molecules such as myeloperoxidase (MPO) and ECP. At the same time, mucosal release of nitric oxide (NO) was measured. Compared to healthy controls, patients with CD showed significant increases in rectal NO and MPO concentrations measured 15 h after challenge with both CM and gluten, while ECP was increased to a similar extent in the two groups. Therefore, a rectal challenge with CM protein seems to induce a local inflammatory mucosal reaction in patients with CD but not in healthy controls. Ten out of 20 patients showed abnormal increases in both MPO and NO as a reaction to CM challenge, but no increase in ECP, indicating the absence of eosinophil activation at least 15 h after challenge [157].

More studies are needed to evaluate the correct time of faecal ECP measurement.

Therefore, we can assume that ECP is not a reliable faecal marker, as its values are not constantly increased in CMPA. This is possibly due to the lack of agreement on the correct time to perform ECP measurement in stool samples, as this faecal marker does not seem to increase immediately after CM exposure. Moreover, as previously said, faecal concentration of ECP seems to be enhanced particularly in patients with immediate-type reactions to the cow's milk challenge (IgE mediated). This can help understanding the different results reached in the cited studies.

12. Conclusions

In the light of this, calprotectin and TNF-α seem to be the most useful faecal marker in the management of non-IgE-mediated GI-CMPA. They can be helpful in achieving a diagnosis, as they can easily differentiate constitutive intestinal epithelial disorders, such as MVA or ED, and immune-inflammatory disorders, such as AIE or IC. Moreover, faecal calprotectin and TNF-α are useful for treatment follow-up, as their values in the stools markedly increase after intake of cow's milk protein, due to the reactivation of mucosal inflammation.

The main difference between TNF-α and calprotectin is that the first is highly susceptible to degradation. Therefore, calprotectin measurement might be preferable and more appropriate to the clinical setting.

Beta-defensins seem to provide a first-line defence against infection, promoting interactions between the innate and adaptive immunity in newborn infants. Their values are mainly influenced by microbiota and are up-regulated during NEC. No significant correlations have been reported with GI-CMPA, so far. Therefore, β-defensins are not considered a useful faecal marker in the management of CMPA.

ECP is one of the most important proteins released by activated eosinophil granulocytes, and the determination of its concentration is a considerably more specific indicator of eosinophil inflammation than eosinophil granulocyte count in peripheral blood.

ECP can be measured in several biological samples, serum and stools among them. In individuals with CMPA, serum ECP values increase considerably during the diet with milk, then returning to normal values several months after milk withdrawal. Also faecal ECP values tend to increase after exposure to cow's milk protein, but this does not happen immediately. ECP values usually increase for several hours, or even days, after CM exposure. Unfortunately, the lack of agreement on when to measure ECP in stool samples after CM exposure, leads ECP not to be considered a reliable faecal marker in CMPA nowadays, as further researches are still needed in order to understand the correct time to perform faecal ECP measurement.

Author details

Maria Elisabetta Baldassarre*, Raffaella Panza and Nicola Laforgia

*Address all correspondence to: mariaelisabetta.baldassarre@uniba.it

Department of Biomedical Science and Human Oncology, Neonatology and NICU Section, University "ALDO MORO", Bari, Italy

References

[1] Deleves PJ. Food Allergy—immunology; allergic disorders—Merck Manuals Professional Edition. http://www.merckmanuals.com/professional/immunology;-allergic-disorders/allergic,-autoimmune,-and-other-hypersensitivity-disorders/food-allergy.

[2] Morita H, Nomura I, Matsuda A, Saito H, Matsumoto K. Gastrointestinal food allergy in infants. *Allergol Int*. 2013;62(3):297–307. doi:10.2332/allergolint.13-RA-0542.

[3] Medicine TA of B. ABM Clinical Protocol #24: Allergic proctocolitis in the exclusively breastfed infant. *Breastfeed Med*. 2011;6(6):435–440. doi:10.1089/bfm.2011.9977.

[4] Arvola T, Ruuska T, Keränen J, Hyöty H, Salminen S, Isolauri E. Rectal bleeding in infancy: clinical, allergological, and microbiological examination. *Pediatrics*. 2006;117(4):e760–e768. doi:10.1542/peds.2005-1069.

[5] Hwang J-B, Park MH, Kang YN, Kim SP, Suh S-I, Kam S. Advanced criteria for clinicopathological diagnosis of food protein-induced proctocolitis. *J Korean Med Sci*. 2007;22(2):213. doi:10.3346/jkms.2007.22.2.213.

[6] Eigenmann PA. Mechanisms of food allergy. *Pediatr Allergy Immunol*. 2009;20(1):5–11. doi:10.1111/j.1399-3038.2008.00847.x.

[7] BeşerÖF, Sancak S, Erkan T, Kutlu T, Çoku◎raşH, Coku◎raşFÇ. Can fecal calprotectin level be used as a markers of inflammation in the diagnosis and follow-up of cow's milk protein allergy? *Allergy Asthma Immunol Res*. 2014. doi:10.4168/aair.2014.6.1.33.

[8] Vasan RS. Biomarkers of cardiovascular disease: molecular basis and practical considerations. *Circulation*. 2006;113(19):2335–2362. doi:10.1161/CIRCULATIONAHA.104.482570.

[9] Kapel N, Campeotto F, Kalach N, Baldassare M, Butel M-J, Dupont C. Faecal calprotectin in term and preterm neonates. *J Pediatr Gastroenterol Nutr*. 2010. doi:10.1097/MPG.0b013e3181e2ad72.

[10] Dale I, Fagerhol MK, Naesgaard I. Purification and partial characterization of a highly immunogenic human leukocyte protein, the L1 antigen. *Eur J Biochem*. 1983;134(1):1–6. http://www.ncbi.nlm.nih.gov/pubmed/6861753. Accessed June 10, 2015.

[11] Brandtzaeg P, Dale I, Fagerhol MK. Distribution of a formalin-resistant myelomonocytic antigen (L1) in human tissues. II. Normal and aberrant occurrence in various epithelia. *Am J Clin Pathol*. 1987;87(6):700–707. http://europepmc.org/abstract/med/3296737. Accessed June 10, 2015.

[12] Striz I, Trebichavsky I. Calprotectin: a pleiotropic molecule in acute and chronic inflammation. *Physiol Res*. 53(3):245–253. http://cat.inist.fr/?a Modele=affiche N&cpsidt=15946642. Accessed June 10, 2015.

[13] Bhardwaj RS, Zotz C, Zwadlo-Klarwasser G, et al. The calcium-binding proteins MRP8 and MRP14 form a membrane-associated heterodimer in a subset of monocytes/ macrophages present in acute but absent in chronic inflammatory lesions. *Eur J Immunol*. 1992;22(7):1891–1897. doi:10.1002/eji.1830220732.

[14] Kligman D, Hilt DC. The S100 protein family. *Trends Biochem Sci*. 1988;13(11):437–443. doi:10.1016/0968-0004(88)90218-6.

[15] Loomans HJ, Hahn BL, Li Q, Phadnis SH, Sohnle PG. Histidine-based zinc-binding sequences and the antimicrobial activity of calprotectin. *J Infect Dis*. 1998;177(3):812–814. doi:10.1086/517816.

[16] Steinbakk M, Naess-Andresen C-F, Fagerhol MK, Lingaas E, Dale I, Brandtzaeg P. Antimicrobial actions of calcium binding leucocyte L1 protein, calprotectin. *Lancet*. 1990;336(8718):763–765. doi:10.1016/0140-6736(90)93237-J.

[17] Berntzen HB, Olmez U, Fagerhol MK, Munthe E. The leukocyte protein L1 in plasma and synovial fluid from patients with rheumatoid arthritis and osteoarthritis. *Scand J Rheumatol*. 1991;20(2):74–82. http://www.ncbi.nlm.nih.gov/pubmed/1709519. Accessed June 11, 2015.

[18] Dunlop O, Bruun JN, Myrvang B, Fagerhol MK. Calprotectin in cerebrospinal fluid of the HIV infected: a diagnostic marker of opportunistic central nervous system infection? *Scand J Infect Dis*. 1991;23(6):687–689. http://www.ncbi.nlm.nih.gov/pubmed/1815329. Accessed June 11, 2015.

[19] Cuida M, Brun JG, Tynning T, Jonsson R. Calprotectin levels in oral fluids: the importance of collection site. *Eur J Oral Sci*. 1995;103(1):8–10. http://www.ncbi.nlm.nih.gov/pubmed/7600253. Accessed June 11, 2015.

[20] Holt J, Fagerhol MK, Dale I. Quantitation of a leukocyte protein (L1) in urine. *Acta Paediatr Scand*. 1983;72(4):615–616. http://www.ncbi.nlm.nih.gov/pubmed/6624438. Accessed June 11, 2015.

[21] Isaksen B, Fagerhol MK. Calprotectin inhibits matrix metalloproteinases by sequestration of zinc. *Mol Pathol*. 2001;54(5):289–292. http://www.pubmedcentral.nih.gov/articlerender.fcgi?artid=1187084&tool=pmcentrez&rendertype=abstract. Accessed June 2, 2015.

[22] Boussac M, Garin J. Calcium-dependent secretion in human neutrophils: a proteomic approach. *Electrophoresis*. 2000;21(3):665–672. doi: 10.1002/(SICI)1522-2683(20000201)21:3<665::AID-ELPS665>3.0.CO;2-U.

[23] Roth J. Myeloid-related protein (MRP) 8and MRP14, calcium-binding proteins of the S100 Family, are secreted by activated monocytes via a novel, tubulin-dependent pathway. *J Biol Chem*. 1997;272(14):9496–9502. doi: 10.1074/jbc.272.14.9496.

[24] Baldassarre ME, Altomare MA, Fanelli M, et al. Does calprotectin represent a regula-

tory factor in host defense or a drug target in inflammatory disease? *Endocr Metab Immune Disord Drug Targets*. March 2007;7(1): 1–5(5). doi: 10.2174/187153007780059441.

[25] Wilkinson M, Busuttil A, Hayward C, Brock D, Dorin J, Van Heyningen V. Expression pattern of two related cystic fibrosis-associated calcium-binding proteins in normal and abnormal tissues. *J Cell Sci*. 1988;91(2):221–230. http://jcs.biologists.org/content/91/2/221.short. Accessed June 11, 2015.

[26] Roth J, Burwinkel F, van den Bos C, Goebeler M, Vollmer E, Sorg C. MRP8 and MRP14, S-100-like proteins associated with myeloid differentiation, are translocated to plasma membrane and intermediate filaments in a calcium-dependent manner. *Blood*. 1993;82(6):1875–1883. http://www.bloodjournal.org/content/82/6/1875.abstract. Accessed June 11, 2015.

[27] Akiyama H, Ikeda K, Katoh M, Mc Geer EG, Mc Geer PL. Expression of MRP14, 27E10, interferon-α and leukocyte common antigen by reactive microglia in postmortem human brain tissue. *J Neuroimmunol*. 1994;50(2):195–201. doi: 10.1016/0165-5728(94)90046-9.

[28] Yui S, Nakatani Y, Mikami M. Calprotectin (S100A8/S100A9), an inflammatory protein complex from neutrophils with a broad apoptosis-inducing activity. *Biol Pharm Bull*. 2003;26(6):753–760. doi: 10.1248/bpb.26.753.

[29] Fagerhol MK. Calprotectin, a faecal marker of organic gastrointestinal abnormality. *Lancet*. 2000;356(9244):1783–1784. doi: 10.1016/S0140-6736(00)03224-4.

[30] Poullis A, Foster R, Northfield TC, Mendall MA. Review article: faecal markers in the assessment of activity in inflammatory bowel disease. *Aliment Pharmacol Ther*. 2002;16(4):675–681. http://www.ncbi.nlm.nih.gov/pubmed/11929384. Accessed June 10, 2015.

[31] Saverymuttu SH, Peters AM, Crofton ME, et al. 111Indium autologous granulocytes in the detection of inflammatory bowel disease. *Gut*. 1985;26(9):955–960. http://www.pubmedcentral.nih.gov/articlerender.fcgi?artid=1432869&tool=pmcentrez&rendertype=abstract. Accessed June 10, 2015.

[32] Naess-Andresen CF, Egelandsdal B, Fagerhol MK. Calcium binding and concomitant changes in the structure and heat stability of calprotectin (L1 protein). *Clin Mol Pathol*. 1995;48(5):M278–M284. http://www.pubmedcentral.nih.gov/articlerender.fcgi?artid=407985&tool=pmcentrez&rendertype=abstract. Accessed June 10, 2015.

[33] Tøn H, Brandsnes, Dale S, et al. Improved assay for fecal calprotectin. *Clin Chim Acta*. 2000;292(1–2):41–54. http://www.ncbi.nlm.nih.gov/pubmed/10686275. Accessed June 10, 2015.

[34] Røseth AG, Fagerhol MK, Aadland E, Schjønsby H. Assessment of the neutrophil dominating protein calprotectin in feces. A methodologic study. *Scand J Gastroenterol*.

1992;27(9):793–798. http://www.ncbi.nlm.nih.gov/pubmed/1411288. Accessed May 18, 2015.

[35] Fagerberg UL, Lööf L, Merzoug RD, Hansson L-O, Finkel Y. Fecal calprotectin levels in healthy children studied with an improved assay. *J Pediatr Gastroenterol Nutr.* 2003;37(4):468–472. http://www.ncbi.nlm.nih.gov/pubmed/14508218. Accessed June 10, 2015.

[36] Laforgia N, Baldassarre M, Pontrelli G, et al. Calprotectin levels in meconium. *Acta Paediatr.* 2007;92(4):463–466. doi:10.1111/j.1651-2227.2003.tb00579.x.

[37] Josefsson S, Bunn SK, Domellöf M. Fecal calprotectin in very low birth weight infants. *J Pediatr Gastroenterol Nutr.* 2007;44(4):407–413. doi:10.1097/MPG.0b013e3180320643.

[38] Nissen, Annemieke C.; van Gils, Carlijn E.; Menheere, Paul P.; Van den Neucker, Anita M.; van der Hoeven, Mark A.; Forget, Pierre-Philippe.

[39] Rugtveit J, Fagerhol MK. Age-dependent variations in fecal calprotectin concentrations in children. *J Pediatr Gastroenterol Nutr.* 2002;34(3):323–324; author reply 324–325. http://www.ncbi.nlm.nih.gov/pubmed/11964964. Accessed June 1, 2015.

[40] Campeotto F, Butel MJ, Kalach N, et al. High faecal calprotectin concentrations in newborn infants. *Arch Dis Child Fetal Neonatal Ed.* 2004;89(4):F353–F355. doi:10.1136/adc.2002.022368.

[41] Campeotto F, Kalach N, Lapillonne A, Butel MJ, Dupont C, Kapel N. Time course of faecal calprotectin in preterm newborns during the first month of life. *Acta Paediatr.* 2007;96(10):1531–1533. doi:10.1111/j.1651-2227.2007.00457.x.

[42] Yang Q, Smith PB, Goldberg RN, Cotten CM. Dynamic change of fecal calprotectin in very low birth weight infants during the first month of life. *Neonatology.* 2008;94(4):267–271. doi:10.1159/000151645.

[43] Golden B, Bunn S, Main M. Age-dependent variations in fecal calprotectin concentrations in children. *J Pediatr Gastroenterol Nutr.* 34(3):324–325. http://cat.inist.fr/?a Modele=affiche N&cpsidt=13566983. Accessed July 4, 2015.

[44] Dorosko SM, Mackenzie T, Connor RI. Fecal calprotectin concentrations are higher in exclusively breastfed infants compared to those who are mixed-fed. *Breastfeed Med.* 2008;3(2):117–119. doi:10.1089/bfm.2007.0036.

[45] Savino F, Castagno E, Calabrese R, Viola S, Oggero R, Miniero R. High faecal calprotectin levels in healthy, exclusively breast-fed infants. *Neonatology.* 2010;97(4):299–304. doi:10.1159/000255161.

[46] Mohan R, Koebnick C, Schildt J, Mueller M, Radke M, Blaut M. Effects of Bifidobacterium lactis Bb12 supplementation on body weight, fecal p H, acetate, lactate, calprotectin, and Ig A in preterm infants. *Pediatr Res.* 2008;64(4):418–422. doi:10.1203/PDR.0b013e318181b7fa.

[47] Savino F, Castagno E, Palumeri E, Oggero R, Mussa GC. Faecal calprotectin levels at two months of age in healthy infants and in infants with atopic and gastrointestinal disorders. *Pediatr Res*. 2004;56(3):503–503. doi:10.1203/00006450-200409000-00256.

[48] Baldassarre ME, Fanelli M, Lasorella ML, et al. Fecal calprotectin (FC) in newborns: is it a predictive marker of gastrointestinal and/or allergic disease? *Immunopharmacol Immunotoxicol*. 2011;33(1):220–223. http://www.tandfonline.com/doi/abs/10.3109/08923973.2010.486035. Accessed June 13, 2015.

[49] Baldassarre ME, Laforgia N, Fanelli M, Laneve A, Grosso R, Lifschitz C. Lactobacillus GG improves recovery in infants with blood in the stools and presumptive allergic colitis compared with extensively hydrolyzed formula alone. *J Pediatr*. 2010;156(3):397–401. doi:10.1016/j.jpeds.2009.09.012.

[50] Rosenfeldt V, Benfeldt E, Nielsen SD, et al. Effect of probiotic Lactobacillus strains in children with atopic dermatitis. *J Allergy Clin Immunol*. 2003;111(2):389–395. doi: 10.1067/mai.2003.389.

[51] Viljanen M, Savilahti E, Haahtela T, et al. Probiotics in the treatment of atopic eczema/dermatitis syndrome in infants: a double-blind placebo-controlled trial. *Allergy*. 2005;60(4):494–500. doi:10.1111/j.1398-9995.2004.00514.x.

[52] Majamaa H, Isolauri E. Probiotics: a novel approach in the management of food allergy. *J Allergy Clin Immunol*. 1997;99(2):179–185. doi:10.1016/S0091-6749(97)70093-9.

[53] Weston S, Halbert A, Richmond P, Prescott SL. Effects of probiotics on atopic dermatitis: a randomised controlled trial. *Arch Dis Child*. 2005;90(9):892–897. doi:10.1136/adc.2004.060673.

[54] Newburg DS, Walker WA. Protection of the neonate by the innate immune system of developing gut and of human milk. *Pediatr Res*. 2007;61(1):2–8. doi:10.1203/01.pdr.0000250274.68571.18.

[55] Kai-Larsen Y, Bergsson G, Gudmundsson GH, et al. Antimicrobial components of the neonatal gut affected upon colonization. *Pediatr Res*. 2007;61(5 Pt 1):530–536. doi: 10.1203/pdr.0b013e318045be83.

[56] Selsted ME. Enteric defensins: antibiotic peptide components of intestinal host defense. *J Cell Biol*. 1992;118(4):929–936. doi:10.1083/jcb.118.4.929.

[57] Schutte BC, Mitros JP, Bartlett JA, et al. Discovery of five conserved beta-defensin gene clusters using a computational search strategy. *Proc Natl Acad Sci U S A*. 2002;99(4):2129–2133. doi:10.1073/pnas.042692699.

[58] Selsted ME, Ouellette AJ. Mammalian defensins in the antimicrobial immune response. *Nat Immunol*. 2005;6(6):551–557. doi:10.1038/ni1206.

[59] Pazgier M, Li X, Lu W, Lubkowski J. Human defensins: synthesis and structural properties. *Curr Pharm Des*. 2007;13(30):3096–3118. doi:10.2174/138161207782110381.

[60] Ouellette AJ. Paneth cell α-defensins in enteric innate immunity. *Cell Mol Life Sci.* 2011;68(13):2215–2229. doi:10.1007/s00018-011-0714-6.

[61] Bevins CL, Salzman NH. Paneth cells, antimicrobial peptides and maintenance of intestinal homeostasis. *Nat Rev Microbiol.* 2011;9(5):356–368. doi:10.1038/nrmicro2546.

[62] Spencer JD, Hains DS, Porter E, et al. Human alpha defensin 5 expression in the human kidney and urinary tract. *PLo S One.* 2012;7(2):e31712. doi:10.1371/journal.pone.0031712.

[63] Quayle AJ, Porter EM, Nussbaum AA, et al. Gene expression, immunolocalization, and secretion of human defensin-5 in human female reproductive tract. *Am J Pathol.* 1998;152(5):1247–1258. http://www.pubmedcentral.nih.gov/articlerender.fcgi?artid=1858596&tool=pmcentrez&rendertype=abstract. Accessed June 25, 2015.

[64] Com E. Expression of antimicrobial defensins in the male reproductive tract of rats, mice, and humans. *Biol Reprod.* 2002;68(1):95–104. doi:10.1095/biolreprod.102.005389.

[65] Svinarich DM, Wolf NA, Gomez R, Gonik B, Romero R. Detection of human defensin 5 in reproductive tissues. *Am J Obstet Gynecol.* 1997;176(2):470–475. doi:10.1016/S0002-9378(97)70517-9.

[66] Schibli DJ, Hunter HN, Aseyev V, et al. The solution structures of the human beta-defensins lead to a better understanding of the potent bactericidal activity of HBD3 against *Staphylococcus aureus*. *J Biol Chem.* 2002;277(10):8279–8289. doi:10.1074/jbc.M108830200.

[67] Wilson SS, Wiens ME, Smith JG. Antiviral mechanisms of human defensins. *J Mol Biol.* 2013;425(24):4965–4980. doi:10.1016/j.jmb.2013.09.038.

[68] Yamaguchi Y, Nagase T, Makita R, et al. Identification of multiple novel epididymis-specific β -defensin isoforms in humans and mice. *J Immunol.* 2002;169(5):2516–2523. doi:10.4049/jimmunol.169.5.2516.

[69] Pazgier M, Hoover DM, Yang D, Lu W, Lubkowski J. Human beta-defensins. *Cell Mol Life Sci.* 2006;63(11):1294–1313. doi:10.1007/s00018-005-5540-2.

[70] Zasloff M. Antibiotic peptides as mediators of innate immunity. *Curr Opin Immunol.* 1992;4(1):3–7. doi:10.1016/0952-7915(92)90115-U.

[71] Ramasundara M, Leach ST, Lemberg DA, Day AS. Defensins and inflammation: the role of defensins in inflammatory bowel disease. *J Gastroenterol Hepatol.* 2009;24(2):202–208. doi:10.1111/j.1440-1746.2008.05772.x.

[72] Liu, Zhanju, and Yurong Yang. Defensins in Ulcerative Colitis. INTECH Open Access Publisher, 2011.

[73] Salzman NH, Hung K, Haribhai D, et al. Enteric defensins are essential regulators of intestinal microbial ecology. *Nat Immunol.* 2009;11(1):76–82. doi:10.1038/ni.1825.

[74] Wehkamp J, Schmid M, Stange EF. Defensins and other antimicrobial peptides in inflammatory bowel disease. *Curr Opin Gastroenterol.* 2007;23(4):370–378. doi:10.1097/MOG.0b013e328136c580.

[75] Salzman NH, Underwood MA, Bevins CL. Paneth cells, defensins, and the commensal microbiota: a hypothesis on intimate interplay at the intestinal mucosa. *Semin Immunol.* 2007;19(2):70–83. doi:10.1016/j.smim.2007.04.002.

[76] Fellermann K, Stange DE, Schaeffeler E, et al. A chromosome 8 gene-cluster polymorphism with low human beta-defensin 2 gene copy number predisposes to Crohn disease of the colon. *Am J Hum Genet.* 2006;79(3):439–448. doi:10.1086/505915.

[77] Wehkamp J, Schmid M, Fellermann K, Stange EF. Defensin deficiency, intestinal microbes, and the clinical phenotypes of Crohn's disease. *J Leukoc Biol.* 2005;77(4):460–465. doi:10.1189/jlb.0904543.

[78] Yoshio H, Lagercrantz H, Gudmundsson GH, Agerberth B. First line of defense in early human life. *Semin Perinatol.* 2004;28(4):304–311. doi:10.1053/j.semperi.2004.08.008.

[79] Mallow EB, Harris A, Salzman N, et al. Human enteric defensins: gene structure and developmental expression. *J Biol Chem.* 1996;271(8):4038–4045. http://www.jbc.org/content/271/8/4038.short. Accessed June 25, 2015.

[80] Salzman NH, Polin RA, Harris MC, et al. Enteric defensin expression in necrotizing enterocolitis. *Pediatr Res.* 1998;44(1):20–26. doi:10.1203/00006450-199807000-00003.

[81] Wehkamp J, Fellermann K, Herrlinger KR, et al. Human β-defensin 2 but not β-defensin 1 is expressed preferentially in colonic mucosa of inflammatory bowel disease. *Eur J Gastroenterol Hepatol.* 2002;14(7):745–752. doi:10.1097/00042737-200207000-00006.

[82] Langhorst J, Wieder A, Michalsen A, Musial F, Dobos GJ, Rueffer A. Activated innate immune system in irritable bowel syndrome? *Gut.* 2007;56(9):1325–1326. doi:10.1136/gut.2007.125005.

[83] Langhorst J, Junge A, Rueffer A, et al. Elevated human beta-defensin-2 levels indicate an activation of the innate immune system in patients with irritable bowel syndrome. *Am J Gastroenterol.* 2009;104(2):404–410. doi:10.1038/ajg.2008.86.

[84] Kapel N, Benahmed N, Morali A, et al. Fecal beta-defensin-2 in children with inflammatory bowel diseases. *J Pediatr Gastroenterol Nutr.* 2009;48(1):117–120. doi:10.1097/MPG.0b013e318174e872.

[85] Richter M, Topf H-G, Gröschl M, et al. Influence of gestational age, cesarean section, and type of feeding on fecal human beta-defensin 2 and tumor necrosis factor-alpha. *J Pediatr Gastroenterol Nutr.* 2010;51(1):103–105. doi:10.1097/MPG.0b013e3181cd26f9.

[86] Underwood MA, Bevins CL. Defensin-barbed innate immunity: clinical associations in the pediatric population. *Pediatrics.* 2010;125(6):1237–1247. doi:10.1542/peds.2009-3289.

[87] Campeotto F, Baldassarre M, Laforgia N, et al. Fecal expression of human β-defensin-2 following birth. *Neonatology*. 2010. doi:10.1159/000315872.

[88] Clinical and Translational Allergy, December 2014, 4(1):8

[89] Swardfager W, Lanctôt K, Rothenburg L, Wong A, Cappell J, Herrmann N. A meta-analysis of cytokines in Alzheimer's disease. *Biol Psychiatry*. 2010;68(10):930–941. doi: 10.1016/j.biopsych.2010.06.012.

[90] Locksley RM, Killeen N, Lenardo MJ. The TNF and TNF receptor superfamilies. *Cell*. 2001;104(4):487–501. doi:10.1016/S0092-8674(01)00237-9.

[91] Dowlati Y, Herrmann N, Swardfager W, et al. A meta-analysis of cytokines in major depression. *Biol Psychiatry*. 2010;67(5):446–457. doi:10.1016/j.biopsych.2009.09.033.

[92] Brynskov J, Foegh P, Pedersen G, et al. Tumour necrosis factor alpha converting enzyme (TACE) activity in the colonic mucosa of patients with inflammatory bowel disease. *Gut*. 2002;51(1):37–43. http://www.pubmedcentral.nih.gov/articlerender.fcgi?artid=1773288&tool=pmcentrez&rendertype=abstract. Accessed July 3, 2015.

[93] Nicholls S, Stephens S, Braegger CP, Walker-Smith JA, Mac Donald TT. Cytokines in stools of children with inflammatory bowel disease or infective diarrhoea. *J Clin Pathol*. 1993;46(8):757–760. doi:10.1136/jcp.46.8.757.

[94] Braegger CP, Nicholls S, Murch SH, Mac Donald TT, Stephens S. Tumour necrosis factor alpha in stool as a marker of intestinal inflammation. *Lancet*. 1992;339(8785):89–91. doi:10.1016/0140-6736(92)90999-J.

[95] Kriegler M, Perez C, De Fay K, Albert I, Lu SD. A novel form of TNF/cachectin is a cell surface cytotoxic transmembrane protein: Ramifications for the complex physiology of TNF. *Cell*. 1988;53(1):45–53. doi:10.1016/0092-8674(88)90486-2.

[96] Tang P, Hung M-C, Klostergaard J. Human pro-tumor necrosis factor is a homotrimer. *Biochemistry*. 1996;35(25):8216–8225. doi:10.1021/bi952182t.

[97] Black RA, Rauch CT, Kozlosky CJ, et al. A metalloproteinase disintegrin that releases tumour-necrosis factor-alpha from cells. *Nature*. 1997;385(6618):729–733. doi: 10.1038/385729a0.

[98] Olszewski MB, Groot AJ, Dastych J, Knol EF. TNF Trafficking to human mast cell granules: mature chain-dependent endocytosis. *J Immunol*. 2007;178(9):5701–5709. doi: 10.4049/jimmunol.178.9.5701.

[99] RougéC, Butel MJ, Piloquet H, et al. Fecal calprotectin excretion in preterm infants during the neonatal period. *PLo S One*. 2010. doi:10.1371/journal.pone.0011083.

[100] Rodes L, Khan A, Paul A, Coussa Charley M, Marinescu D, Tomaro Duchesneau C, Shao W, Kahouli I, Prakash S. Effect of probiotics Lactobacillus and Bifidobacterium on gut-derived lipopolysaccharides and inflammatory cytokines: an in vitro study

using a human colonic microbiota model. *J Microbiol Biotechnol.* 23(4):518–526. http://www.papersearch.net/view/detail.asp?detail_key=05212249. Accessed July 3, 2015.

[101] Zilbauer M, Jenke A, Wenzel G, et al. Expression of human beta-defensins in children with chronic inflammatory bowel disease. *PLo S One.* 2010;5(10):e15389. doi:10.1371/journal.pone.0015389.

[102] Kapel N, Roman C, Caldari D, et al. Fecal tumor necrosis factor-alpha and calprotectin as differential diagnostic markers for severe diarrhea of small infants. *J Pediatr Gastroenterol Nutr.* 2005;41(4):396–400. http://www.ncbi.nlm.nih.gov/pubmed/16205505. Accessed June 1, 2015.

[103] Heyman M, Darmon N, Dupont C, et al. Mononuclear cells from infants allergic to cow's milk secrete tumor necrosis factor α, altering intestinal function. *Gastroenterology.* 1994;106(6):1514–1523. doi:10.5555/uri:pii:0016508594904057.

[104] Rodriguez P, Heyman M, Candalh C, Blaton MA, Bouchaud C. Tumour necrosis factor-alpha induces morphological and functional alterations of intestinal HT29 cl.19A cell monolayers. *Cytokine.* 1995;7(5):441–448. doi:10.1006/cyto.1995.0060.

[105] Morita H, Nomura I, Orihara K, et al. Antigen-specific T-cell responses in patients with non-Ig E-mediated gastrointestinal food allergy are predominantly skewed to T(H)2. *J Allergy Clin Immunol.* 2013;131(2):590–592.e1–e6. doi:10.1016/j.jaci.2012.09.005.

[106] Benlounes N, Dupont C, Candalh C, et al. The threshold for immune cell reactivity to milk antigens decreases in cow's milk allergy with intestinal symptoms. *J Allergy Clin Immunol.* 1996;98(4):781–789. doi:10.1016/S0091-6749(96)70127-6.

[107] Kapel N, Matarazzo P, Haouchine D, et al. Fecal tumor necrosis factor alpha, eosinophil cationic protein and Ig E levels in infants with cow's milk allergy and gastrointestinal manifestations. *Clin Chem Lab Med.* 1999;37(1):29–32. doi:10.1515/CCLM.1999.004.

[108] Wada H, Horisawa T, Inoue M, Yoshida T, Toma T, Yachie A. Sequential measurement of fecal parameters in a case of non-immunoglobulin E-mediated milk allergy. *Pediatr Int.* 2007;49(1):109–111. doi:10.1111/j.1442-200X.2007.02294.x.

[109] [109] Majamaa H, Miettinen A, Laine S, Isolauri E. Intestinal inflammation in children with atopic eczema: faecal eosinophil cationic protein and tumour necrosis factor-alpha as non-invasive indicators of food allergy. *Clin Exp Allergy.* 1996;26(2):181–187. doi:10.1111/j.1365-2222.1996.tb00078.x.

[110] Ruemmele FM, Brousse N, Goulet O. Autoimmune enteropathy: molecular concepts. *Curr Opin Gastroenterol.* 2004;20(6):587–591. doi:10.1097/00001574-200411000-00014.

[111] Goulet O, Kedinger M, Brousse N, et al. Intractable diarrhea of infancy with epithelial and basement membrane abnormalities. *J Pediatr.* 1995;127(2):212–219. doi:10.1016/S0022-3476(95)70297-0.

[112] Walker-Smith JA. Intractable diarrhoea in infancy: a continuing challenge for the

paediatric gastroenterologist. *Acta Paediatr.* 1994;83(s395):6–9. doi:10.1111/j.1651-2227.1994.tb13220.x.

[113] SouléJC, Andant C, Kapel N, Gobert JC. Evaluation of fecal tumor necrosis factor alpha (TNFα) as a marker of Crohn's disease (CD) activity. *Gastroenterology.* 1998;114(114):A1089. doi:10.1016/S0016-5085(98)84428-4.

[114] Bystrom J, Garcia RC, Hakansson L, et al. Eosinophil cationic protein is stored in, but not produced by, peripheral blood neutrophils. *Clin Exp Allergy.* 2002;32(7):1082–1091. doi:10.1046/j.1365-2222.2002.01408.x.

[115] Byström J, Tenno T, Håkansson L, et al. Monocytes, but not macrophages, produce the eosinophil cationic protein. *APMIS.* 2008;109(7–8):507–516. doi:10.1111/j.1600-0463.2001.907804.x.

[116] Koh GC-H, Shek LP-C, Goh DY-T, Van Bever H, Koh DS-Q. Eosinophil cationic protein: is it useful in asthma? A systematic review. *Respir Med.* 2007;101(4):696–705. doi:10.1016/j.rmed.2006.08.012.

[117] Noguchi E, Iwama A, Takeda K, et al. The promoter polymorphism in the eosinophil cationic protein gene and its influence on the serum eosinophil cationic protein level. *Am J Respir Crit Care Med.* 2003;167(2):180–184. doi:10.1164/rccm.200204-292OC.

[118] Egesten A, Calafat J, Weller PF, et al. Localization of granule proteins in human eosinophil bone marrow progenitors. *Int Arch Allergy Immunol.* 1997;114(2):130–138. doi:10.1159/000237657.

[119] Peters MS, Rodriguez M, Gleich GJ. Localization of human eosinophil granule major basic protein, eosinophil cationic protein, and eosinophil-derived neurotoxin by immunoelectron microscopy. *Lab Invest.* 1986;54(6):656–662. http://europepmc.org/abstract/med/3520144. Accessed June 26, 2015.

[120] Bystrom J, Amin K, Bishop-Bailey D. Analysing the eosinophil cationic protein--a clue to the function of the eosinophil granulocyte. *Respir Res.* 2011;12(1):10. doi:10.1186/1465-9921-12-10.

[121] Dodig, Slavic. "Eosinophil cationic protein-current concepts and controversies." Biochemia medica 21.2 (2011).

[122] Monteseirin J, Vega A, Chacon P, et al. Neutrophils as a novel source of eosinophil cationic protein in Ig E-mediated processes. *J Immunol.* 2007;179(4):2634–2641. doi:10.4049/jimmunol.179.4.2634.

[123] Tomassini M, Tsicopoulos A, Tai PC, et al. Release of granule proteins by eosinophils from allergic and nonallergic patients with eosinophilia on immunoglobulin-dependent activation. *J Allergy Clin Immunol.* 1991;88(3):365–375. doi:10.1016/0091-6749(91)90099-A.

[124] Carlson M, Peterson C, Venge P. The influence of IL-3, IL-5, and GM-CSF on normal human eosinophil and neutrophil C3b-induced degranulation. *Allergy*. 1993;48(6):437–442. http://europepmc.org/abstract/med/8238799. Accessed July 23, 2015.

[125] Woschnagg C, Rubin J, Venge P. Eosinophil cationic protein (ECP) is processed during secretion. *J Immunol*. 2009;183(6):3949–3954. doi:10.4049/jimmunol.0900509.

[126] Meng Q, Ying S, Corrigan CJ, et al. Effects of rapamycin, cyclosporin A, and dexamethasone on interleukin 5-induced eosinophil degranulation and prolonged survival. *Allergy*. 1997;52(11):1095–1101. doi:10.1111/j.1398-9995.1997.tb00181.x.

[127] Eda R, Sugiyama H, Hopp RJ, Okada C, Bewtra AK, Townley RG. Inhibitory effects of formoterol on platelet-activating factor induced eosinophil chemotaxis and degranulation. *Int Arch Allergy Immunol*. 1993;102(4):391–398. doi:10.1159/000236588.

[128] Kawashima T, Iwamoto I, Nakagawa N, Tomioka H, Yoshida S. Inhibitory effect of pemirolast, a novel antiallergic drug, on leukotriene C4 and granule protein release from human eosinophils. *Int Arch Allergy Immunol*. 1994;103(4):405–409. doi: 10.1159/000236662.

[129] Mc Laren DJ, Peterson CG, Venge P. Schistosoma mansoni: further studies of the interaction between schistosomula and granulocyte-derived cationic proteins in vitro. *Parasitology*. 1984;88 (Pt 3):491–503. http://www.ncbi.nlm.nih.gov/pubmed/6739134. Accessed July 23, 2015.

[130] Lehrer RI, Szklarek D, Barton A, Ganz T, Hamann KJ, Gleich GJ. Antibacterial properties of eosinophil major basic protein and eosinophil cationic protein. *J Immunol*. 1989;142(12):4428–4434. http://www.ncbi.nlm.nih.gov/pubmed/2656865. Accessed June 14, 2015.

[131] Domachowske JB, Dyer KD, Adams AG, Leto TL, Rosenberg HF. Eosinophil cationic protein/RNase 3 is another RNase A-family ribonuclease with direct antiviral activity. *Nucleic Acids Res*. 1998;26(14):3358–3363. doi:10.1093/nar/26.14.3358.

[132] Boix, E., Torrent, M., Sanchez, D., & Nogues, M. V. (2008). The antipathogen activities of eosinophil cationic protein. Current pharmaceutical biotechnology, 9(3), 141–152.

[133] Young JD, Peterson CG, Venge P, Cohn ZA. Mechanism of membrane damage mediated by human eosinophil cationic protein. *Nature*. 321(6070):613–616. doi: 10.1038/321613a0.

[134] Miner JH. Laminins and their roles in mammals. *Microsc Res Tech*. 2008;71(5):349–356. doi:10.1002/jemt.20563.

[135] Torrent M, Odorizzi F, Nogués MV, Boix E. Eosinophil cationic protein aggregation: identification of an N-terminus amyloid prone region. *Biomacromolecules*. 2010;11(8): 1983–1990. doi:10.1021/bm100334u.

[136] Torrent M, Badia M, Moussaoui M, Sanchez D, Nogués MV, Boix E. Comparison of human RNase 3 and RNase 7 bactericidal action at the Gram-negative and Gram-positive bacterial cell wall. *FEBS J.* 2010;277(7):1713–1725. doi:10.1111/j.1742-4658.2010.07595.x.

[137] Zagai U, Lundahl J, Klominek J, Venge P, Sköld CM. Eosinophil cationic protein stimulates migration of human lung fibroblasts in vitro. *Scand J Immunol.* 2009;69(4):381–386. doi:10.1111/j.1365-3083.2009.02233.x.

[138] Rosenberg HF, Domachowske JB. Eosinophils, eosinophil ribonucleases, and their role in host defense against respiratory virus pathogens. *J Leukoc Biol.* 2001;70(5):691–698. http://www.jleukbio.org/content/70/5/691.short. Accessed July 23, 2015.

[139] Garofalo R, Kimpen JL, Welliver RC, Ogra PL. Eosinophil degranulation in the respiratory tract during naturally acquired respiratory syncytial virus infection. *J Pediatr.* 1992;120(1):28–32. http://www.ncbi.nlm.nih.gov/pubmed/1731020. Accessed July 23, 2015.

[140] Domachowske JB, Rosenberg HF. Respiratory syncytial virus infection: immune response, immunopathogenesis, and treatment. *Clin Microbiol Rev.* 1999;12(2):298–309. http://cmr.asm.org/content/12/2/298.short. Accessed July 23, 2015.

[141] Trautmann A, Schmid-Grendelmeier P, Krüger K, et al. T cells and eosinophils cooperate in the induction of bronchial epithelial cell apoptosis in asthma. *J Allergy Clin Immunol.* 2002;109(2):329–337. doi:10.1067/mai.2002.121460.

[142] Navarro S, Boix E, Cuchillo CM, Nogués MV. Eosinophil-induced neurotoxicity: the role of eosinophil cationic protein/RNase 3. *J Neuroimmunol.* 2010;227(1–2):60–70. doi:10.1016/j.jneuroim.2010.06.012.

[143] Chang K-C, Lo C-W, Fan T-C, et al. TNF-alpha mediates eosinophil cationic protein-induced apoptosis in BEAS-2B cells. *BMC Cell Biol.* 2010;11(1):6. doi:10.1186/1471-2121-11-6.

[144] Plager DA, Davis MDP, Andrews AG, et al. Eosinophil ribonucleases and their cutaneous lesion-forming activity. *J Immunol.* 2009;183(6):4013–4020. doi:10.4049/jimmunol.0900055.

[145] Venge P, Bystrom J, Carlson M, et al. Eosinophil cationic protein (ECP): A review on molecular and biological properites and the use of ECP as a marker of eosinophil activation in disease. *Clin Exp Allergy.* September 1999;29(9):1172–1186 http://swepub.kb.se/bib/swepub:oai:Di VA.org:uu-55915. Accessed July 23, 2015.

[146] Gomes I, Mathur SK, Espenshade BM, Mori Y, Varga J, Ackerman SJ. Eosinophil-fibroblast interactions induce fibroblast IL-6 secretion and extracellular matrix gene expression: implications in fibrogenesis. *J Allergy Clin Immunol.* 2005;116(4):796–804. doi:10.1016/j.jaci.2005.06.031.

[147] Peona V, Amici M De, Quaglini S, et al. Serum eosinophilic cationic protein: is there a

role in respiratory disorders? *J Asthma.* 2010;47(2): 131–134 http://www.tandfonline.com/doi/abs/10.3109/02770900903497170#.Vb Za Tkvs8sw. Accessed July 27, 2015.

[148] Kato M, Yamada Y, Maruyama K, Hayashi Y. Serum eosinophil cationic protein and 27 cytokines/chemokines in acute exacerbation of childhood asthma. *Int Arch Allergy Immunol.* 2010;152 (Suppl 1):62–66. doi:10.1159/000312127.

[149] Kämpe M, Stolt I, Lampinen M, Janson C, Stålenheim G, Carlson M. Patients with allergic rhinitis and allergic asthma share the same pattern of eosinophil and neutrophil degranulation after allergen challenge. *Clin Mol Allergy.* 2011;9(1):3. doi: 10.1186/1476-7961-9-3.

[150] Moneret-Vautrin D-A. [Is the seric eosinophil cationic protein level a valuable tool of diagnosis in clinical practice?]. *Rev Med Interne.* 2006;27(9):679–683. doi:10.1016/j.revmed.2006.02.007.

[151] Park Y-J, Oh E-J, Park J-W, Kim M, Han K. Plasma eosinophil cationic protein, interleukin-5, and ECP/Eo count ratio in patients with various eosinophilic diseases. *Ann Clin Lab Sci.* 2006;36(3):262–266. http://www.annclinlabsci.org/content/36/3/262.short. Accessed July 27, 2015.

[152] Hogan SP, Rothenberg ME. Eosinophil function in eosinophil-associated gastrointestinal disorders. *Curr Allergy Asthma Rep.* 2006;6(1):65–71. doi:10.1007/s11882-006-0013-8.

[153] Peterson CGB, Eklund E, Taha Y, Raab Y, Carlson M. A new method for the quantification of neutrophil and eosinophil cationic proteins in feces: establishment of normal levels and clinical application in patients with inflammatory bowel disease. *Am J Gastroenterol.* 2002;97(7):1755–1762. doi:10.1111/j.1572-0241.2002.05837.x.

[154] Silva AC, Levy L, Trindade JC, Mendonça P, Silva C, Lopes AI. Faecal and serum levels of eosinophil cationic protein in a healthy paediatric population. *Scand J Clin Lab Investig.* 2007;67(7): 757–766 http://www.tandfonline.com/doi/abs/10.1080/00365510701308337#.Vb AIzkvs8sw. Accessed July 22, 2015.

[155] Rodríguez Jiménez B, Domínguez Ortega J, González García JM, Kindelan Recarte C. Eosinophilic gastroenteritis due to allergy to cow's milk. *J Investig Allergol Clin Immunol.* 2011;21(2):150–152. http://www.ncbi.nlm.nih.gov/pubmed/21462806. Accessed July 15, 2015.

[156] Saarinen KM, Sarnesto A, Savilahti E. Markers of inflammation in the feces of infants with cow's milk allergy. *Pediatr Allergy Immunol.* 2002;13(3):188–194. http://www.ncbi.nlm.nih.gov/pubmed/12144641. Accessed July 15, 2015.

[157] Kristjánsson G, Venge P, Hällgren R. Mucosal reactivity to cow's milk protein in coeliac disease. *Clin Exp Immunol.* 2007;147(3):449–455. doi:10.1111/j.1365-2249.2007.03298.x.

7

Casein Proteins: Structural and Functional Aspects

Mohd Younus Bhat, Tanveer Ali Dar and Laishram Rajendrakumar Singh

Abstract

Mammalian milk is a complex fluid mixture of various proteins, minerals, and lipids, which play an important role in providing nutrition and immunity to the newborn. Casein proteins, which form about 80% of the bovine milk proteins, form large colloidal particles with calcium phosphate to form casein micelles, which for many years have been an important subject of interest. Casein micelles are composed of four main types of proteins: α_{S1}-casein, α_{S2}-casein, β-casein, and k-casein. These constituent casein proteins lack well-defined secondary and tertiary structure due to large amount of propyl residues. These micelles are being extensively studied because of their importance in functional behavior of milk and various milk products. However, the exact structure and nature of these casein micelles are still under debate. These different casein proteins possess different functional properties due to their primary amino acid sequence.

Keywords: Milk, Casein, Phosphoproteins, calcium

1. Introduction

It is now widely known that milk is a complex biological fluid secreted by mammals whose most important biological function is to supply nutrients for the nourishment of the off-spring. The term "micelle" has been applied to the dispersed phase of milk, that is the casein-protein complex. Casein protein component of milk is made up of different proteins, which possess different functions despite having no well-defined secondary and tertiary structure. These proteins, which include α_{S1}-, α_{S2}-, β-, and k-casein, have a primary amino acid sequence different from each other and occupy different positions in micelle and perform specific functions. Some of the proteins are involved in calcium phosphate transport while others in stability of other caseins and micelle. The structure and properties of casein micelle as a whole and individual casein proteins, which constitute the micelle, are discussed.

2. Milk proteins and casein micelle

Milk is a complex biological fluid with high content of proteins, minerals, and lipids secreted by mammals to supply nutrition and to provide immunological protection to the newborn. The differences in the metabolic processes of the lactating mother and the nutritive requirements of the newborn are thought to be responsible for the interspecies differences in the composition of milk [1]. The main function of milk is to provide essential amino acids and minerals that are vital for the development and therefore function of muscular and other tissues in new born mammals. It also includes active proteins providing antibodies, metal and vitamin-binding proteins, and several protein hormones [2]. Milk proteins coagulate very rapidly in the stomach of newborn as they are structurally built in a way that they form large complexes with calcium phosphate. Normal bovine milk contains almost 3.2–3.7% protein which varies in composition and concentration during different stages of lactation [3, 4]. Milk proteins are divided into two classes and are no more thought to be a homogeneous protein [2]. Caseins constitute about 75–80% of total protein and precipitate at pH 4.6 at 30°C. The remaining fraction, serum or whey protein, is soluble under similar conditions [5]. The rest of proteins found in milk are trace fractions of glycoprotein [6]. Casein proteins and calcium phosphate form large colloidal particles called casein micelles, which have been the subject of interest for many years [7]. The main function of the casein micelle is to provide fluidity to casein molecules and solubilize phosphate and calcium. There is a very large flow of calcium through the mammary epithelial tissue, and despite this, there is rarely any formation of calcium stones in the mammary gland. It has been suggested that the calcification of the mammary gland is prevented by the formation of casein-micelle complex with calcium phosphate. The primary amino acid sequence of casein proteins and their conformation in solution are therefore thought to prevent calcification of the mammary gland in addition to providing nutrition [8]. In addition to their biological role, which is to provide nutrition, caseins are also studied for their role in human health and other malfunctions such as stone-forming diseases in bovine animals [9–12].

The casein-micelle structure is being studied extensively because of its importance in the functional behavior of milk and some milk products [5]. However, the exact structure of casein and its micelles is still under debate. Various physical and biochemical studies of these colloids have mainly focused on the properties, size of the colloids, protein composition, micelle reconstitution, etc. Due to the large size of the casein-micelles, which interfere with absolute structure determination, different models have been proposed. Models can be classified into three categories: coat–core model, subunit or submicelle model, and internal structure model. Waugh and Nobel in 1965, Payens in 1966, Parry and Carroll in 1969, and Paquin and others in 1987 proposed coat-core models. Coat-core model dictates that micelle is an aggregate of caseins with outer and interior of micelle having different composition, and there is an inaccuracy in the identification of inner part of the structure [13–16]. This actually contains two diametrically opposite theories. Waugh and Nobel in 1965 were the first who proposed a model which fits in this class which is based primarily upon the solubilities of various caseins in Ca^{2+} solutions. According to them, α_{S1}- and k-caseins form low weight ratio complexes in the absence of calcium. Monomers of α_{S1}- or β-caseins with charged phosphate loop form caseinate core due to addition of calcium ions. The α_{S1}- or -β-caseins in their monomeric form with charged phosphate loops form limiting size aggregates/caseinate core. The formation of low weight α_{S1}-k-complex monolayer leads to the prevention of caseinate precipitation. The k-casein monomers spread out entirely on the surface of coat/complex formed, and therefore, its amount dictate the size of casein micelle. This model explains the lyophilic nature of the colloidal casein complex and also the ready accessibility of k-casein to chymosin and therefore is quite appealing [17]. According to Payens (1966) model based on his experimental data on the association of caseins, the densely folded α_{S1}-caseins remain adhered to loose network of β-caseins to form micelle core. Unlike the Waugh and Nobel models, colloidal calcium phosphate is present both on the outer surface and in the inner side of the micelle, while k-casein is confined to the surface of the micelle [7]. In 1969, Parry and Carroll used electron microscopy to locate k-casein on the surface of micelle as proposed by Waugh. They suggested that k-casein is present at the interior and acts as a nucleating agent to which calcium-insoluble caseins might cluster and gets stabilized the colloidal calcium phosphate. They used ferritin-labeled anti-k-casein antibodies to localize kappa casein at the outer surface of casein micelles. They found very little or no concentration of k-casein protein on the outer surface of the casein micelles as was suggested by previous workers. According to this model of casein micelle, the surface of the micelle comprises α_{S1}- and β-caseins with some colloidal calcium phosphate [18]. Paquin et al. in 1987 proposed a model based on results obtained from experiments using gel chromatography of EDTA-dissociated casein micelles for identifying two protein fractions by monolayer methods. This model describes the micelle core as a scaffold of colloidal calcium phosphate and α_{S1}-caseins, while β-caseins are held by hydrophobic interactions. These models predict a precise distribution of k-casein and are based upon nucleation around a core which is k-casein in case of Parry and Carroll and αS1, β-calcium caseinate in case of Waugh [7].

The submicelle models that were proposed by Shimmin and Hill (1964), Morr (1967), Slattery and Evard (1973), Schmidt (1980), Walstra (1984), and Ono and Obata (1989) considered that casein micelles are composed of uniform subunits that are roughly spherical in shape [19–23]. Shimmin and Hill (1964) were the first who postulated a submicellar structure for the casein micelle [24]. They used electron microscopy to study the ultrathin cross sections of embedded casein micelles and measured a diameter of 10 nm for the submicelles [24]. Another model proposed by Morr (1967) which was based on results obtained from study of oxalate and urea treatment on the disruption of casein micelles and proposed that α_{S1}-, β-, and k-monomers formed small uniform submicelles. These casein micelles are composed of numerous, loosely packed, calcium caseinate complex units, joined in association by a combination of calcium and colloidal calcium phosphate and citrate linkages between casein phosphoserine and carboxyl groups. Hydrophobic bonding and calcium caseinate bridges stabilize the submicelles, while colloidal calcium phosphate helps to aggregate the submicelles into micellar structure [19]. Each of these calcium caseinate complex units is probably composed of an inner core consisting of a α_{S1}- and β-casein, surrounded by an outer layer rich in α_{S1}- and k-casein, as suggested by Waugh and noble [14, 25]. According to Morr, these submicelles have a diameter of ~30 nm studied by using sedimentation velocity which is somewhat larger than that postulated by Shimmin and Hill. Additional β-casein could become associated with the outer surfaces of the micelle under appropriate conditions which favor conversion of soluble casein (mainly β-casein) to micellar casein. A model was proposed for the native casein micelle which consists of numerous loosely packed calcium caseinate complex units joined in association by a combination of calcium and colloidal calcium phosphate–citrate linkages. The colloidal calcium phosphate–citrate is considered to be distributed throughout the micelle rather than as a layer on its outer surface. α_{S1}-, α_{S2}-, and β-casein precipitate when calcium binds to their phosphoserine residues. k-casein at the other end is not only calcium insoluble, but it also interacts with other calcium-sensitive caseins and stabilizes them thereby initiates the formation of the stable colloidal state. Various enzymatic, immunological, and chemical techniques usually recognized that while majority of the k-casein must reside on the outer surface of the casein micelles, other caseins might also occur there as well [26, 27]. k-casein is thought to be predominantly present on the outer surface of the casein micelle as shown by almost all researchers working in this field till date. Various methods for disruption of casein micelles have been used by several other researchers to study the nature of submicelle. Carroll et al. (1971) used urea, EDTA, sodium fluoride, and sodium lauryl sulfate for the disruption of micelles and found particles ~8 to 12 nm in diameter [28]. Submicelles of 10 nm diameter were also found by Schmidt and Buchheim (1970) after they dialyzed milk free calcium in cold and using high pressure to disrupt casein micelles [29, 30]. These results were confirmed by Buchheim and Welsch in 1973. Pepper and Farrell (1982) used gel chromatography to study interaction of concentration-dependent interactions of EDTA dissociated whole-casein micelles. It was found that with increasing protein concentration at 37°C and pH 6.6, the individual components of casein formed polymers which approached a molecular radius of ~9.5 nm [31]. These submicelles were thought to be formed by the interaction of SH-k-casein monomers with those of αS- and β-caseins as seen by analyzing concentration elution profiles. Carroll et al. (1970) and Farrell and Thompson (1971) also observed particles of ~10-nm diameter in the Golgi vacuoles of lactating rat mammary gland and therefore supported the hypothesis of Shimmin and Hill (1964) [28, 32]. Another model for casein micelle structure is based on the results of various experiments on the effect of calcium on the sedimentation behavior of those particles which are formed in mixtures of caseins was proposed by Slattery and Evard in 1973. This model based upon casein interactions combines the best features of most casein micelle models. According to this model, submicelles which are rich in k-casein are found predominantly on the outer surface of the casein micelle, while those poor in k-casein content are internalized. They suggested that casein monomers interact to form submicelles of variable composition depending upon their casein content. This model also suggests an inverse relationship between k-casein content and micelle size. This model predicts large casein micelles which are poor in k-casein content, k-casein will occupy position on surface, while in smaller micelles which are rich in k-casein, k-casein is uniformly distributed [22]. Walstra (1984) proposed the submicelle model for casein which is the most accepted model for casein. According to this model, spherical subunits or submicelles are the building blocks of casein micelles. Each submicelle is variable in composition with 20–25 casein molecules per submicelle, and the diameter of submicelle is 12–15 nm. Hydrophobic interactions between the constituent proteins and the calcium phosphate linkages keep the submicelle together. According to this model, there are two types of submicelles one consisting of αS- and β-caseins and another αS- and k-caseins, the former is has hydrophobic regions buried in the center while latter is more hydrophilic because of the presence of sugar residues on k-caseins. Further aggregation of submicelles is avoided by the steric and electrostatic repulsions by the hydrophilic part of the C-terminal end of k-casein located near outside of micelle, protruding from the micelle surface as a hairy layer [23]. Carroll and Farrell in 1983 also found that the location of k-casein is indeed related to casein micelle size using ferritin-labeled double-antibody technique coupled with electron microscopy [33]. These results confirm the inverse relationship between micelle size and k-casein content and also that larger casein micelles contain higher polymers of k-casein, indicating that k-k interactions are greater in k-poor micelles. Since according to this model, k-casein is not totally precisely localized in the micelles this model therefore contradicts with models proposed by Parry, Waugh, Garnier, and Ribadeau-Dumas and supports the more flexible model of Slattery and Evard [22, 25, 34].

The internal structure models, which are the last models, were proposed by Rose (1969), Garnier and Ribadeau-Dumas (1970), Holt (1992), and Horne (1998) indicate the manner in which different caseins aggregate [34–37]. The internal structure model of casein micelle is based upon the properties of isolated protein components which are involved in the formation of internal structure of the micelle. Rose (1969) was the first to propose internal structure model by using the endothermic polymerization of β-casein as the basis for his casein micelle model. According to this model, α_{S1}-monomers attach to chain like polymers of β-casein which are self-associated from β-casein monomers. k-casein interacts with α_{S1}-monomers. The β-casein is directed inward, while k-casein is directed outwards and a small amount of k-casein is placed in an internal position as these two segments associate. Colloidal calcium phosphate is incorporated as a stabilizing during the formation of micelle. The occurrence of some overall stoichiometry of the various casein components and the role of colloidal calcium phosphate in stabilizing micelle make this model appealing [37]. However, synthetic micelles can be formed from simple k- and α_{S1}-casein complexes in the complete absence of β-casein which makes β-casein as the basis of micelle formation questionable. Waugh et al. (1970) have also shown that the α_{S1}- and β-caseins tend to form mixed polymers randomly and β-casein is structure less in solution. It also forms micellar-like complexes rather than linear polymers [38]. Garnier and Ribadeau-Dumas (1970) who proposed another model emphasize on k-casein as the foundation of micelle structure. According to this model, three chains of α_{S1}- and β-casein are linked to the trimers of k-casein which radiate from the k-casein node which is present as a Y-like structure. There is a formation of loosely packed network when these α_{S1}- and β-caseins connect to other k-nodes. This model places steric restraints upon k-casein which posses few secondary structures. This model provides demonstrated porosity and explains a uniform distribution of k-casein regardless of micelle size. The model assigns no role to calcium caseinate interactions and ignores the role of colloidal calcium phosphate involvement in stabilization of the micelle. Although the submicelle casein model proposed by Walstra in 1999 has been widely accepted, there are two alternative models proposed by Holt in 1992 and

Horne in 1998 which fall into internal structure model [6, 35, 36]. According to model proposed by Holt, the casein micelle forms a tangled web of flexible casein networks forming a gel-like structure with C-terminal region of k-casein extending to form a hairy layer and microgranules of colloidal calcium phosphate at center. The surface location of k-casein and the cementing role of colloidal calcium phosphate are the two main features of this model. The caseins micelles according to this model are stabilized by two main factors one of which is steric stabilization by protruding k-casein layer hairs and another is by surface potential of approximately -20mV at pH 6.7. In 1998, Horne proposed dual bonding model which suggests that it is a balance between electrostatic repulsions and attractive hydrophobic interactions which held the proteins in casein micelles together. According to this model, hydrophobic interaction is the driving force for the formation of casein micelles and electrostatic repulsions are responsible for limiting the growth of polymers [36]. α_{S1}- and β-caseins self-associate by hydrophobic interactions as a result of formation of train–loop–train and tail–train like structures, respectively, upon adsorption at hydrophobic interfaces. There occurs a reduction in electrostatic repulsion because of colloidal calcium phosphate which form linkages between casein micelles and neutralizing agents of the negative charge of phosphoserine residues which makes the hydrophobic interaction between caseins a dominant force for the association of proteins. The lack of phosphoserine cluster to bind calcium in k-casein makes it to interact hydrophobically and act as a propagation terminator.

3. Forces responsible for the stability of the casein micelle

Linderstrom-Lang in 1929 postulated that mixture of calcium-insoluble proteins stabilized by calcium-soluble protein form the colloidal milk complex [39]. The calcium-soluble protein would be readily split by chymosin which leads to its coagulation due to destabilization of colloid. Such fractions exist as α_{S1}-, α_{S2}-, and β-caseins which are insoluble in calcium and k-casein which is soluble in presence of calcium and is split readily by chymosin. Sedimentation velocity experiments performed by Waugh et al., in 1971, demonstrated that α_{S1}- and k-casein complexes can be reformed from already isolated fractions [40]. A brief summary of the various types of bonding forces responsible for the stabilization of protein structure will be discussed.

3.1. Hydrophobic interactions

There is presence of large number of hydrophobic residues clustered together in α_{S1}-, β-, and k-casein as found by amino acid sequence analysis of these proteins. Since these are among the most hydrophobic proteins, role of hydrophobic bonding in the stabilization of casein cannot be ignored. The ability of β-casein to self-associate was reduced after removal of isoleucine and valine at C-terminal end of protein which normally self-associate in the absence of calcium [41]. Additionally, the ability of β-casein to form polymers was destroyed completely after removal of 20 amino acids at C-terminal which are mainly hydrophobic in nature [41]. Various investigators have found that α_{S1}-, β- and k-caseins diffuse out of the micelle at low temperature due to decrease in hydrophobic interactions [42–44]. The micelles containing rare α_{S1}-A genetic variant which possesses similar physical and solubility properties like that of β-casein is also less stable in cold. On the basis of light scattering and electron microscopy, it has been found that increased pressure disrupts casein micelle structure which also acts primarily on hydrophobic interactions [42–46]. The dependence of hydrophobic interactions on temperature and pressure also explains the resistance of skim milk to withstand higher temperature which is otherwise destabilized at extremely low temperatures.

3.2. Electrostatic interactions

There are many potential sites for strong ion bonding in apolar environment that might play a role in the stabilization of casein micelles. It is not possible to exactly assess the role of various inter- and intramolecular ionic bonds present in αs-, β-, and k-casein in stabilization of casein micelle structure. The ability of k-casein to stabilize the α_{S1}-casein is abolished when there is carbamylation of lysine residues in k-casein which further demonstrate that ionic interactions play a role in the casein micelle structure [47]. Modification of arginine side chains also affects the casein micelle stability and chymosin-induced coagulation [48].

3.3. Hydrogen bonding

The α-helical and β-pleated structures in many globular and fibrous structures are stabilized by hydrogen bonding along the polypeptide chain. Since casein proteins posses very little secondary structure and 72–76% of protein exists in aperiodic form, the degree of stabilization by α-helix and β-structure is very low [49, 50]. Hydrogen bonds between the various components of casein during the formation of highly aggregated casein micelle are possible but in case of ionizable side chains of monomeric proteins which are accessible to water, their contribution to the stability of these monomeric proteins is very less. For the formation of a residue-residue hydrogen bond in case of these monomeric proteins, there must be breakage of water-residue hydrogen bond which has already formed. During the interaction of two subunits of a protein, there are chances of formation of hydrogen bonds between individual monomers as the surface groups are no longer fully hydrated. Hydrogen bond may exist during the formation of aggregated casein micelles and self-association of α_{S1}-casein.

3.4. Disulfide bonds

Disulfide bonds between cysteine residues during folding of pleated sheet structures, helical segments, and unordered structures leads to the formation of tertiary structure. Both α_{S2}- and k-casein contain cysteine but the degree of disulfide cross-linkages which are normally present in the casein micelle is controversial [51–53]. It has been reported by many investigators that disulfide cross-linkages contribute to the overall stability of the casein micelle but they are not the driving force for the formation of casein micelle. Slattery in 1978 found that larger micelles have higher molecular weight disulfide-bonded polymers of k-casein. These k-casein molecules are thought to be contiguous with each other and form disulfide-linked aggregates which compose the casein micelle structure [54].

4. Casein proteins as internally disordered calcium-binding phosphoproteins

Casein proteins are phosphoproteins which comprise approximately 80% of the total protein present in bovine milk [55]. They were defined as phosphoproteins which precipitate from raw milk upon acidification at pH 4.6 at 20°C [56]. Casein proteins belong to one of the larger family of secretory calcium-binding phosphoproteins as has been found by the analysis of structure of human genome. Casein proteins provide one of the best example of intrinsically disordered or natively disordered or natively unfolded proteins [57]. The previous assumption that only those proteins which possess a well-defined folded conformation is able to perform a specific biological function is not valid in case of many intrinsically disordered proteins as they have specific biological functions even in their unfolded state [58]. Furthermore, it has also been found that in case of many of the eukaryotic proteins involved mainly in signaling pathways, there is presence of regions with disordered backbone conformations. The presence of disordered region in a protein involved in signaling provides larger surface area for interactions with other proteins. This property can therefore also help these proteins to interact with multiple proteins or target molecules at one time [59]. All of the casein proteins possess very little secondary and tertiary structure but are still able to perform their function in their disordered state. Caseins belong to the scavenger family of secretory calcium-binding phosphoproteins as they are involved in trapping of calcium phosphate. The intrinsic disorder of these proteins not only help in forming a thermodynamically stable complex with calcium phosphate but also allow these proteins to form a more tightly packed complex than a globular structure [58]. These casein proteins are post-translationally phosphorylated at seryl and very less frequently at threonyl residues which is one of their unique characteristic. k-casein contains only one or two phosphoseryl residues and is only casein which is glycosylated [53]. These calcium-sensitive caseins are not only able to bind to calcium phosphate crystal surface but are also able to form calcium phosphate nanoclusters which are thermodynamically stable chemical complexes by sequestering amorphous calcium phosphate. Calcium phosphate sequestration also depends upon the formation of phosphate centers in the primary sequence by clustering of phosphorylated residues [60].

5. Properties and functions of different protein components of casein

Earlier principle protein of bovine milk was considered to be homogenous protein casein. Later on it was found that casein proteins are heterogeneous and are composed of distinct fractions like α-, β-, and k-casein [61]. Casein in milk in its native state exists as large associate of macromolecules in colloidal dispersion with a mass of ~ 10^8 Da and ~200 nm in size [62]. The major protein of the casein complex is α_{S1}-casein which almost 38% followed by β-casein 36%, k-casein 13%, and α_{S2}-casein 10% [63]. Proline which is known to disrupt alpha-helical and β-structures is present in higher amount in α_{S1}-casein. It has been found that 70% of α_{S1}-casein is in unordered form with only a small amount of α-helical and β-structure. α_{S1}-Casein plays an important role in the ability of milk to transport calcium phosphate. It has also been found that one of its antioxidant peptide has 2,2-diphenyl-1-picrylhydrazyl (DPPH) radical scavenging activity. All casein proteins in their native states do not possess a well-defined tertiary or secondary structure [49]. Casein is classified as intrinsically disordered proteins implying that the protein is functional in its unstructured form. Further, it has been found by various investigators that α_{S1}- and β-casein possess chaperonic activity and are responsible for the stabilization of micelle by preventing aggregation of α_{S2}- and k-casein, respectively. Various investigators have also found that αs-casein prevents the stress-induced aggregation of various proteins like bovine serum albumin, whey proteins, β-lactoglobulin, carbonic anhydrase, and alcohol dehydrogenase by forming soluble, high molecular weight complexes [64]. The self-association of α_{S1}-casein monomers in aqueous solution is attributed to the high degree of hydrophobicity and small amount of structural content [38, 45]. There are ~43% hydrophobic amino acids in case of α_{S1}-casein, ~33% in case of α_{S2}-casein, ~52% in β-casein, and ~43% in k-casein. Under the normal pH, ionic strength, and temperature of milk, the major protein components of milk are insoluble. The second most abundant milk protein is β-casein with five phosphoserine residues and a molecular weight of 23,980 [65]. All the protein's net charge, phosphoserine content, and α-helical residues are restricted to the first 40 amino acid residues present at N-terminal portion of β-casein, while C-terminal contains many apolar residues responsible for its high hydrophobicity [49]. β-Casein possesses very little secondary structure and is present as a random coil in aqueous solution. The lack of secondary structure is due to the evenly distributed proline found during complete amino acid sequence of β-casein [66]. β-Casein plays an important role in determining the surface property of casein micelle. One of its peptide acts as a macrophage activator thereby increase the phagocytic activity of macrophages and their peroxide release. β-casein also possesses an antioxidant peptide which has antioxidant activity. The last casein sequenced was α_{S2}-casein which possesses most unique primary structure of all the caseins with a molecular weight of 25,150 [67]. This protein has two cysteine residues with no known carbohydrate. α_{S2}-Casein exists as a dimer or may have some intrachain disulfide. α_{S2}-Casein is least susceptible to aggregation because of alternating negatively charged and hydrophobic areas [68]. It also plays important role in the transport of calcium phosphate. Its anti-microbial peptide casocidin-I has the ability to inhibit growth of E. coli and other bacteria. k-casein, which is soluble over a very broad range of calcium ion concentrations unlike other forms of caseins like α_{S1}-, α_{S2}-, and β-casein, is the fourth major component of the milk-protein complex [69]. Calcium solubility of k-casein has led workers to assign to it the role of casein micelle stabilization whose other components are insoluble in calcium. It stabilizes micelle formation thereby prevent precipitation of casein in milk. Casoxins peptide possesses opoid antagonist, whereas casoplatelin inhibits platelet aggregation. The molecular weight of reduced k-casein is about 19,000 [51]. It exists as a high molecular weight mixtures of polymers. It is the only major component of casein which possesses carbohydrates bound to the highly soluble macropeptide portion formed after chymosin hydrolysis. There are only one or two phosphate residues per k-casein casein monomer which makes it soluble in calcium [70].

There are several genetic variants of casein components with variable numbers of phosphoseryl residues especially in case of α_{S2}-casein which exhibits a large variability in the extent of phosphorylation [71]. Another unique feature of caseins is the large amount of propyl residues especially in case of β-casein which greatly affect the secondary and tertiary structure of caseins [53]. In addition, all casein proteins possess different hydrophilic and hydrophobic regions along the protein chain [46]. αS-Caseins are the major casein proteins with 8–12 seryl phosphate groups, while β-casein contains about 5 phosphoserine residues and is more hydrophobic than αS-caseins and k-casein [49]. Highly phosphorylated αS-caseins and β-caseins are very sensitive to the of calcium salt concentration, that is, these proteins precipitate in presence of high Ca^{2+} ions [40, 72]. Unlike other types of caseins, k-caseins are glycoproteins [56, 71] with only one phosphoserine group. This makes them stable in the presence of Ca^{2+} ions thereby playing an important role in protecting other caseins which are calcium sensitive from precipitation and makes casein micelle stable [69]. Casein is insensitive to heat, and it is only temperature above 120°C that causes the casein proteins to become insoluble, whereas it is sensitive to pH and will precipitate at its isoelectric pH [73]. The individual families of casein proteins were identified by alkaline urea gel electrophoresis. Each of the four different caseins may have a variety of numbers of phosphate groups attached through their serine or threonine residues. In terms of the extent of phosphorylation, α_{S1}-casein may have 8 or 9, α_{S2}-casein 10–13, β-casein may have 4 and k-casein, 1–3. α_{S1}- and β-Casein contain no disulfide bonds, and α_{S2}- and k-casein contain two cysteine residues which form inter- or intramolecular disulfide bonds [74, 75]. α_{S2}-Casein exists as a dimer, and k-casein can exist from dimer to decamer depending upon the pattern of intermolecular disulfide bonding [68]. There are 10 different molecular forms of k-casein on the basis of degree of glycosylation and is the only casein which is glycosylated [56, 70, 71]. Another source of variability in caseins is genetic polymorphism. α_{S1}-Casein has been shown to be present in bovine milk as α_{S1}-casein A-D [71]. Caseins are structurally classified as natively or intrinsically disordered proteins which is different from random coil conformation found in some globular proteins [76, 77]. Due to the lack of well-defined structure, crystallization of casein proteins to provide a three-dimensional crystal structure is not possible, but at the other end, this lack of structure helps to facilitate proteolysis and therefore ready absorption of amino acids and small peptides in the intestine [2, 78]. Caseins proteins possess very little three dimensional structure but possess some secondary structure [79]. The high number of proline residues which distort protein folding into α-helices and β-sheets is responsible for inhibition of higher proportions of secondary and tertiary structure. Casein proteins contain 32–42% non-polar amino acids which makes them highly hydrophobic but due to the presence of high number of phosphate and sulfur-containing amino acids and carbohydrates in case of k-casein, they are quite soluble in aqueous solvents [2]. Casein proteins are homologous in all the species as has been found by various protein and gene sequencing studies [80]. The proportion of various caseins varies widely. All species form colloidal casein micelles for the transport of calcium and phosphate. Casein micelles of most species appear quite similar at the ultra structural level. Despite the variations in casein components, the α_{S1}- and α_{S2}-caseins are calcium sensitive, whereas β- and k-casein are not sensitive to calcium. Casein proteins are important nutritionally because of their high phosphate content due to which they bind significant quantities of calcium and also are rich in lysine which is an essential amino acid in humans. α_{S1}- and α_{S2}-casein possess 14 and 24 lysines, respectively [2]. Each of the caseins possesses significant variability due to extent of their post-translational modification, disulfide bonding, genetic polymorphism [81]. The properties of milk and various milk products mainly depend on proteins present in it and to some extent on other components like salts, fat, and lactose. Caseins which possess an extraordinary high heat stability make the milk and other milk products highly stable even at higher temperature [61].

6. Conclusion

Mammalian milk contains casein micelles that help to provide adequate nutrients to the neonate and at the same time prevent any risk of pathological calcification or amyloidosis. Interestingly, all caseins exhibit a disordered conformation and many have chaperonic activity (α_{S1}- and β-casein) which might be an attribute that help these casein proteins to self-associate and assembly into functional micelle. The structural disorderness and the chaperonic property would have been evolutionarily selected to make these molecules ideally suitable to thrive under various environmental insults since the milk is secretory product. It might be possible that that in addition to casein, many other milk proteins may also have chaperonic function. Identifying chaperonic function of other proteins present in milk will have many industrial and clinical insights.

Author details

Mohd Younus Bhat[1], Tanveer Ali Dar[1] and Laishram Rajendrakumar Singh[2*]

*Address all correspondence to: lairksingh@gmail.com

1 Department of Clinical Biochemistry, University of Kashmir, Srinagar, J&K, India

2 Dr. B.R. Ambedkar Centre for Biomedical Research, University of Delhi, North Campus, New Delhi, India

References

[1] R. Jenness, R.E. Sloan, The composition of milks of various species: a review, Dairy Science Abstract, vol. 32, 1970, pp. 599–612.

[2] P. F. Fox, T. Uniacke-Lowe, P.L.H. McSweeney, J.A. O'Mahony, Heat-induced changes in milk, Dairy Chemistry and Biochemistry, Springer International Publishing (2015) 345–375.

[3] R. Aschaffenburg, Imherited casein variants in cow's milk, Nature 192 (1961) 431.

[4] A. Author, Variations in the composition of milk, vol. 51, Royal Society of Chemistry, 1926, pp. 146–149.

[5] D.G. Dalgleish, Milk proteins. Chemistry and physics, Food proteins, Applied Sciences Publisher, London (1982) 155–178.

[6] P. Walstra, Casein sub-micelles: do they exist? International Dairy Journal 9 (1999) 189–192.

[7] H.S. Rollema, P.F. Fox, Casein association and micelle formation, Advanced Dairy Chemistry-1: Proteins (1992) 111–140.

[8] T.M. Doherty, K. Asotra, L.A. Fitzpatrick, J.-H. Qiao, D.J. Wilkin, R.C. Detrano, C.R. Dunstan, P.K. Shah, T.B. Rajavashisth, Calcification in atherosclerosis: bone biology and chronic inflammation at the arterial crossroads, Proceedings of the National Academy of Sciences 100 (2003) 11201–11206.

[9] R.B. Beems, E. Gruys, B.J. Spit, Amyloid in the corpora amylacea of the rat mammary gland, Veterinary Pathology Online 15 (1978) 347–352.

[10] S.C. Nickerson, L.M. Sordillo, N.T. Boddie, A.M. Saxton, Prevalence and ultrastructural characteristics of bovine mammary corpora amylacea during the lactation cycle, Journal of Dairy Science 68 (1985) 709–717.

[11] I.M. Reid, Corpora amylacea of the bovine mammary gland: histochemical and electron microscopic evidence for their amyloid nature, Journal of comparative pathology 82 (1972) 409–413.

[12] H. Taniyama, A. Kitamura, Y. Kagawa, K. Hirayama, T. Yoshino, S. Kamiya, Localized amyloidosis in canine mammary tumors, Veterinary Pathology Online 37 (2000) 104–107.

[13] H.M. Farrell Jr, Physical equilibria: proteins, Fundamentals of dairy chemistry, Springer, 1988, pp. 461–510.

[14] R.W. Noble Jr, D.F. Waugh, Casein Micelles. Formation and Structure. I1, Journal of the American Chemical Society 87 (1965) 2236–2245.

[15] P. Paquin, M. Britten, M.F. Laliberte, M. Boulet, Interfacial properties of milk casein proteins, ACS Symposium series-American Chemical Society (USA), 1987.

[16] T.A.J. Payens, D.G. Schmidt, Boundary spreading of rapidly polymerizing α_{S1}-casein B and C during sedimentation: Numerical solutions of the Lamm-Gilbert-Fujita equation, Archives of Biochemistry and Biophysics 115 (1966) 136–145.

[17] J. Garnier, Models of casein micelle structure, Netherlands milk and dairy journal 27 (1973) 240–248.

[18] P. Walstra, The voluminosity of bovine casein micelles and some of its implications, Journal of Dairy Research 46 (1979) 317–323.

[19] C.V. Morr, Effect of oxalate and urea upon ultracentrifugation properties of raw and heated skimmilk casein micelles, Journal of Dairy Science 50 (1967) 1744–1751.

[20] T. Ono, T. Obata, A model for the assembly of bovine casein micelles from F2 and F3 subunits, Journal of Dairy Research 56 (1989) 453–461.

[21] D.G. Schmidt, Collodial aspects of casein [in milk products] [A review article], Netherlands Milk and Dairy Journal (1980) 47, 34–42.

[22] C.W. Slattery, R. Evard, A model for the formation and structure of casein micelles from subunits of variable composition, Biochimica et Biophysica Acta (BBA)-Protein Structure 317 (1973) 529–538.

[23] P. Walstra, R. Jenness, Dairy chemistry & physics, John Wiley & Sons, USA, 1984.

[24] P.D. Shimmin, R.D. Hill, An electron microscope study of the internal structure of casein micelles, Journal of Dairy Research 31 (1964) 121–123.

[25] D.F. Waugh, Formation and structure of casein micelles, Milk Proteins: Chemistry and Molecular Biology 2 (1971) 3–85.

[26] S.H. Ashoor, R.A. Sair, N.F. Olson, T. Richardson, Use of a papain superpolymer to elucidate the structure of bovine casein micelles, Biochimica et Biophysica Acta (BBA)-Protein Structure 229 (1971) 423–430.

[27] R.J. Carroll, H.M. Farrell, Immunological approach to location of k-casein in the casein micelle by electron microscopy, Journal of Dairy Science 66 (1983) 679–686.

[28] R.J. Carroll, M.P. Thompson, H.M. Farrell Jr, Formation and structure of casein micelles in lactating mammary tissue, 28th Annual Proceedings EMSA 150 (1970).

[29] W. Buchheim, U. Welsch, Evidence for the submicellar composition of casein micelles on the basis of electron microscopical studies, Netherlands milk and dairy journal (1973) 27–163.

[30] D.G. Schmidt, Association of caseins and casein micelle structure, Developments in Dairy Chemistry 1 (1982) 61–86.

[31] L. Pepper, H.M. Farrell, Interactions leading to formation of casein submicelles, Journal of Dairy Science 65 (1982) 2259–2266.

[32] H.M. Farrell, M.P. Thompson, Biological significance of milk protein polymorphism, Journal of Dairy Science 54 (1971) 1219–1228.

[33] H.M. Farrell, F.W. Douglas, Effects of ultra-high temperatures pasteurization on the functional and nutritional properties of milk proteins, Kieler dairy research reports 35 (1983) 345–356.

[34] B.R. Dumas, J. Garnier, Structure of the casein micelle. The accessibility of the subunits to various reagents, Journal of Dairy Research 37 (1970) 269–278.

[35] C. Holt, Structure and stability of bovine casein micelles, Advances in Protein Chemistry 43 (1992) 63–151.

[36] D.S. Horne, Casein interactions: casting light on the black boxes, the structure in dairy products, International Dairy Journal 8 (1998) 171–177.

[37] D. Rose, A proposed model of micelle structure in bovine milk, Dairy Science Abstract (1969) 35, 1083–1093.

[38] D.F. Waugh, L.K. Creamer, C.W. Slattery, G.W. Dresdner, Core polymers of casein micelles, Biochemistry 9 (1970) 786–795.

[39] K. Linderstørm-Lang, S. Kodama, Studies on casein. III. On the fractionation of casein, compt. Rend Trav. lab. Carlsberg. ser. chem 17 (1929) 1–116

[40] D.F. Waugh, C.W. Slattery, L.K. Creamer, Binding of cations to caseins. Site binding, Donnan binding, and system characteristics, Biochemistry 10 (1971) 817–823.

[41] G.P. Berry, L.K. Creamer, Association of bovine β-casein. Importance of the C-terminal region, Biochemistry 14 (1975) 3542–3545.

[42] A.E. Ali, A.T. Andrews, G.C. Cheeseman, Influence of storage of milk on casein distribution between the micellar and soluble phases and its relationship to cheese-making parameters, Journal of Dairy Research 47 (1980) 371–382.

[43] D.T. Davies, A.J.R. Law, Variation in the protein composition of bovine casein micelles and serum casein in relation to micellar size and milk temperature, Journal of Dairy Research 50 (1983) 67–75.

[44] W.K. Downey, R.F. Murphy, The temperature dependent dissociation of b-casein from bovine casein micelles and complexes, Journal of Dairy Research 37 (1970) 361.

[45] D.G. Schmidt, W. Buchheim, Electron microscope investigation of the substructure of casein micelles in cows' milk, Milchwissenschaft 25 (1970) 596–600.

[46] D.G. Schmidt, T.A.J. Payens, E. Matijevic, Surface and Colloid Science, John Wiley, New York 165 (1976) 199.

[47] L. Pepper, N.J. Hipp, W.G. Gordon, Effects of modification of ε-amino groups on the interaction of k-and α_{S1}-caseins, Biochimica et Biophysica Acta (BBA)-Protein Structure 207 (1970) 340–346.

[48] R.D. Hill, The effect of the modification of arginine side chains in casein on the coagulation of rennin-altered casein, Journal of Dairy Research 37 (1970) 187–192.

[49] L.K. Creamer, T. Richardson, D.A.D. Parry, Secondary structure of bovine α_{S1}-and β-casein in solution, Archives of Biochemistry and Biophysics 211 (1981) 689–696.

[50] J. Raap, K.E.T. Kerling, H.J. Vreeman, S. Visser, Peptide substrates for chymosin (rennin): conformational studies of k-casein and some k-casein-related oligopeptides by circular dichroism and secondary structure prediction, Archives of Biochemistry and Biophysics 221 (1983) 117–124.

[51] H.E. Swaisgood, J.R. Brunner, Characteristics of kappa-casein in the presence of various dissociating agents, Biochemical and Biophysical Research Communications 12 (1963) 148–151.

[52] H.E. Swaisgood, J.R. Brunner, H.A. Lillevik, Physical parameters of K-casein from cow's milk*, Biochemistry 3 (1964) 1616–1623.

[53] H.E. Swaisgood, P.F. Fox, Chemistry of the caseins, Advanced Dairy Chemistry-1: Proteins. (1992) 63–110.

[54] C.W. Slattery, Variation in the glycosylation pattern of bovine k-casein with micelle size and its relationship to a micelle model, Biochemistry 17 (1978) 1100–1104.

[55] J.R. Brunner, Milk proteins, Food proteins (1977) 175–208.

[56] R.M. Whitney, J.R. Brunner, K.E. Ebner, H.M. Farrell, R.V. Josephson, C.V. Morr, H.E. Swaisgood, Nomenclature of the proteins of cow's milk: fourth revision, Journal of Dairy Science 59 (1976) 795–815.

[57] L.W. Fisher, D.A. Torchia, B. Fohr, M.F. Young, N.S. Fedarko, Flexible structures of SIBLING proteins, bone sialoprotein, and osteopontin, Biochemical and Biophysical Research Communications 280 (2001) 460–465.

[58] P. Tompa, Intrinsically unstructured proteins, Trends in Biochemical Sciences 27 (2002) 527–533.

[59] H.J. Dyson, P.E. Wright, Coupling of folding and binding for unstructured proteins, Current Opinion in Structural Biology 12 (2002) 54–60.

[60] C. Holt, P.A. Timmins, N. Errington, J. Leaver, A core-shell model of calcium phosphate nanoclusters stabilized by beta-casein phosphopeptides, derived from sedimentation equilibrium and small-angle X-ray and neutron-scattering measurements, European Journal of Biochemistry 252 (1998) 73–78.

[61] C. Holt, L. Sawyer, Caseins as rheomorphic proteins: interpretation of primary and secondary structures of the αS1-, β- and κ-caseins, Journal of the Chemical Society, Faraday Transactions 89 (1993) 2683–2692.

[62] D.G. Dalgleish, M. Corredig, The structure of the casein micelle of milk and its changes during processing, Annual Review of Food Science and Technology 3 449–467.

[63] D.T. Davies, A.J.R. Law, The content and composition of protein in creamery milks in south-west Scotland, Journal of Dairy Research 47 (1980) 83–90.

[64] J. Bhattacharyya, K.P. Das, Molecular chaperone-like properties of an unfolded protein, αs-casein, Journal of Biological Chemistry 274 (1999) 15505–15509.

[65] B.R. Dumas, J.C. Mercier, G. Brignon, F. Grosclau, Primary structure of bovine-casein-complete-sequence., European Journal of Biochemistry 25 (1972) 505.

[66] M. Noelken, M. Reibstein, Conformation of β-casein B, Archives of Biochemistry and Biophysics 123 (1968) 397–402.

[67] G. Brignon, B.R. Dumas, J.-C. Mercier, J.-P. Pelissier, B.C. Das, Complete amino acid sequence of bovine α_{S2}-casein, FEBS Letters 76 (1977) 274–279.

[68] T.H.M. Snoeren, B. Van Markwijk, R. Van Montfort, Some physicochemical properties of bovine α_{S2}-casein, Biochimica et Biophysica Acta (BBA)-Protein Structure 622 (1980) 268–276.

[69] D.F. Waugh, P.H. Von Hippel, k-casein and the stabilization of casein micelles, Journal of the American Chemical Society 78 (1956) 4576–4582.

[70] I.R. Dickson, D.J. Perkins, Studies on the interactions between purified bovine caseins and alkaline-earth-metal ions, Biochemical Journal 124 (1971) 235–240.

[71] W.N. Eigel, J.E. Butler, C.A. Ernstrom, H.M. Farrell, V.R. Harwalkar, R. Jenness, R.M. Whitney, Nomenclature of proteins of cow's milk: fifth revision, Journal of Dairy Science 67 (1984) 1599–1631.

[72] M.P. Thompson, W.G. Gordon, R.T. Boswell, H.M. Farrell, Solubility solvation, and stabilization of α_{S1}-and β-caseins, Journal of Dairy Science 52 (1969) 1166–1173.

[73] P. Walstra, T.J. Geurts, A. Noomen, A. Jellema, M. Van Boekel, Heat treatment, Dairy Technology-Principles of Milk Properties and Processes, Marcel Dekker Inc., New York (1999) 189–234.

[74] R. Beeby, The presence of sulphydryl groups in k-casein, Biochimica et Biophysica Acta (BBA)-General Subjects 82 (1964) 418–419.

[75] P. JollÃ¨s, C. Alais, Amino acid composition of k-casein and terminal amino acids of k- and para-k-casein, Archives of Biochemistry and Biophysics 98 (1962) 56–57.

[76] A.K. Dunker, C.J. Brown, J.D. Lawson, L.M. Iakoucheva, Z. Obradovic, Intrinsic disorder and protein function, Biochemistry 41 (2002) 6573–6582.

[77] V.N. Uversky, What does it mean to be natively unfolded?, European Journal of Biochemistry 269 (2002) 2–12.

[78] M.H. Alaimo, H.M. Farrell, M.W. Germann, Conformational analysis of the hydrophobic peptide α_{S1}-casein (136–196), Biochimica et biophysica Acta (BBA)-Protein Structure and Molecular Enzymology 1431 (1999) 410–420.

[79] K. Kuwajima, The molten globule state of alpha-lactalbumin, The FASEB Journal 10 (1996) 102–109.

[80] H.J. Vogel, R.D. Brokx, H. Ouyang, Calcium-binding proteins, Springer, USA, 2002.

[81] P.F. Fox, Milk proteins: general and historical aspects, Advanced Dairy Chemistry-1 Proteins, Springer, USA, 2003, pp. 1–48.

8

The Protein Component of Sow Colostrum and Milk

P.K. Theil and W.L. Hurley

Abstract

The production of colostrum and milk by the sow are primary limiting factors affecting survival, growth and development of the piglets. The proteins of colostrum and milk provide not only a supply of amino acids to the neonate but also a wide range of bioactive factors. Proteins in sow mammary secretions include those associated with the milk fat membranes, caseins, mammary-derived whey proteins, immunoglobulins, hormones and growth factors, enzymes, and a wide range of other proteins. Concentrations of most milk-specific proteins typically are lower in colostrum than in milk, while concentrations of immunoglobulins and other bioactive proteins often are enriched in colostrum compared with mature milk. Dietary protein is utilized for milk protein production with approximately 50% efficiency. During both the colostrum period and at peak lactation as much as 700–800 g of protein is secreted daily by today's highly prolific sows. Estimates of daily milk protein secretion during lactation suggest that sows are not able to consume sufficient dietary protein and energy to account for output of solids in milk and therefore must mobilize body protein and body fat to support their milk production. Milk protein content typically is not affected by dietary treatment, indicating that the sow mobilizes her body reserves to maintain milk production and milk protein production. These observations are of particular interest for today's highly prolific sows, which may require more dietary protein than previous genotypes.

Keywords: Colostrum, milk, protein, sow, lactation

1. Introduction

The newborn piglet is completely reliant on the sow's colostrum and milk for a nutrient supply as well as for a range of bioactive factors that impact its immunological and physiological development. Production of colostrum and milk by the sow are primary limiting factors affecting survival, growth and development of the piglets [1–6]. Milk production by sows is impacted by numerous factors, including litter size, piglet size, nursing frequency, stage of lactation, sow age, body condition and parity, maternal behavior, and nutrition [7–10].

General composition of sow colostrum and milk has been reviewed previously [11–16]. Composition of mammary secretions after giving birth changes rapidly. For the sow, colostrum is considered as the secretions during the first 24 h after giving birth [17]. The period between 24 and 96 h after parturition is considered as transition milk [18, 19].

Recently, the lactation curve of sows was described [20], which showed that milk yield increases day by day in the first two weeks of lactation, depending on the number of suckling piglets. Indeed, genetic selection for large litters has increased the milk yield of sows and at peak lactation highly prolific sows (14 suckling piglets) produce on average 13–15 kg of milk daily. The milk protein production increases in a pattern similar to the lactation curve and peaks at 700–800 g of protein secreted daily.

The proteins of colostrum and milk provide not only a supply of amino acids to the neonate but also a wide range of bioactive factors. This chapter provides an overview of the proteins found in sow colostrum and milk, and a description on how the milk protein changes over time during the colostrum and lactation periods. In addition, estimates of milk protein production by the sow are discussed, and some of the factors that may impact milk protein production.

2. Proteins in milk fractions

The proteins of milk generally are characterized according to how they segregate during some fundamental methods of fractionating milk components. These methods originally were developed to fractionate bovine milk. One basic method of separating the fractions of milk is to separate the lower density lipid-containing cream fraction from the aqueous component, typically by centrifugation. The residual aqueous or defatted fraction then is referred to as the skim milk fraction. This type of fractionation has most often been used as a starting point for identifying milk proteins.

Milk is a partially stable emulsion of fat globules. In addition to the lipid component of the fat globules (primarily triacylglycerides), they also contain numerous proteins, which are associated with the milk fat globule membrane (MFGM). The MFGM proteins have been extensively studied in cow milk [21–23]; however, they have received relatively less attention in sow milk. Seventeen proteins have been identified in association with sow MFGM, including xanthine dehydrogenase, lactadherin, butyrophilin, adipophilin, acyl-CoA synthetase, fatty acid-binding protein 3, and others [24].

The skim milk fraction will contain the major milk protein components, including the caseins and whey proteins. Caseins are a special set of proteins found only in milk and are the major proteins found in sow milk. The caseins in sow milk include αS1-casein (*CSN1S1*), αS2-casein (*CSN1S2*), β-casein (*CSN2*), and κ-casein (*CSN3*). The whey proteins are defined as all of the skim milk proteins excluding the caseins. The whey protein fraction encompasses a wide range of proteins with an extensive array of functions. The functionality of the various whey proteins includes direct involvement in the process of milk synthesis, provision of immune protection to the neonate, and hormones and growth factors that may impact neonatal development as well as enzymes and other proteins associated with cellular functions arising from epithelia, leukocytes, and other mammary tissue cells.

3. Major sow milk proteins

The concentration of total casein as a proportion of total milk protein generally increases from a low of 9–32% at parturition to 30–45% at 24 h postpartum [25, 26] and accounts for 50–55% of total milk protein in mature milk [25–28]. Total whey protein content of sow colostrum may be as high as 90% at parturition, followed by a decline to 70% by 24 postpartum [18, 26]. Total whey protein as a percentage of total protein is approximately 50% in mature milk [26].

The major whey proteins in sow milk include β-lactoglobulin (*BLG*), α-lactalbumin (*LALBA*), whey acidic protein (*WAP*), lactoferrin (also referred to as lactotransferrin; *LTF*), serum albumin (*ALB*), and immunoglobulins. *BLG* is a member of the lipocalin family, which transports small hydrophobic molecules; however, the specific function of this protein in milk or in the neonate is not understood. *BLG* has significant sequence similarity with glycodelin [29], also known as progestin-associated endometrial protein (*PAEP*). Isolated bovine *BLG* has also been shown to have antimicrobial activity against some mastitis causing bacteria [30]. Concentrations of *BLG*, the major whey protein in sow milk, are between approximately 8 and 15 g/L during the initial week postpartum [31, 32]. *LALBA* is a small molecular weight protein that is the regulatory subunit of the lactose synthase enzyme activity. *LALBA* binds with β-1,4-galactosyltransferase in the Golgi apparatus of the mammary epithelial cell resulting in the synthesis of the disaccharide lactose from glucose and galactose. The synthesis of lactose, the primary osmole in sow milk, is responsible for drawing water into the Golgi apparatus and secretory vesicles during milk synthesis. Concentrations of *LALBA* are lower in colostrum (0.8–1.9 g/L) than in transition or mature milk (approximately 2–3 g/L) [31, 32], consistent with the observed lower concentrations of lactose in colostrum compared with mature sow milk. The function of the *WAP* in sow milk also is not understood. Concentrations of *WAP* in sow mammary secretions increase from 0.3 g/L at parturition to approximately 1 g/L at day 7 [31, 32]. Lactoferrin is an iron-binding protein with antimicrobial properties. Concentrations of lactoferrin in colostrum are relatively high (1.2 g/L) until day 3 postpartum and then decline to 0.3 g/L by one week postpartum [33]. Colostrum concentrations of albumin decline rapidly between parturition and 12 h postpartum (from 19 down to 8 g/L) and then gradually decline to 3.0 g/L or less [34].

4. Protective and bioactive proteins

Colostrum and milk contain an array of protective factors that can impact the neonate. The role of colostrum and milk immunoglobulins in the neonate has been extensively studied. The immunoglobulins are transported through the mammary epithelial cells by a receptor-mediated process, thought to involve the neonatal Fc-receptor (FcRn) in the case of IgG1 and IgG2, and the polymeric immunoglobulin receptor (pIgR) for IgA and IgM (see Ref. [35] for review).

Colostrum during or immediately after parturition contains the highest concentrations of each immunoglobulin, with average concentrations of total IgG, IgA, and IgM of approximately 65, 13, and 8 g/L, respectively [16]. These immunoglobulin concentrations remain elevated during the initial 6 h postpartum, decline by approximately 30–45% by 12 h, and continue declining through the next several days. Concentrations of IgG and IgM in mature milk are approximately 1 and 1.6 g/L, respectively. Interestingly, IgA is the major immunoglobulin in mature milk of the sow with a concentration of approximately 4 g/L [16]. The changes in proportion and total amount of immunoglobulins in sow mammary secretions are shown in **Figure 1**.

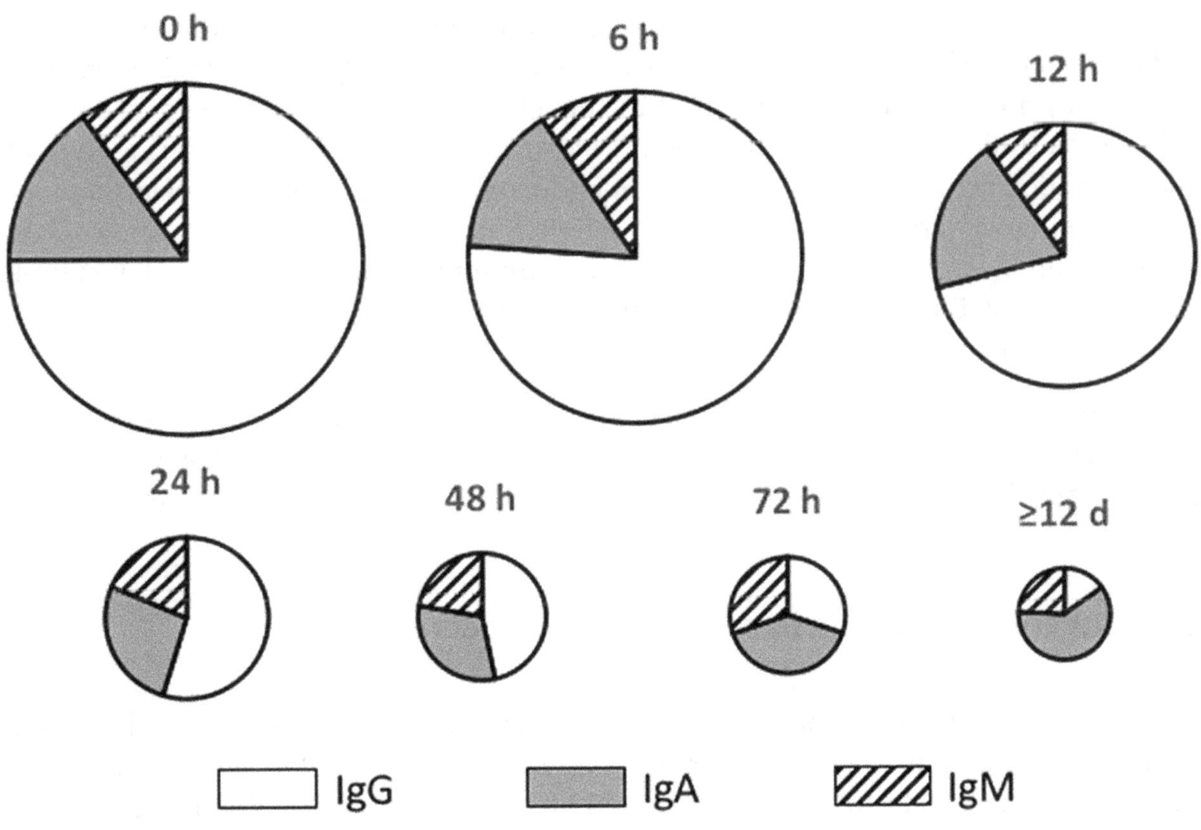

Figure 1. Relative total amounts and proportions of immunoglobulins G, A, and M in sow mammary secretions during lactation. Circles for 6 h through ≥12 days represent the total immunoglobulin content in mammary secretions compared with the content at 0 h. Adapted from Ref. [16].

Milk contains a wide array of antimicrobial and immunomodulatory proteins and peptides. Certainly the immunoglobulins, especially IgA, lactoferrin and lysozyme activity individually and synergistically probably provide for some of the antimicrobial function that has been found in sow milk [36]. Lysozyme activity has been identified in sow colostrum immediately after parturition [37] but not in mature milk [38]. Even some of the proteins associated with the milk fat globule may have bactericidal properties, such as the mucins [39, 40], lactadherin, butyrophilin, adipophilin, acyl-CoA synthetase, and fatty acid-binding protein 3 [24]. The specific role of these proteins in modulating the immune health of the sow's mammary gland and the health of her litter deserve additional evaluation. Several cytokines have been identified in sow colostrum and milk, including IL-4, IL-6, IL-10, IL-12, IFN-γ, TNF-α, and TGF-β [41]. Concentrations of these cytokines are greatest in colostrum and then decline as lactation progresses [41]. Butyrophilin has also been implicated as having an immunomodulatory function [42].

Milk contains many hormones and growth factors, which may have biological activity in the neonate [43, 44]. Prolactin in sow colostrum is at its highest concentration just prior to parturition and then declines rapidly during the first 24 h postpartum [45], followed by a slower decline through the remainder of lactation [46]. Similarly, concentrations of relaxin are greatest in colostrum and then decline over the first week of lactation [47]. This milk-borne relaxin has been associated with maternal programming of neonatal development in what is called the lactocrine hypothesis [48]. Specifically, relaxin absorbed into the circulatory system in the piglet prior to gut closure can affect the developing reproductive tract of the neonate. Leptin concentrations in the skim fraction of colostrum and milk decrease during the initial week of lactation before stabilizing for the remainder of lactation, while stage of lactation does not seem to impact leptin concentrations in whole milk [49]. Sow colostrum and milk also contain insulin, neurotensin, bombesin [50], triiodothyronine and thyroxine [51], insulin-like growth factor-I (IGF-I; [52]), epidermal growth factor-like peptide [53, 54], and prostaglandin-like activity [55].

Proteins in sow colostrum and milk include many enzyme activities, such as lipase and ribonuclease activity [38], trypsin inhibitor activity [56], ceruloplasmin [57], and enolase and glycogen phosphorylase [58]. A high molecular weight glycoprotein of unknown origin and function has been identified, which may be specific to sow milk [59].

5. Protein concentrations in sow mammary secretions

Sow colostrum, i.e. the mammary secretions during the first 24 h after onset of parturition, is rich in protein, especially colostrum from the earliest phase. The protein concentration drops from approximately 16 to 10% during the first 12 h after onset of parturition (**Figure 2**; [60]). Recent studies indicate that the protein content continues to decrease to reach a plateau at 6–7% protein at 36 h after parturition [61]. However, according to Ref. [60], the protein content remains rather stable in transient milk, i.e. in mammary secretions released from 24 h post parturition and the next couple of days (**Figure 2**). The low protein content of mammary secretions 24–36 h post parturition observed in Ref. [61] may indicate that late colostrum/

transient milk produced by modern hyperprolific sows, with an average of 18 piglets born, may be compromised in terms of protein content. In Ref. [60], published in 1995, sows had an average of 9.7 piglets born.

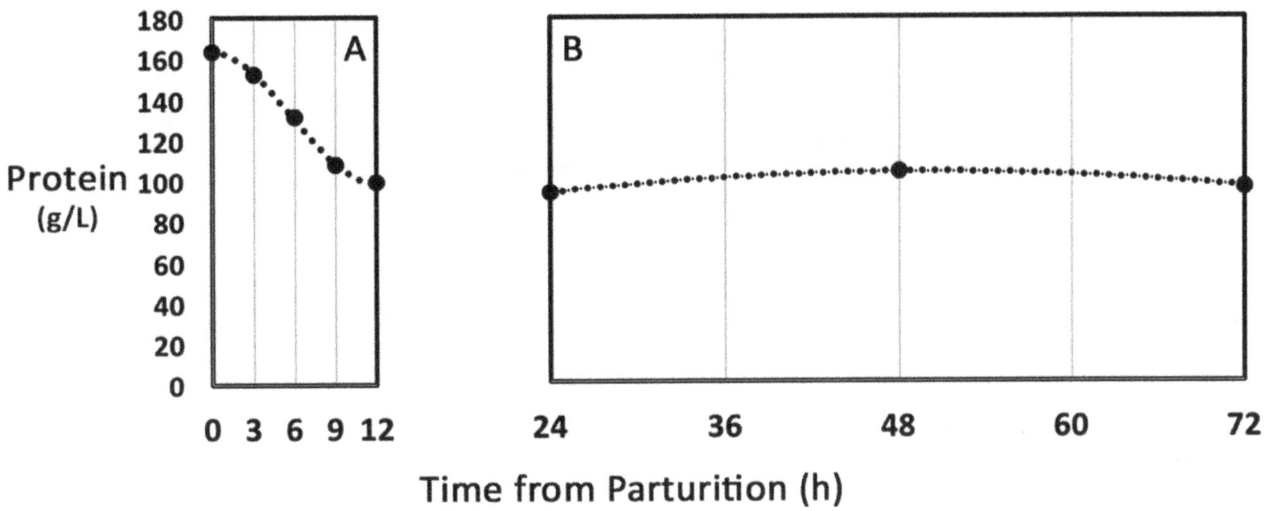

Figure 2. Examples of mammary secretion protein concentration change during the initial 12 h of the colostrum period and during the transition milk period. (A) Changes in colostrum protein concentration from parturition to 12 h can be described by $y = 0.0757x^3 - 1.3411x^2 - 0.1717x + 163.25$ ($R^2 = 0.99931$). (B) Changes in protein concentration from 24 to 72 h can be described by $y = -0.0151x^2 + 1.5054x + 66.15$ ($R^2 = 1$). Adapted from Ref. [60].

Changes in the protein concentration of sow milk during lactation can be estimated using the mathematical model described in Ref. [20]. Milk protein concentrations decline from 7% at day 2 to reach a plateau at approximately 5%, which is then maintained through at least four weeks after parturition (**Figure 3**; [20]).

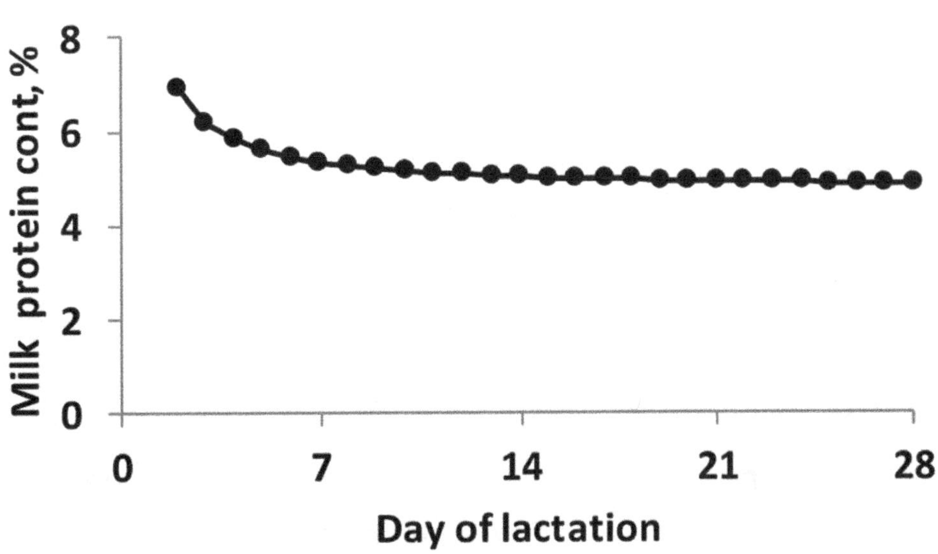

Figure 3. Milk protein concentration during lactation. The profile was calculated using the mathematical model developed in Ref. [20] with the following input: Dietary protein (16%).

As lactation progresses, other milk constituents than the protein fraction also change. Indeed, the milk lactose concentration increases slightly from 5.2 to 5.5%, while the milk fat concentration decreases from 8.0 to 6.3% from day 2 to 28, respectively [20]. As a consequence, the energy content of milk decreases in a curvilinear manner throughout lactation (**Figure 4**).

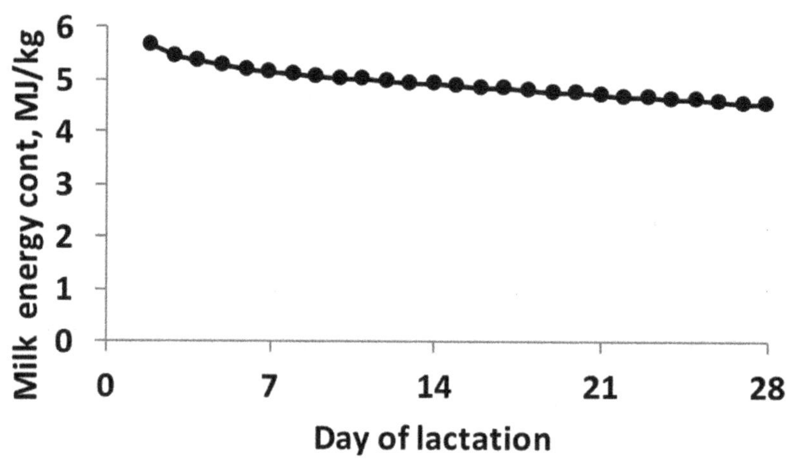

Figure 4. Milk energy content during lactation. The profile was calculated using the mathematical model developed in Ref. [20] with the following input: Dietary protein (16%).

The protein-to-energy ratio of the milk declines during the first week of lactation; however, there is a gradual increase thereafter until weaning (**Figure 5**). The protein-to-energy ratio may be indicative of the nutritive quality of sow milk to support piglet growth [19]. However, the relative consistency of the protein-to-energy ratio of sow milk through lactation may indicate that milk intake of individual piglets is the major determinant of piglet growth rather than milk composition per se.

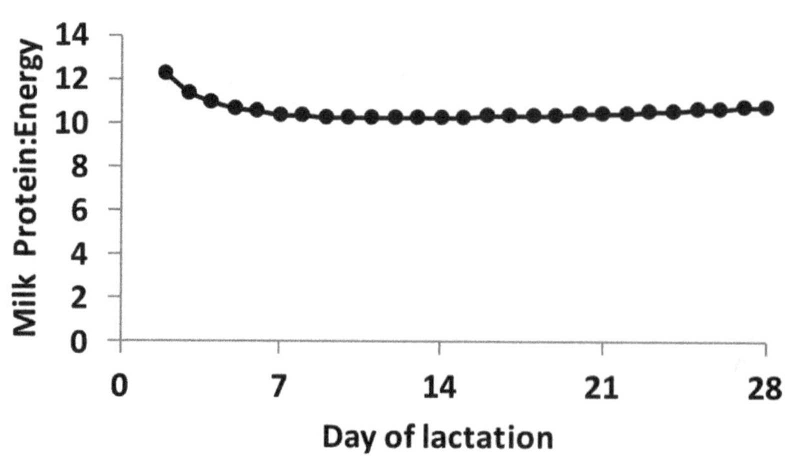

Figure 5. Protein-to-energy ratio of sow milk during lactation. The profile was calculated using the mathematical model developed in Ref. [20] with the following input: Dietary protein (16%).

6. Daily protein yields in sow colostrum and milk

Recently, it was demonstrated that the yield of sow colostrum was approximately 50% higher than previously believed [4, 5, 17]. With the high protein content in colostrum secreted in the early phase (described above), the daily protein secretion through colostrum amounts to 700–800 g [61], which is similar to the daily secretion of protein in milk at peak lactation (see below; [20]). While recent studies have shed light on the amounts of nutrients being secreted through colostrum and when they are secreted [4, 61], it is still unknown when the colostral proteins are being synthesized [5]. The high concentration of protein in colostrum at parturition suggests that this synthesis occurs in late gestation, allowing for accumulation of the proteins in colostrum. Milk-specific proteins, such as the caseins, *BLG*, *LALBA*, and *WAP*, are synthesized by mammary epithelial cells, while most of the immunoglobulin and the serum albumin are synthesized elsewhere in the sow.

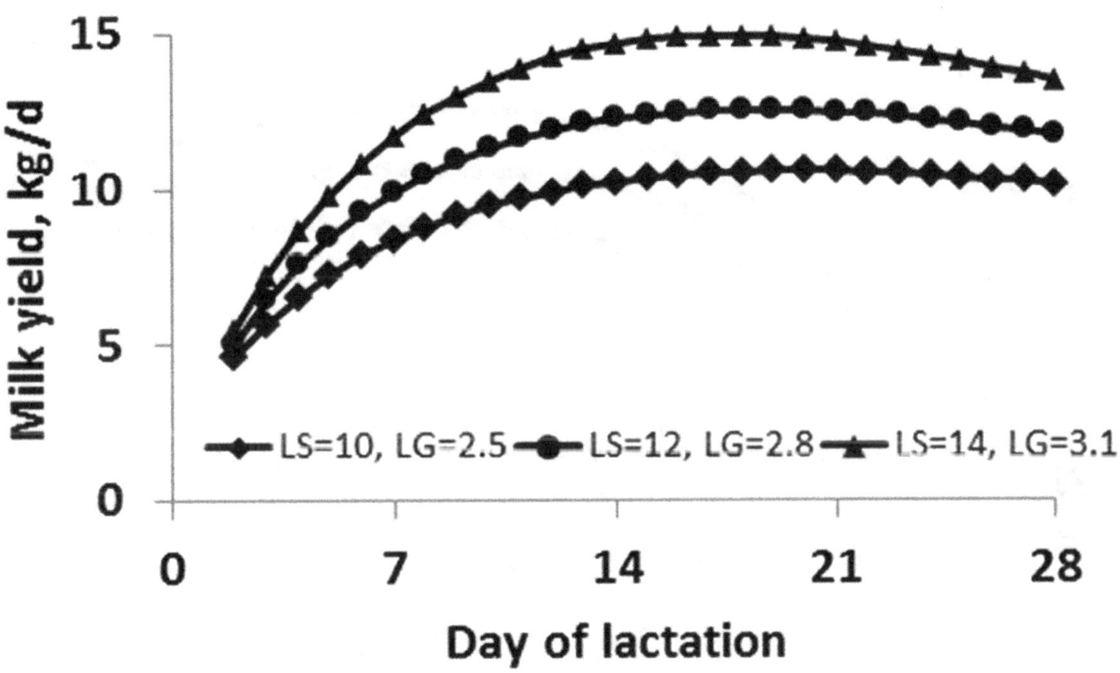

Figure 6. Daily milk yield of sows with litter size (LS) of 10, 12, or 14 piglets. Note that the profile was calculated using the mathematical model developed in Ref. [20], assuming that sows with 10, 12, and 14 piglets on average produced litter gain (LG) of 2.5, 2.8, and 3.1 kg/d, respectively.

From a nutritional perspective, the timing of synthesis of the colostral proteins is important because fetal growth, mammary growth, and colostrum protein production compete for the dietary proteins supplied during late gestation. Factorial calculations of the daily protein requirement in late gestation suggest that dietary protein likely is scarcely supplied [62]. In a recent study, colostrum yield and colostrum protein content were studied across five different experiments to reveal potential limitations in colostrum synthesis [63]. In that study, it was found that the colostrum yield was positively correlated with litter size (including total born, live born, and 24 h litter size), and it is therefore logical to suggest that highly prolific sows may require more dietary protein than previous genotypes. However, no relationship between colostrum yield and colostrum protein concentration was found, indicating that colostrum protein production likely is not compromised, in spite of high colostrum production due to their large litters [63].

Daily milk production increases during the first two weeks of lactation and peaks around day 17 of lactation [20], although it seems to peak a few days later if the production level is lower. Milk yield is greatly affected by the number of suckling piglets [64] because each piglet takes ownership of a single mammary gland and stimulates mammary growth [65, 66] and maintains milk production within that gland [67]. Milk yield is around 13-15 kg/d at peak lactation in modern highly prolific sows (with 14 piglets; [20]), whereas sows with a more moderate litter of 10 suckling piglets and consequently less litter gain per day produce around 10 kg/d (**Figure 6**). Consequently, sows produce around or slightly above 1.0 kg of milk daily for each piglet at peak lactation. After peak lactation, the milk yield decreases; however, it remains nearly maximal if piglets are weaned after three to four weeks.

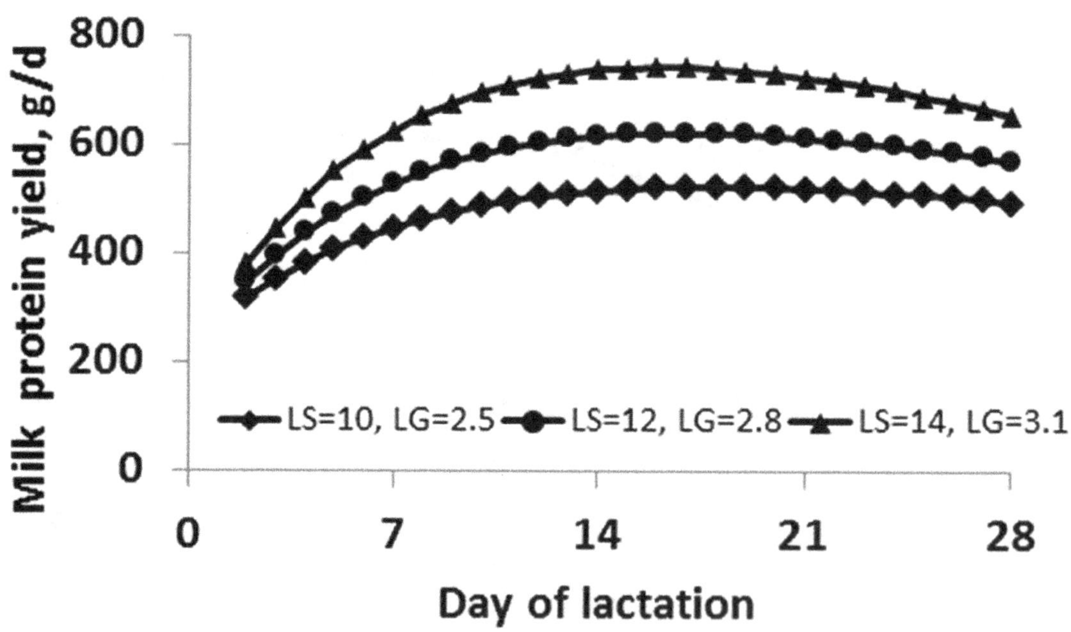

Figure 7. Daily milk protein yields of sows with litter size (LS) of 10, 12, or 14 piglets. Note that the profile was calculated using the mathematical model developed in Ref. [20], assuming that sows with 10, 12, and 14 piglets on average produced litter gain (LG) of 2.5, 2.8, and 3.1 kg/d, respectively.

Daily milk protein production increases in a fashion similar to the lactation curves of sows (**Figure 7**). Thus, the daily milk protein yield also peaks around day 17 (slightly later for sows with a lower yield). For highly prolific sows, as much as 750 g of milk protein is secreted daily, whereas the daily milk protein yield is around 500 g when sows have 10 suckling piglets and have a moderate level of productivity (i.e. litter gain of 2.5 kg/d). In a recent study, it was found that milk protein concentration was negatively related to the milk yield of highly prolific sows for all four weeks of lactation [63]. The negative relationship suggests that the mammary capacity for milk protein synthesis has been reached or, alternatively, that the dietary protein supply is inadequate to support the milk production of high prolific sows. In favor of the latter explanation, it was demonstrated that dietary protein is utilized for milk protein production with approximately 50% efficiency [68]. At peak lactation, sows ingest around 1300–1400 g of dietary protein daily, but if high producing sows secrete 750 g of milk protein daily, they should ingest 1500 g/d of dietary protein (assuming 50% utilization efficiency). Most sows are not able to ingest enough feed to obtain a zero balance of protein nor a zero balance of energy, and consequently, sows typically mobilize body protein and body fat to support their milk production [19, 62, 68].

Milk production and milk protein production are highly prioritized by lactating sows [19]. If the dietary supply of protein or energy is insufficient, the sow mobilizes amino acids from muscle tissues or fat from adipose tissues, respectively. Mobilization of muscle tissues can be substantial and result in sow weight loss that in extreme cases may amount to 75–80 kg during three weeks of lactation [69]. However, in a pioneer experiment [70] where lactating sows were provided an insufficient nutrient supply, a linear decrease in weight loss was reported in sows in response to feed intake that was gradually increased from 1.5 to 4.8 kg/d (over six different feeding levels). In each case, the nitrogen balance of the sows was negative but was clearly responsive to feed intake. Interestingly, the milk protein content was not affected by dietary treatment, and the piglet growth rate, which may be seen as an indicator for the milk yield, was not compromised until the fourth week of lactation. These findings are consistent with the sow's physiology during lactation being set to maintain milk production and milk protein production as a high priority, and she compensates for insufficient nutrient intake by increasing the body mobilization. The reduced piglet weight gain in the fourth week of lactation observed in Ref. [70] could be interpreted as a result of the depletion of readily available body pools of amino acids after three weeks of insufficient nutrient supply.

7. Conclusion

Colostrum is much richer in protein than in mature sow milk. The protein component changes dramatically during the colostrum period, where the decline in contents of bioactive proteins, including immunoglobulins, hormones, and growth factors, is pronounced. The concentrations of milk-specific proteins, such as the caseins and the major mammary-derived whey proteins, increase during the early lactation period. The amount of protein secreted in colostrum and milk is extremely high. During both the colostrum period and at peak lactation as much as 700–800 g of protein is secreted daily. Currently, it is not known which factors are most limiting the sow milk production or the sow milk protein production. Without doubt, daily feed intake and dietary content of essential amino acids are important factors. However, sow live weight, body pools of amino acids in muscle tissues, and amino acid mobilization capacity may be equally important to sustain this demanding daily production of milk and milk protein. Current research suggests that today's highly prolific sows may require more dietary protein than previous genotypes.

Author details

P.K. Theil[1*] and W.L. Hurley[2]

*Address all correspondence to: Peter.Theil@anis.au.dk

1 Aarhus University, Tjele, Denmark

2 University of Illinois, Urbana-Champaign, USA

References

[1] Boyd RD, Kensinger RS. Metabolic precursors for milk synthesis. In: Verstegen MWA, Moughan PJ, Schrama JW, editors. The Lactating Sow. Wageningen: Wageningen Press, 1998. p. 71–95.

[2] Le Dividich J, Rooke JA, Herpin P. Nutrition and immunological importance of colostrum for the new-born pig. Journal of Agricultural Science. 2005;143:469–485. DOI: 10.1017/S0021859605005642

[3] Quesnel H, Farmer C, Devillers N. Colostrum intake: Influence on piglet performance and factors of variation. Livestock Science. 2012;146:105–114. DOI: 10.1016/j.livsci.2012.03.010

[4] Theil PK, Flummer C, Hurley WL, Kristensen NB, Labouriau RL, Sørensen MT. Mechanistic model to predict colostrum intake based on deuterium oxide dilution technique data and impact of gestation and pre-farrowing diets on piglet intake and sow yield of colostrum. Journal of Animal Science. 2014;92:5507–5519. DOI: 10.2527/jas.2014-7841

[5] Theil PK, Lauridsen C, Quesnel H. Neonatal piglet survival: impact of sow nutrition around parturition on fetal glycogen deposition and production and composition of colostrum and transient milk. Animal. 2014;8:1021–1030. DOI: 10.1017/S1751731114000950

[6] Quesnel H, Farmer C, Theil PK. Colostrum and milk production. In: Farmer C, editor. The Gestating and Lactating Sow. Wageningen: Wageningen Academic Publishers, 2015. p. 173–192. DOI: 10.3920/978-90-8686-803-2_8

[7] Fraser D. Behavioural perspectives on piglet survival. In: Cole DJA, Foxcroft GR, Weir BJ, editors. Control of Pig Reproduction III. Essex, U.K.: Journal of Reproduction and Fertility Ltd., 1990. p. 355–370.

[8] Brooks PH, Burke J. Behaviour of sows and piglets during lactation. In: Verstegen MWA, Moughan PJ, Scharma JW, editors. The Lactating Sow. Wageningen: Wageningen Press, 1998. p. 301–338.

[9] King RH. Factors that influence milk production in well-fed sows. Journal of Animal Science. 2000;78(Suppl. 3):19–25.

[10] Hurley WL. Mammary gland growth in the lactating sow. Livestock Production Science. 2001;70:149–157. DOI: 10.1016/S0301-6226(01)00208-1

[11] Oftedal OT. Milk composition, milk yield and energy output at peak lactation: a comparative review. Symposium of the Zoological Society of London. 1984;51:33–85.

[12] Hartmann PE, Holmes MA. Sow Lactation. In: Barnett JL, Hennessy DP, editors. Manipulating Piglet Production II. Proc. Australasian Pig Science Association, Werribee, Victoria, Australia, 1989. p. 72–97.

[13] Oftedal OT, Iversen SJ. Comparative analysis of nonhuman milks. A. Phylogenetic variation in the gross composition of milks. In: Jensen RG, editor. Handbook of Milk Composition. San Diego: Academic Press, Inc., 1995. p. 749–789.

[14] Darragh AJ, Moughan PJ. The composition of colostrum and milk. In: Verstegen MWA, Moughan PJ, Schrama JW, editors. The Lactating Sow. Wageningen: Wageningen Press, 1998. p. 3–21.

[15] Park YW. Milks of other domesticated mammals (Pigs, Yaks, Reindeer, etc). In: Fuquay JW, Fox PF, McSweeney PLH, editors. Encyclopedia of Dairy Sciences, Volume 3. 2nd ed. San Diego: Academic Press, 2011. p. 530–537.

[16] Hurley WL. Composition of sow colostrum and milk. In: Farmer C, editor. The Gestating and Lactating Sow. Wageningen: Wageningen Academic Publishers, 2015. p. 193–229. DOI: 10.3920/978-90-8686-803-2_9

[17] Devillers N, van Milgen J, Prunier A, Le Dividich J. Estimation of colostrum intake in the neonatal pig. Animal Science. 2004;78:305–313.

[18] Klobasa F, Werhahn E, Butler JE. Composition of sow milk during lactation. Journal of Animal Science. 1987;64:1458–1466.

[19] Theil PK, Nielsen MO, Sørensen MT, Lauridsen C. Lactation, milk and suckling. In: Bach Knudsen KE, Kjeldsen NJ, Poulsen HD, Jensen BB, editors. Nutritional Physiology of Pigs. Chapter 17, 2012. Available at: http://vsp.lf.dk/Viden/Laerebog_fysiologi/Chapter%2017.aspx

[20] Hansen AV, Strathe AB, Kebreab E, France J, Theil PK. Predicting milk yield and composition in lactating sows – A Bayesian approach. Journal of Animal Science. 2012;90:2285–2298. DOI: 10.2527/jas.2011-4788

[21] Heid HW, Keenan TW. Intracellular origin and secretion of milk fat globules. European Journal of Cell Biology. 2004;84:245–258. DOI: 10.1016/j.ejcb.2004.12.002

[22] Fong BY, Norris CS, MacGibbon AKH. Protein and lipid composition of bovine milk-fat-globule membrane. International Dairy Journal. 2007;17:275–288. DOI: 10.1016/j.idairyj.2006.05.004

[23] Cavaletto M, Giuffrida MG, Conti A. Milk fat globule membrane components – a proteomic approach. Adv Exp Med Biol. 2008;606:129–141. DOI: 10.1007/978-0-387-74087-4_4

[24] Novakovic P, Huang YY, Lockerbie B, Shahriar F, Kelly J, Gordon JR, Midleton DM, Loewen ME, Kidney BA, Simko E. Identification of *Escherichia coli* F4ac-binding proteins in porcine milk fat globule membrane. Canadian Journal of Veterinary Research. 2015;79:120–128.

[25] Brent BE, Miller ER, Ullrey DE, Kemp KE. Postpartum changes in nitrogenous constituents of sow milk. Journal of Animal Science. 1973;36:73–78.

[26] Csapo J, Martin TG, Csapo-Kiss ZS, Hazas Z. Protein, fats, vitamin and mineral concentrations in porcine colostrum and milk from parturition to 60 days. International Dairy Journal. 1996;6:881–902. DOI: 10.1016/0958-6946(95)00072-0

[27] Mahan DC, Becker DE, Harmon BG, Jensen AH. Effect of protein levels and *opaque-2* corn on sow milk composition. Journal of Animal Science. 1971;32:482–496.

[28] Richert BT, Goodband RD, Tokach MD, Nelssen JL. Increasing valine, isoleucine, and total branched-chain amino acids for lactating sows. Journal of Animal Science. 1997;75:2117–2128. DOI: 10.1997.7582117x

[29] Schiefner A, Rodewald F, Neumaier I, Skerra A. The dimeric crystal structure of the human fertility lipocalin glycodelin reveals a protein scaffold for the presentation of complex glycans. Biochemistry Journal. 2015;466:95–104. DOI: 10.1042/BJ20141003

[30] Chaneton L, Perez Saez JM, Bussman LE. Antimicrobial activity of bovine β-lactoglobulin against mastitis-causing bacteria. Journal of Dairy Science. 2011;94:138–145. DOI: 10.3168/jds.2010-3319

[31] Jackson JR. Effect of gestation length and supplemental dietary fat on colostrum and milk composition in swine [thesis]. Urbana: University of Illinois, 1990.

[32] Hurley WL, Bryson JM. Enhancing sow productivity through an understanding of mammary gland biology and lactation physiology. Pig News and Information. 1999;20:125N–130N.

[33] Elliot JL, Senft B, Erhardt G, Fraser D. Isolation of lactoferrin and its concentration in sows' colostrum and milk during a 21-day lactation. Journal of Animal Science. 1984;59:1080–1084.

[34] Klobasa F, Butler JE. Absolute and relative concentrations of immunoglobulins G, M, and A, and albumin in lacteal secretion of sows of different lactation numbers. American Journal of Veterinary Research. 1987;48:176–182. DOI: 10.1038/icb.1974.6

[35] Hurley WL, Theil PK. Immunoglobulins in mammary secretions. In: McSweeney PLH, Fox PF, editors. Advanced Dairy Chemistry: Volume 1A: Proteins: Basic Aspects. 4th ed. New York: Springer Science + Business Media, 2013. p. 275–294. DOI: 10.1007/978-1-4614-4714-6_9

[36] Nagy LK, Mackenzie T, Bharucha Z. In vitro studies on the antimicrobial effects of colostrum and milk from vaccinated and unvaccinated pigs on *Escherichia coli*. Research in Veterinary Science. 1976;21:132–140.

[37] Krakowski L, Krzyzanowski J, Wrona Z, Kostro K, Siwicki AK. The influence of nonspecific immunostimulation of pregnant sows on the immunological value of colostrum. Veterinary Immunology and Immunopathology. 2002;87:89–95. DOI: 10.1016/S0165-2427(02)00004-1

[38] Chandan RC, Parry RMJr, Shahani KM. Lysozyme, lipase, and ribonuclease in milk of various species. Journal of Dairy Science. 1968;51:606–607.

[39] Patton S. MUC1 and MUC-X, epithelial mucins of breast and milk. In: Newberg DS, editor. Bioactive Components of Human Milk. New York: Kluwer Academic/Plenum Publishers, 2001. p. 35–45. DOI: 10.1007/978-1-4615-1371-1_4

[40] Schroten H. Chemistry of milk mucins and their anti-microbial action. Advances in Nutritional Research. 2001;10:231–245. DOI: 10.1007/978-1-4615-0661-4

[41] Nguyen TV, Yuan L, Azevedo SP, Jeong K-I, Gonzalez A-M, Saif LJ. Transfer of maternal cytokines to suckling piglets: In vivo and in vitro models with implications for immunomodulation of neonatal immunity. Veterinary Immunology and Immunopathology. 2007;117:236–248. DOI: 10.1016/j.vetimm.2007.02.013

[42] Arnett HA, Viney JL. Immune modulation by butyrophilins. Nature Reviews Immunology. 2014;14:559–569. DOI: 10.1038/nri3715

[43] Pakkanen R, Aalto J. Growth factors and antimicrobial factors in bovine colostrum. International Dairy Journal. 1997;7:285–297. DOI: 10.1016/S0958-6946(97)00022-8

[44] Baumrucker CR, Magliaro-Macrina AL. Hormones in milk. In: Fuquay JW, Fox PF, McSweeney PLH, editors. Encyclopedia of Dairy Sciences. 2nd ed. San Diego: Academic Press, 2011. p. 765–771. DOI: 10.1016/B978-0-12-374407-4.00228-4

[45] Devillers N, Farmer C, Mounier A-M, Le Dividich J, Prunier A. Hormones, IgG and lactose changes around parturition in plasma, and colostrum or saliva of multiparous sows. Reproduction Nutrition Development. 2004;44:381–396. DOI: 10.1051/rnd:2004043

[46] Mulloy AL, Malven PV. Relationships between concentrations of porcine prolactin in blood serum and milk of lactating sows. Journal of Animal Science. 1979;48:76–81.

[47] Yan W, Wile AA, Bathgate AD, Frankshun A-L, Lasano S, Crean BD, Steinetz BG, Bagnell CA, Bartol FF. Expression of LGR7 and LGR8 by neonatal porcine uterine

tissues and transmission of milk-borne relaxin into the neonatal circulation by suckling. Endocrinology. 2006;147:4303–4310. DOI: 10.1210/en.2006-0397

[48] Bagnell CA, Steinetz BG, Bartol FF. Milk-borne relaxin and the lactocrine hypothesis for maternal programming of neonatal tissues. Annals of the New York Academy of Science. 2009;1160:152–157. DOI: 10.1111/j.1749-6632.2009.03834.x

[49] Estienne MJ, Harper AF, Barb CR, Azain MJ. Concentrations of leptin in serum and milk collected from lactating sows differing in body condition. Domestic Animal Endocrinology. 2000;19:275–280. DOI: 10.1016/S0739-7240(00)00082-5

[50] Westrom BR, Ekman R, Svendsen L, Svendsen J, Karlsson BW. Levels of immunoreactive insulin, neurotensin, and bombesin in porcine colostrum and milk. Journal of Pediatric Gastroenterology and Nutrition. 1987;6:460–465. DOI: 10.1097/00005176-198705000-00025

[51] Mostyn A, Sebert S, Litten JC, Perkins KS, Laws J, Symonds ME, Clarke L. Influence of porcine genotype on the abundance of thyroid hormones and leptin in sow milk and its impact on growth, metabolism and expression of key adipose tissue genes in offspring. Journal of Endocrinology. 2006;190:631–639. DOI: 10.1677/joe.1.06731

[52] Monaco MH, Gronlund DE, Bleck GT, Hurley WL, Wheeler MB, Donovan SM. Mammary specific transgenic over-expression of insulin-like growth factor-I (IGF-I) increases pig milk IGF-I and IGF binding proteins, with no effect on milk composition or yield. Transgenic Research. 2005;14:761–773. DOI: 10.1007/s11248-005-7219-8

[53] Jaeger LA, Lamar CH, Bottoms GD, Cline TR. Growth-stimulating substances in porcine milk. American Journal of Veterinary Research. 1987;48:1531–1533.

[54] Tan TJ, Schober DA, Simmen FA. Fibroblast mitogens in swine milk include an epidermal growth factor-related peptide. Regulatory Peptides. 1990;27:61–74. DOI: 10.1016/0167-0115(90)90205-B

[55] Maffeo G, Damasio M, Balabio R, Jochle W. Detection of prostaglandin-like substances in sow's milk. Zuchthygiene. 1987;22:209–214. DOI: 10.1111/j.1439-0531.1988.tb00808.x

[56] Jensen PT. Tryspin inhibitor in sow colostrum and its function. Annales De Recherches Veterinaires. 1978;9:225–228.

[57] Cerveza PJ, Mehrbod F, Cotton SJ, Lomeli N, Linder MC, Fonda EG, Wickler SJ. Milk ceruloplasmin and its expression by mammary gland and liver in pigs. Archives of Biochemistry and Biophysics. 2000;373:451–461. DOI: 10.1006/abbi.1999.1572

[58] Danielsen M, Pedersen LJ, Bendixen E. An in vivo characterization of colostrum protein uptake in porcine gut during early lactation. Journal of Proteomics. 2011;74:101–109. DOI: 10.1016/j.jprot.2010.08.011

[59] Qin Y, Qi N, Tang Y, He J, Li X, Gu F, Zou S. Isolation and identification of a high

molecular weight protein in sow milk. Animal. 2015;9:847–854. DOI: 10.1017/S1751731114003280

[60] Jackson JR, Hurley WL, Easter RA, Jensen AH, Odle J. Effects of induced or delayed parturition and supplemental dietary fat on colostrum and milk composition in sows. Journal of Animal Science. 1995;73:1906–1913.

[61] Krogh U, Bruun TS, Amdi C, Flummer C, Poulsen J, Theil PK. Colostrum production in sows fed different sources of fiber and fat during late gestation. Canadian Journal of Animal Science. 2015;95:211–223. DOI: 10.4141/CJAS-2014-060

[62] Theil PK. Transition feeding of sows. In: Farmer C, editor. The Gestating and Lactating Sow. Wageningen: Wageningen Academic Publishers, 2015. p. 147–172. DOI: 10.3920/978-90-8686-803-2_7

[63] Vadmand CN, Krogh U, Hansen CF, Theil PK. 2015. Impact of sow and litter characteristics on colostrum yield, time for onset of lactation, and milk yield of sows. Journal of Animal Science. 2015;93:2488–2500. DOI: 10.2527/jas.2014-8659

[64] Auldist DE, Morrish L, Eason P, King RH. The influence of litter size on milk production of sows. Animal Science. 1998;67:333–337. DOI: 10.1017/S1357729800010109

[65] Kim SW, Hurley WL, Han IK, Easter RA. Growth of nursing pigs related to the characteristics of nursed mammary glands. Journal of Animal Science. 2000;78:1313–1318.

[66] Nielsen OL, Pedersen AR, Sørensen MT. Relationships between piglet growth rate and mammary gland size of the sow. Livestock Production Science. 2001;67:273–279. DOI: 10.1016/S0301-6226(00)00197-4

[67] Theil PK, Sejrsen K, Hurley WL, Labouriau R, Thomsen B, Sørensen MT. Role of suckling in regulating cell turnover and onset and maintenance of lactation in individual mammary glands of sows. Journal of Animal Science. 2006;84:1691–1698. DOI: 10.2527/jas.2005-518

[68] Theil PK, Jørgensen H, Jakobsen K. Energy and protein metabolism in lactating sows fed two levels of dietary fat. Livestock Production Science. 2004;89:265–276. DOI: 10.1016/j.livprodsci.2004.01.001

[69] Kim SW, Hurley WL, Wu G, Ji F. Ideal amino acid balance for sows during gestation and lactation. Journal of Animal Science. 2009;87(E. Suppl):E123–E132. DOI: 10.2527/jas.2008-1452

[70] King RH, Dunkin AC. The effect of nutrition on the reproductive performance of first-litter sows. 3. The response to graded increases in food intake during lactation. Animal Production. 1986;42:119–126.

9

Human Milk Lactoferrin and Antibodies: Catalytic Activities, Complexes, and Other Features

Sergey E. Sedykh, Valentina N. Buneva and Georgy A. Nevinsky

Abstract

Human milk is a source of biologically active proteins, including lactoferrin (LF) and antibodies (Abs). These proteins are considered as the most polyfunctional proteins of human milk. Apparently, human milk is not a simple mixture of proteins and peptides: recently it was shown that human milk contains stable supramolecular protein complex, composed of LF, α-lactalbumin, milk albumin, β-casein, IgG, and sIgA molecules. We believe that the whole set of different biological functions of the individual milk proteins is significantly supplemented by features of their complexes.

Keywords: antibodies, catalytic antibodies, Fab arm exchange, lactoferrin, protein complexes

1. Introduction

Human milk not only has nutritional value but also is a source of biologically active proteins, including lactoferrin (LF) and antibodies (Abs). LF and Abs are often considered as the most polyfunctional proteins of human milk. LF is an iron-binding glycoprotein and is assumed to play a key role in iron uptake by the intestinal mucosa; it also keeps iron from iron-requiring bacteria acting as a bacteriostatic agent. Major immunoglobulin of human milk is sIgA; besides that milk contains significant amounts of IgG, along with conventional Abs, human milk contains anti-idiotypic Abs capable of enhancing infant antibody response. Maternal milk Abs coat infant mucosal surfaces, play protective role, and also catalyze various chemical reactions (so-called abzymes). Recently it was shown that human milk contains stable supramolecular protein complex, which is composed of LF, α-lactalbumin, milk albumin, β-casein, IgG, and sIgA molecules. Here we describe the biological roles of LF and Abs as the major multifunctional proteins of human milk.

2. Human milk lactoferrin and its known functions

LF is a single polypeptide chain of 76–80 kDa containing two lobes [1] either of which binds one Fe^{3+} ion and contains one glycan chain [2]. LF was first isolated and purified from bovine and human milk [3–5]; then LF was described in human epithelial secretions and barrier body fluids [6–9]. The concentration of LF in human milk (~1 mg/ml) is much higher comparing to cow's milk (up to 0.01 mg/ml) [10]. Considering that LF binds iron [4], it was hypothesized that LF transports iron in the infant gut. In 1991 the LF receptor specifically binding and transporting LF was identified on the gut mucosa cells [11, 12]. Since there is no evidence that iron then enters the bloodstream, it can be concluded that LF does not facilitate iron absorption in the infant gut [13].

Many different functions have been attributed to LF: immunomodulation and cell growth regulation [9, 14], protection from iron-induced lipid peroxidation, DNA binding [9], RNA hydrolysis [15, 16], and transcriptional activation of specific DNA sequences [17, 18]. LF also activates natural killer cells [19] and may be effective in antitumor defense, which is not dependent on iron [14, 20]. It was shown that LF induces granulopoiesis [21], cytotoxicity mediated by Abs [22], production of cytokines [23], and growth of some cells in vitro [24]. The role of LF in physiology of these activities is not yet clear, but there are suggestions that LF is effective for primary defense against viral and microbial infections [6, 8]. The highest concentration of LF is usually detected in the inflammatory sites since LF is known as a protein of the acute phase. The blood of newborn babies contains LF during several hours after feeding and it was shown that LF easily penetrates plasmatic and nuclear membranes of the cells [25]. Protective activity of LF against viral and microbial diseases may increase the newborns' passive immune system. Since the removal of iron from bacteria eliminates this important microelement which is ultimately needed for the proliferation of microflora, it was initially suggested that the antimicrobial properties might be attributed to LF iron-binding capacity [26]. Surface receptors for LF are expressed on many microorganisms that may explain different iron-independent antimicrobial and antiviral properties of LF [27, 28] in which mechanisms are still unknown.

It has been proposed that the existence of several oligomeric forms of LF may provide polyfunctional properties of this relatively small protein. The oligomerization and dissociation of LF may be under control of specific ligands such as ATP and others [29, 30]. There were shown some peculiarities of LF molecule ATP-binding site: interaction of the protein with ATP leads to changes of LF interaction with DNA, proteins, and polysaccharides [30]. Further it was demonstrated that LF possesses two DNA-binding sites which interact with specific and nonspecific DNAs in an anti-cooperative manner and may coincide or overlap with the known polyanion-binding and antimicrobial domains of the protein [31].

3. Oligomerization of human milk lactoferrin

It has been reported that at physiological concentrations of NaCl and KCl LF in solution predominantly exist in two forms—monomeric and tetrameric [19, 32]. However, LF oligomerization has been so far studied by gel filtration using polysaccharide sorbents [19, 32], but it is known that LF can efficiently interact with polysaccharides [33]. Therefore, LF can bind to such polysaccharide sorbents and as a result may be eluted only with buffers with high ionic strength which promote dissociation of LF oligomers [34]. Thus, gel filtration is inadequate for studying LF and may lead to improper conclusions about the state of LF in solution in the presence or absence of salts.

The analysis of the LF oligomerization by gel filtration [35] has shown that the LF monomer and oligomer interactions with resins are remarkably weaker in case of Sepharose 4B. Prior to gel filtration, the solutions of lyophilized LF (3 mg/ml) were incubated in the absence or in the presence of different ligands and 0.1–0.15 M KCl for 10 days at 20°C. Gel filtration of LF preincubated in 50 mM Tris-HCl pH 7.5 on the Sepharose 4B column resolves the LF sample into several fractions (**Figure 1A**, curve 1) demonstrating small reproducible peak at ~800 kDa presumably containing LF decamers, major wide peak at ~310–275 kDa (a tetramer), and small peaks at ~150 kDa (a dimer) and at ~70 kDa (a monomer).

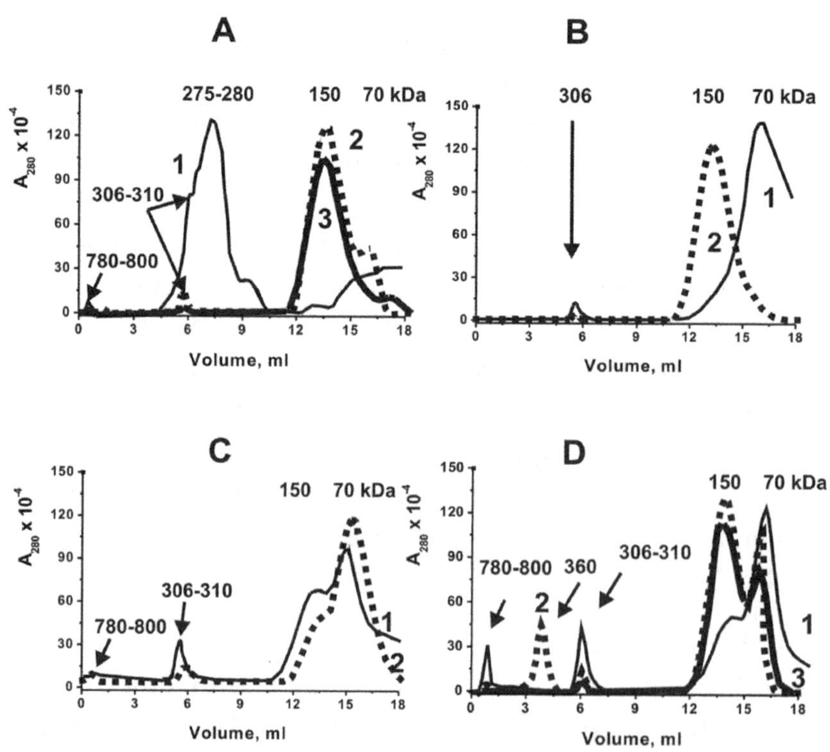

Figure 1. Gel filtration analysis of LF oligomeric forms on Sepharose 4B. Prior to gel filtration, LF was incubated for 20 days at 22°C in 50 mM Tris-HCl (pH 7.5) in the absence of other ligands (A) or in the presence of 5 mM oligosaccharide (B), 50 mM ATP (C), or 50 mM AMP (D). The reaction mixtures containing no salt (curve 1), containing 0.1 M (curve 2), or 0.15 M KCl (curve 3) were used. The column was equilibrated and LF was eluted with the buffer used for the LF preincubation.

The addition of 0.1–0.15 M KCl to the buffers leads to the significant decrease of the LF oligomer forms (**Figure 1A**, curves 2 and 3). One can suppose that absence of KCl leads to nonspecific absorbing of dimer and monomer LF forms on the Sepharose 4B resin [35]. Preincubation of LF in the presence of 0.1 M KCl prior to gel filtration leads to complete dissociation of the tetramer form (**Figure 1B**, curve 2) with the most of LF fraction being a dimer. The gel filtration profile of LF sample incubated with 50 mM ATP in the presence and in the absence of KCl indicates that such conditions are favorable either for decamer (~800 kDa) and for tetramer (~306 kDa) forms and especially favorable for the dimer and monomer forms of LF (**Figure 1C**). As it can be seen from **Figure 1D**, 50 mM AMP allows formation of even more stable complexes: in this condition the most expressed peaks of decamer and tetramer forms are observed, and again, addition of 0.1 M KCl leads to dissociation of decamer form, but increase of KCl to 0.15 M stabilizes dimer LF forms. Hereby, oligosaccharide, ATP, and AMP stimulate the formation of LF oligomers [35].

Denaturing PAGE with 0.1% SDS of the LF preparations preheated with 1% SDS at 100°C reveals on the gel only one band corresponding to the 80 kDa LF monomer [29–31, 35]. Mild treatment with 0.1% SDS without heating leads to the detection of oligomeric forms containing up to six monomers [36]. This leads us to suggestion that LF forms oligomers with a higher monomer number in solution and that these oligomers are quite stable under mild concentrations of SDS. The native electrophoresis allowed us to estimate the initial degree of LF oligomerization and changes of the oligomerization degree due to the presence of various ligands. The gel after the native electrophoresis revealed several weak bands with different electrophoretic mobility, including two major bands that could be attributed to the monomer and the tetramer. After the preincubation with ATP or AMP, most of the proteins remain at the start of the gel, which leads to suggestion that generation of LF oligomer forms with high molecular weight is taking place [36]. The results demonstrated that gel filtration and electrophoresis are not the optimal methods for the detection of the LF oligomers existing before and after binding with different ligands. Therefore, the methods allowing direct detection of protein oligomer forms in solutions were used: small-angle X-ray scattering (SAXS) and light scattering (LS) [35].

According to the data of 2.8 Å X-ray analysis and calculations of LF hydration [1], the volume LF globule is ~132 Å^3. The shape of molecule is close to the ellipsoid and the average radius of gyration (Rg) and is ~26.7 Å. The SAXS method allows detecting only small particles, which are comparable with LF oligomers containing one to four monomers. SAXS spectra from a standard LF solution (5 mg/ml) were obtained and the average Rg was calculated as 43.7 Å [35]. LF oligomers dissociate in the presence of 1 M NaCl; therefore the changes in the SAXS spectra after adding NaCl to LF solution to 1 M final concentration were studied. Values of Rg after adding NaCl were 38.6 Å after 1 h, 32.9 Å after 20 h, and 27.4 Å after 42 h; during the further incubation SAXS spectra were stable and the profile matched expected Rg 26.7 Å for LF monomer. Then the LF solutions were analyzed by the LS method: Rg values estimated for the solutions of various LF preparations ranged from 450 to 575 Å and depended on the time of storage and the initial LF concentration. Rg values obtained using LS were one order higher than obtained by SAXS (43.7 Å). LF oligomers analyzed by LS demonstrated approximately the same slow rate of monomer dissociation, and the final values of Rg after long incubation were similar to SAXS data. Interestingly, lyophilized LF preparations that were solubilized in a neutral buffer contained substantially lower quantities of oligomer LF: the Rg values for such solutions were 50–100 Å, which is 5–10 times lower than for the nonlyophilized LF solutions (450–575 Å). Thereby LF exists in neutral solutions in different oligomer forms, the increase of LF concentration promotes oligomer formation, but all the oligomers dissociate in the presence of 1 M NaCl [35].

There may be different possible modes of monomer association into oligomers: a sphere (model 1), a cylinder (model 2), and a plate of the height *h* and the radius *R* (model 3) [35]; the estimation of LF oligomerization type was done taking into account these three different models. The LF oligomers with Rg 30–44 Å (SAXS) or Rg 50–100 Å (LS) can consist of 8–70, 5–25, or 2–5 monomers depending on the proposed shape of molecule. According to the gel filtration data (**Figure 1**), the LF solutions contain at least ~10-mer complexes. These results comply with the model assuming formation of cylindrical structures composed with 10–25 monomers. LS method has shown that the presence of 0.1 M KCl slows the initial rate of LF oligomerization in 1.5- to twofolds compared to one without salts. The presence of the nucleotides, oligonucleotides (ODNs), and oligosaccharides increases the Rg values of LF complexes up to 300–700 Å which corresponds to 5–25 monomers in cylindrical oligomer model. The relatively low correlation level of Rg values can be explained by conformational changes of LF molecules in the presence of 1 M KCl due to high conformational lability of LF. According to the SAXS and LS data, the increase of LF concentration promotes oligomer formation, but all the oligomers dissociate in the presence of 1 M NaCl.

The effect of $d(pT)_{10}$ and maltoheptaose on the oligomerization of 3.5 μM LF was analyzed in detail. In the presence of the oligosaccharide and $d(pT)_{10}$, Rg was increased from the initial Rg = 75–100 Å to 304 ± 30 Å and 260 ± 20 Å, respectively [35]. Interestingly, AMP exerted a greater effect on the efficiency of oligomerization in several LF preparations (Rg = 75–100 Å) to oligomers demonstrating Rg. = 450–480 Å. Addition of 1 M NaCl to the oligomer complexes formed in the presence of $d(pT)_{10}$ and maltoheptaose led to the dissociation into free LF monomers with Rg 27.3 Å. Remarkably, according to the LS data, the LF decamer can be formed in all analyzed reaction mixtures and the best ligand-stimulating decameric complex formation is AMP and then ATP and the last maltoheptaose. It should be mentioned that each of the three LF ligands demonstrated two K_d values characterizing the interactions with LF (see below). Taken together, the results show that interaction of LF with mono-oligonucleotides, ODNs, and oligosaccharides can lead to the formation of complexes with various degrees of oligomerization.

4. Oligonucleotide recognition and catalytic activity of human milk lactoferrin

It was shown that LF interacts with RNA [15, 16, 37, 38]. Later it was shown that LF can be considered as sequence-specific protein, since it interacts with DNA molecules containing three specific sequences including specific d(TAGAAGATCAAA) ODN [17, 29, 30, 39]. Later, it was shown that human milk LF interacts with ATP, NTP, and oligosaccharides [34] and contains two DNA-binding sites, interacting with specific and nonspecific ODNs [31].

Detailed analyses of LF interaction with a number of nucleotides, NTPs, NDPs, NMPs, dNTPs, dNDPs, and dNMPs and their derivatives (e.g., cAMP, cGMP, NAD, ppppA) by fluorescence analysis have shown that LF contains three nucleotide-binding sites, which are characterized by three dissociation constants (K_d). In the case of the first site, the K_d values for various nucleotides and their derivatives varied from 1.4×10^{-5} M for ADP to 3.6×10^{-7} M for dGDP [40]. Similar results were observed for the second site: the K_d values varied from 3.9×10^{-4} M for NAD to 7.3×10^{-6} M for cAMP, while the K_d for the third site were in the range from 1.1×10^{-3} M for NAD to 9.2×10^{-5} M for dCMP [40]. The results obtained by using fluorescence analysis were confirmed with the measurements of the circular dichroism spectra for LF in the absence and in the presence of ATP: the K_d for LF-ATP complex (1.8×10^{-6} M) was comparable with the fluorescence data (8.3×10^{-6} M). It has been shown that LF binding to the different nucleotides leads to a change of the intrinsic LF fluorescence indicating that conformation changes in the protein are caused by LF-nucleotide complex formation [40].

Using affinity modification of LF by chemically active derivatives of ATP, it was shown that ATP-binding site is localized in the C-terminal domain of LF in contrast to the antibacterial and polyanion-binding sites which are located in the N-terminal domain [30]. In addition, it was shown that LF has two anti-cooperative DNA-binding centers which are localized on the N domain [31]. Interaction of LF with three ATP molecules consists with binding one of the ATP molecules (high affinity) with a specific ATP-binding center on C-terminal domain and binding of two other ATP molecules (with a lower affinity) with two sites on N-terminal domain of LF [41].

It was shown that human milk LF possesses two DNA-binding sites with different affinity for specific ODNs: $K_{d1} \approx 8$ nM (high-affinity site), while $K_{d2} \approx 0.1$ mM (low-affinity site) [31]. Considering the wide range of LF biological functions, it was highly interesting to analyze how the high-affinity site of LF recognizes DNA and what are the differences of LF interactions with DNA comparing to other canonical DNA metabolism enzymes. To evaluate the relative contributions of individual DNA elements to the enzyme affinity for long DNA, a new approach, stepwise increase in ligand complexity (or SILC), was developed (reviewed in [42–44]). Using the SILC approach, many DNA-dependent enzymes were studied, including those of not specific for DNA structure or sequence such as *E. coli* RecA; specific for DNA structure but not for sequence, such as DNA polymerases of prokaryotes, eukaryotes, viruses, and human DNA ligase I; specific for DNA damage such as human uracil DNA glycosylase (UDG), *E. coli* Fpg and human 8-oxoguanine DNA glycosylases (OGG1), and human apurinic/apyrimidinic endonuclease (AP endonuclease); and specific for DNA sequence such as *Eco*RI restriction endonuclease, human topoisomerase I (Topo I), and HIV integrase. It was shown that complex formation, including formation of contacts between all these enzymes and specific sequences, cannot provide high enzyme affinity for DNA or substrate specificity. All mentioned enzymes recognize DNA by forming multiple additive contacts with all DNA units covered by the protein globule (7–20 units depending on the enzyme), and the total interaction is a combination of weak electrostatic, hydrophobic, and/or van der Waals interactions of the enzyme with the individual structural elements of DNA. Thus total interaction can be described by the following geometric progression:

$$K_d\left[d(pN)_n\right] = K_d\left[P_i\right] \times e^{-n} \times h_A^{-a} \times h_C^{-c} \times h_G^{-g} \times h_T^{-t}$$

where $K_d[P_i]$ is the constant of dissociation (K_d) for the *ortho*phosphate (P_i); *e* is the electrostatic factor reflecting the interaction with one inter-nucleoside phosphate group and increase in enzyme affinity; and *h* is hydrophobic factor for A, C, G, and T nucleotide bases, corresponding numbers of which are, respectively, *a*, *c*, *g*, and *t*. It might be proposed that the interaction of noncognate single-stranded (ss) and double-stranded (ds) DNA with any enzyme can be described by this equation. The *h*, K_d, and *e* values for *ortho*phosphate were reviewed in [42–44]. The affinity of some DNA-dependent enzymes to nonspecific $d(pN)_n$ does not always depend on the relative hydrophobicity of the bases. In the case of repair enzymes (8-oxoguanine DNA glycosylases FPG and OGG1, apurinic/apyrimidinic endonuclease APE) and *Eco*RI restriction endonuclease, all hydrophobic factors (*a*, *c*, *g*, and *t*) were equal to one. These enzymes interact mainly with internucleotide phosphate groups, and these weak electrostatic interactions provide high affinity of the $d(pN)_n$ for these enzymes (reviewed in [42–44]).

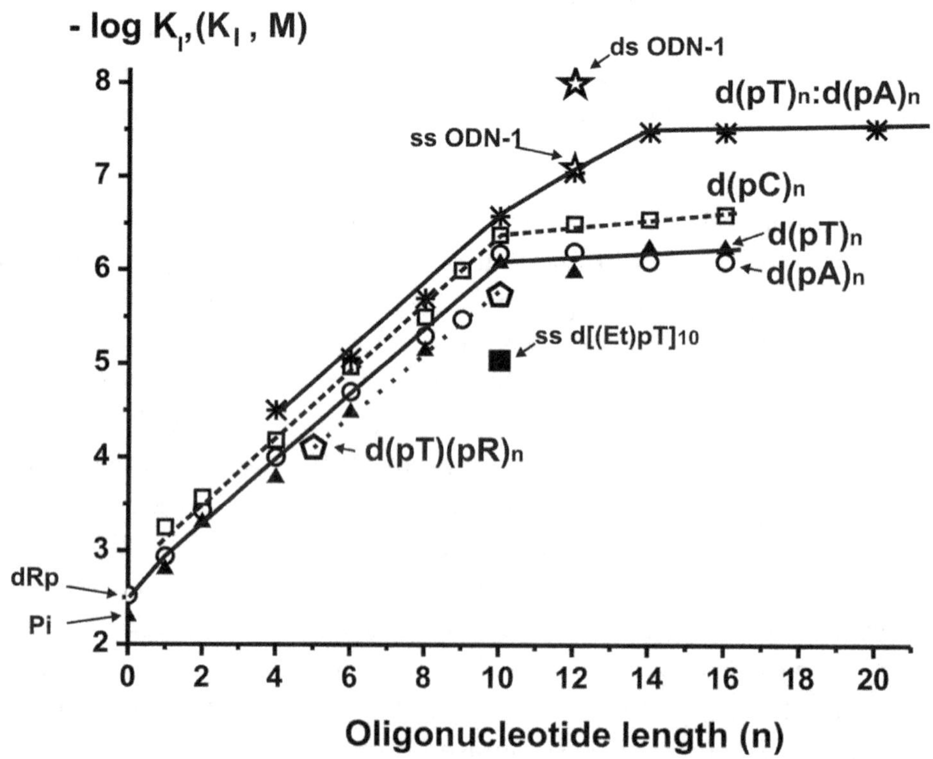

Figure 2. Logarithmic dependencies of LF affinity ($K_i = K_d$ values) for ss and ds; homo- and hetero-ODNs of different lengths (*n*) were calculated by LF complexation analysis with ss [^{32}P]sODN1 and its inhibition. Positions of -log(K_i) values for ethylated $d[(pEt)T)_{10}$, $d[(pT)(pR)_9]$ (pR is a tetrahydrofuran analog of deoxyribose), specific ss and ds sODN1, are shown.

Using the SILC method, the formation of the LF-DNA complexes with $d(pA)_{1-16}$, $d(pT)_{1-16}$, $d(pC)_{1-16}$, $r(pA)_{10}$, $r(pU)_{10}$, $r(pC)_{10}$, specific ODN, and several oligodeoxyribonucleotides was analyzed by electrophoretic mobility shift assay (EMSA) and the Scatchard plot [45]. It was shown that the interaction of the first high-affinity LF DNA-binding site with single-stranded (ss) specific and nonspecific ODNs as well as double-stranded (ds) ODNs is competitive, and K_i ($K_d = K_i$) values were estimated [45]. Affinity of LF (K_i) for homo- and hetero-ss and hetero-ds ODNs of different lengths (n) is determined by using inhibition of LF complexation with ss specific for LF [^{32}P]sODN1. It was shown that like all abovementioned DNA-dependent enzymes [42–44], the minimal ligands of two LF DNA-binding sites are mononucleotides and nucleotide derivatives [45]. The K_d values for minimal ligands of the first LF DNA-binding site were calculated: orthophosphate (5×10^{-6} M), deoxyribose phosphate (3×10^{-6} M), and different dNMPs (($0.56–1.6) \times 10^{-6}$ M). It means that similarly to other analyzed enzymes [42–44], LF recognizes free dNMPs through interaction with all structural elements (base, sugar, phosphate), and the phosphate group makes the major contribution to the affinity for dNMPs [45]. Logarithmic analysis of dependencies the K_d for $d(pN)_n$ versus the number of mononucleotide units ($n = 1–20$) has shown the additivity of free ΔG^0 values for the interaction of 10–11 individual nucleotide units of ODNs with LF and the absence of strong cooperative interactions (**Figure 2**).

Values of factor f (2.28 ± 0.02), the increase in affinity of LF for various $d(pT)_n$ and $d(pA)_n$, were evaluated from the slopes of the linear parts of these curves (**Figure 2**). The factor f (2.36 ± 0.03) for $d(pC)_n$ was slightly higher. The values of $K_d = 1/f = 0.44 \pm 0.004$ M for $d(pT)_n$ and $d(pA)_n$ and 0.42 ± 0.005 M for $d(pC)_n$ correspond to the increase of enzyme K_d values. These are reflecting the interaction between one of ss DNA units and enzyme active center and in turn are equal to the reciprocals of these factors. The affinity of the first LF-binding site for dNMPs (0.56–1.6 mM) is approximately 81–770-fold higher than for other 9–10 nucleotides ($K_d = 0.42–0.44$ M) of an extended ODN [45]. In contrast to DNA polymerases, UDG, AP endonuclease, and Topo I but similarly to *E. coli* Fpg, human 8-oxoG OGG1 glycosylase, and *Eco*RI endonuclease [42–44], the affinity of LF for $d(pC)_n$, $d(pT)_n$, and $d(pA)_n$ did not depend on the relative hydrophobicity of their bases (**Figure 2**). Moreover, LF does not contact 9 of the 10–11 DNA bases but mainly interacts with the sugar-phosphate backbone of the ligands. The internucleoside moieties of ethylated phosphates neutralize their charges and increase the affinity of $d(pT)_{10}$ (0.83 μM) up to 12-fold comparing to ethylated $d[p(Et)T]_{10}$ ($K_i = 10.0$ μM) (**Figure 2**), indicating that negative charges play an important role in DNA internucleoside phosphate group interaction with LF. It might be supposed that the lack of all bases except one nucleotide unit of $d(pN)_n$ interacting with LF active center does not influence significantly the LF interactions with the sugar-phosphate backbone of ss DNA. Similar to other enzymes [42–44], the interactions between LF and sugar-phosphate backbone of DNA may be considered as weak interactions of oppositely charged surfaces by ion-dipole and dipole-dipole interactions. Most probably, DNA- and RNA-binding sites of LF may be the same or at least significantly overlap. It was shown that LF binding to polyanions such as heparin and tRNA inhibits the binding to DNA [31]. Moreover, LF demonstrates higher affinity for $r(pA)_{10}$ (2.9-fold), $r(pC)_{10}$ (2.8-fold), and $r(pU)_{10}$ (7.5-fold) than that corresponding for $d(pN)_{10}$. This can be attributed to extreme biological polyfunctionality of LF, including its ability to activate transcription; also it cannot be excluded that $r(pN)_n$ can be easier adopted to the first LF nucleic acids–binding site [45].

It was shown previously that LF interacts specifically with three sequences of ds DNAs and activates transcription: TAGAAGATCAAA (ODN1), ACTACAGTCTACA (ODN2), and GGCACTTAC (ODN3) [17]. The affinity of these three ODNs for LF was estimated. It was shown that ss and ds ODN3 demonstrate the K_d values of 8.0×10^{-7} M and 2.5×10^{-7} M, respectively. Interestingly, the ss $d(pN)_9$ ($(1–3) \times 10^{-6}$ M)) and ds $d(pT)_{10}$:$d(pA)_{10}$ (2.7×10^{-7} M) values are comparable for those different substrated. It has been proposed that the interaction of LF DNA-binding site's central cytosine bases of GGCACTTAC and in the case of different hetero-ODNs results in equality of f factor for d(pG)-mononucleotide units, d(pT)- and d(pA)-links ($f = 2.28$; see above). Thereby K_d value for ODN3 was calculated from the equation:

$$K_d[ODN3] = K_d[d(CMP)] \cdot \left[\frac{1}{f_{T,A,G}} \right]^6 \cdot \left[\frac{1}{f_C} \right]^2 = 5.6 \cdot 10^{-4} \cdot \left(\frac{1}{2.28} \right)^6 \cdot \left(\frac{1}{2.36} \right)^2 = 7.1 \cdot 10^{-7} M$$

The calculated K_d value $(7.1 \pm 0.7) \times 10^{-7}$ M was comparable with experimental one $(8.0 \pm 0.7) \times 10^{-7}$ M. In a similar way the K_d values for five different nonspecific hetero-ODNs of different lengths and sequences were estimated. The experimental K_d values for all five ODNs were also comparable with the calculated values. The affinity of LF for specific ss ODN2 ($K_d = 2.2 \times 10^{-7}$ M) was ~3.9-fold higher than that for ss $d(pT)_{12–14}$ or ss $d(pA)_{12–14}$ (average $K_d = (8.6 \pm 2.0) \times 10^{-7}$ M). Interestingly, these values for ss $d(pC)_{12–14}$ ($(3.0 \pm 0.2) \times 10^{-7}$ M) are 1.4-fold lower, but ds ODN2 ($K_d = 1.8 \times 10^{-7}$ M) affinity was approximately twofold lower than that for ds $d(pT)_{12}$:$d(pA)_{12}$ ($K_d = 9.0 \times 10^{-8}$ M), which are completely nonspecific. Specific ss ODN1 ($K_d = 8.0 \times 10^{-8}$ M; increase in affinity 4.0–7.9-fold) and ds ODN1 ($K_d = 1.0 \times 10^{-8}$ M; increase in affinity ninefold) but not others, compared to nonspecific ss ($K_d = (3.2–6.3) \times 10^{-7}$ M) and ds ($K_d = 9.0 \times 10^{-8}$ M) homo-$d(pN)_{12}$ of the same length, demonstrated significantly higher affinity for LF. Of all hetero-ODNs only specific ss and ds ODN1 demonstrates significant increase in the affinity compared to nonspecific ODNs and only this ODN contains TCA trinucleotide. Therefore, one cannot exclude that this particular trinucleotide sequence may be important for specific interaction of all nucleotide units of ODN1 with the DNA-binding site of LF [45]. Similar to all analyzed enzymes, the second strand contributes to the LF affinity for nonspecific and specific ds DNA (14–24-fold; $(4–7) \times 10^{-2}$ M) in a less degree than the first strand ($(3–8) \times 10^{-7}$ M) (**Figure 2**) [45].

Interestingly, the relative contributions of specific interactions with specific modified nucleotide units in the case of UDG, AP endonuclease, and Fpg are comparable on the level of minimal ligands (dNMP), ss $d(pN)_n$, and ds ODNs [42–44]. This may explain why total affinity for these repair enzymes is close to additive in case of the specific and nonspecific interactions of specific DNA with different nucleotides. The cooperative effect on the DNA-binding groove of the active site of human OGG1 enzyme may be due to the formation of specific contacts with the oxoG unit [46]. Some nonspecific, previously weak interactions with noncognate ODNs may also greatly strengthen when cognate DNA is bound with LF. Since the ratio in the affinity

(K_d values) for specific and nonspecific ligands significantly increases from mononucleotides (1.8–2.9 ratio) to ss (4.0–7.9 ratio) and ds ODNs (9.0 ratio), it seems likely that the TCA motif of ODN1 may be important for possible cooperative interaction of LF DNA-binding groove with the specific ds ODN1 [45].

Many data indicate that all enzymes acting on extended DNA molecules first bind to any sequence of DNA and then slide to the site containing a specific sequence or a lesion (reviewed in [42–44]). The increase in the affinity (estimated 5 to 790-fold for different enzymes) for specific sequence of DNA can change the DNA backbone conformation in multiple ways. The promotion of the proper fit of specific bases into recognition pockets of the enzymes may be a result of these bonds formation. Similar mechanism was proposed for specific sequence searching and DNA recognition by LF. Moreover it was shown that the interaction of LF with nucleic acids demonstrates some specific features of DNA and RNA recognition compared to other enzymes studied [45].

5. Five catalytic activities of human milk LF

It was previously shown that human LF possesses protease [47] and RNase [15, 16] activities. Later the evidence that homogeneous human milk LF possesses five different enzymatic activities was presented: DNase, RNase, ATPase, phosphatase, and amylase [34]. Several rigid criteria were applied to show that all hydrolyzing activities are intrinsic properties of LF, rather than a result of contamination with canonical enzymes. The most important of these criteria are given below: (a) electrophoretic homogeneity of LF after SDS-PAGE with silver staining, (b) adsorption of LF hydrolyzing activities to the anti-LF Sepharose leading to the disappearance of catalytic activities from solution and the elution by acidic pH of peak possessing five catalytic activities, and (c) FPLC gel filtration of LF under conditions of "acidic shock" (pH 2.6) that does not lead to the disappearance of the activity and to the elution of five in one peak corresponding to ~80 kDa LF.

To exclude possible artifacts of contaminating enzymes, modified with affinity reagents, LF preparations were separated by SDS-PAGE (**Figure 3**). It was shown that only LF contains ^{32}P-label after its affinity modification with [α-^{32}P]oxATP (2′,3′-dialdehyde derivative of ATP), 2′, 3′-dialdehyde derivatives of [5′-^{32}P]-d(pT)$_9$r(pU), and [5′-^{32}P](pU)$_{10}$ (**Figure 3**, lanes 3–5). Two LF catalytic activities were detected after extraction of proteins from the gel slices (**Figure 3A**, lanes 9 and 10) or were detected in situ using gels containing DNA and RNA (lanes 6 and 7; ethidium bromide staining after 16 h at 37°C revealed dark bands corresponding to LF molecular mass on a fluorescing background of copolymerized DNA or RNA).

The same results were obtained using in gel ATPase assay in which enzyme-generated P_i was precipitated by Pb^{2+} resulting in $Pb_2(PO_4)_3$ formation in gel regions containing ATP-hydrolyzing molecules (**Figure 3**, lane 8). Since SDS dissociates any protein complexes, the detection of five activities in the gel regions corresponding to intact LF molecular mass and the absence of any other bands of the activity (**Figure 3**) provide direct evidence that LF possesses five catalytic activities. It was shown that trypsin at pH 8.2 cleaves the LF molecule between Lys283

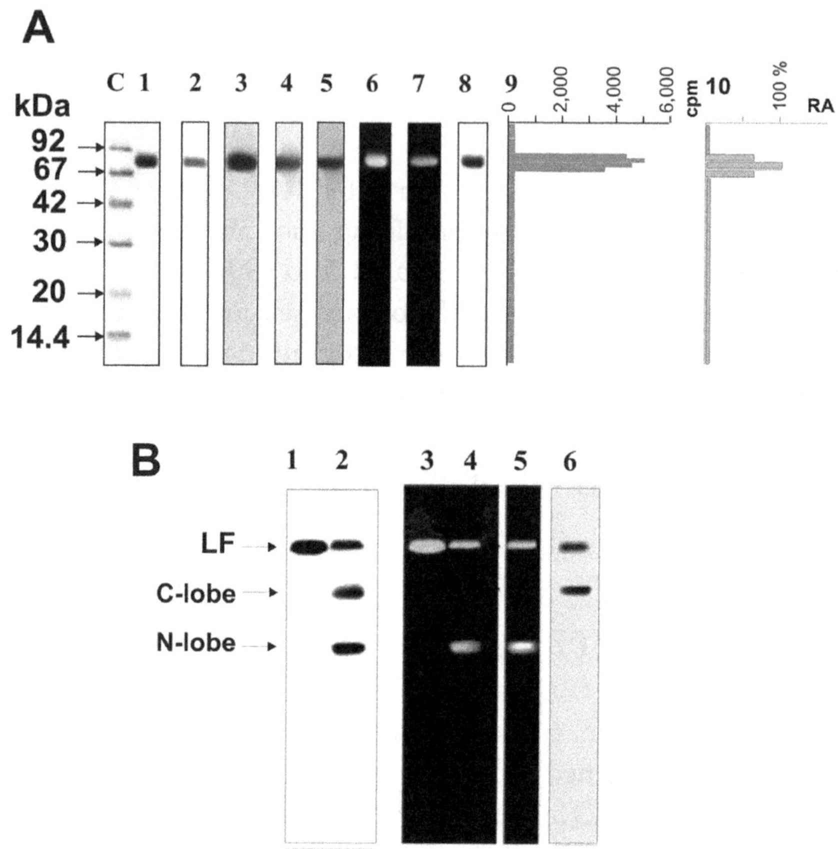

Figure 3. In-gel detection of enzymatic activities of intact LF and tryptic fragments in 12% SDS-PAGE gels. (A) Lane 1, silver stained; lane 2, immunoblotting (alkaline phosphatase-conjugated anti-LF antibodies); lanes 3–5, affinity labeling of LF with periodate-oxidized [α-^{32}P]ATP (3), 5′-[^{32}P](pU)$_{10}$ (4), or 5′-[^{32}P]d(pT)$_9$r(pU) (5) (autoradiographs); lanes 6–7 (negatives), DNase and RNase in gels containing calf thymus DNA (6) or yeast RNA (7); lane 8, ATPase activity; lane 9, phosphatase activity, lane 10, amylolytic activity (RA, relative activity): extracts from 2 to 3 mm gel slices were incubated with 5′-[^{32}P](pT)$_8$ (line 9) or with or with oligosaccharide. (B) Lanes 1 and 2, Coomassie R250-stained LF (1) and LF tryptic fragments (2); lanes 3–5, the negatives of the films corresponding to DNase (3, 4), RNase (5), and ATPase (6; [^{32}P]Pb$_3$(PO$_4$)$_2$ activity of LF (3) and its tryptic fragments (4–6).

and Ser284 residues resulting in N-lobe (30 kDa) and C-lobe (50 kDa) formation [48]. There were compared DNase and ATPase activities of LF before and after trypsinolysis (**Figure 3B**). Interestingly, the high-affinity DNA (lane 4)- and RNA (lane 5)-binding sites are located on the N-fragment (**Figure 3B**), while ATP-binding site is located on the C-fragment (**Figure 3B**, lane 6). Overall, our results demonstrate that all five catalytic activities are intrinsic properties of LF [34].

The above data were obtained using total preparations of LF from different donors. At the same time, it was shown that LF can be separated by chromatography on Cibacron Blue Sepharose into many fractions with different affinities for the sorbent [16]. It was analyzed whether LF fractions exhibiting different affinities for Cibacron Blue Sepharose possess different catalytic activities [34]. The main subfraction of LF (peak 4, LF-5, **Figure 4**) had the highest affinity for this sorbent, as for three additional subfractions (peaks 1–3, **Figure 4**) they represented ~10–20% of the total LF depending on the milk donor. The first protein peak had not demonstrated enzymatic activity; at the same time LF fractions corresponding to three other peaks possessed ODN 5′-phosphatase, DNase, RNase, ATPase, and maltooligosaccharide-hydrolyzing activities, and each activity was eluted in several peaks (**Figure 4A–C**). LF subfraction corresponding to the peak 2 possessed four different activities: phosphatase, DNase, RNase, and ATPase. Eluate corresponding to the protein peak 3 demonstrated three prominent peaks of ODN 5′-phosphatase activity (**Figure 4B**) and two peaks of RNase activity (**Figure 4C**). Thus, LF subfractions LF-1-LF-5 (**Figure 4**) possess different combinations of various catalytic activities.

The nature of the structural variations that gives rise to these profound functional differences is not defined. The LF molecule contains two potential glycosylation sites [2], and the degree of different molecules' glycosylation varies. LF can contain hexose, mannose, hexoseamines, or other saccharides [49] and may also differ in level of phosphorylation. Therefore, one cannot exclude that different affinities of LF fractions to Cibacron Blue Sepharose and specific distribution of various catalytic activities among these fractions may be due to different types of LF glycosylation and/or phosphorylation.

DNase activity of human LF differs significantly from other known canonical DNases [34]. The pH optimum of LF is 7.0–7.5—significantly higher than that (5.0–5.5) of human blood DNase II. The LF activates significantly (100–150%) by 100 mM NaCl, whereas DNase I is 70% inhibited by 50 mM NaCl [50, 51]. LF-mediated cleavage of [^{32}P]-ODNs and DNA is stimulated three- to fivefold by Ca^{2+}, Cu^{2+}, and Zn^{2+} and eight to ninefold by Mn^{2+} or Mg^{2+} ions. In contrast to known human DNases, LF DNase activity is activated by ATP, dATP, and NAD (150 mM) by 1.5–2.5-fold [34]. Human LF hydrolyzes supercoiled plasmid DNA ~30–200 times faster (k_{cat} = 2–9 min^{-1}) than ODNs; the rate is comparable to some DNA restriction endonucleases. It was shown that in case of RNA hydrolysis LF substrate specificity stands out from RNase A and other human milk RNases [34]. LF preparations from different donors hydrolyze ATP with K_m ~3.5 × 10^{-5} M, and the k_{cat} value ~2 × 10^{-3} min^{-1}. As it was shown, maltooligosaccharide is the best substrate for different LF preparations demonstrating K_m ~ 10^{-6} M [34].

Taking into account the relative k_{cat} values for LF in the hydrolysis of DNA, RNA, and ATP and the relative content of LF in human milk comparing to canonical DNases, RNases, and ATPases, it was shown that the LF is the major DNase, RNase, and ATPase of human milk [34, 41, 52]. Unlike the main LF fraction with the highest affinity for Blue Sepharose, all LF fractions with DNase activity are cytotoxic and suppress growth of mouse and human tumor cells [52]. The discovery of LF catalytic activities may contribute to understanding of multiple physiological functions of this extremely polyfunctional protein and explain its protective role against microbial and viral infections.

6. Catalytic antibodies of human milk

Abs are produced by the plasmatic cells. In traditional view they have a unique function binding antigens and elicit the immune response. However, it was shown that Abs raised against chemically stable analogs of the transition states of chemical reaction can possess different enzymatic activities. These artificial catalytic Abs were termed abzymes (*anti*body

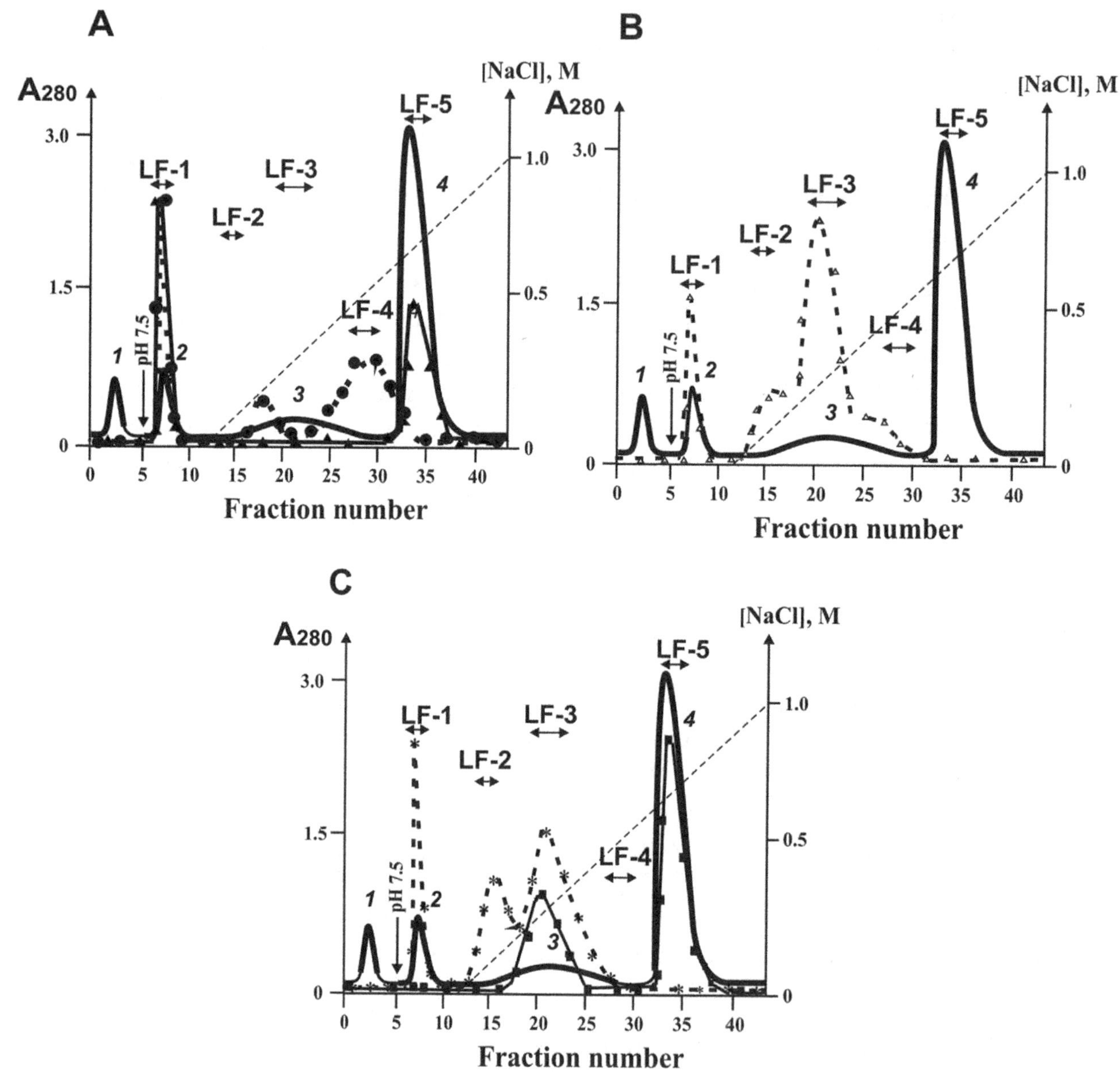

Figure 4. Chromatography of LF on Blue Sepharose and enzymatic activities in five fractions of LF (LF-1-LF-5): (A) solid line (-), A_{280}; DNase (•) and ATPase (▲); (B) 5′- (Δ); (C) RNase (*) and amylase (■). Aliquots (1–3 μl) of chromatography fractions were used for the determination of DNase (supercoiled DNA), RNase (5′-[^{32}P](pU)$_{10}$), phosphatase (5′-[^{32}P](pT)$_8$), ATPase (γ-[^{32}P]ATP), and amylase (4-nitrophenyl-4,6-O-ethylidene-α-D-maltoheptaose) activities.

en*zymes*). Abzymes catalyzing more than 100 distinct chemical reactions are reviewed in Refs. [53–60].

During past two decades, it was shown that the sera of patients with autoimmune diseases contain auto-Abs that possess catalytic activities. Now it is obvious that the occurrence of catalytic Abs is a distinctive feature of AIDs (reviewed in [61, 62]). Naturally occurring abzymes may be Abs raised directly against enzymes acting as haptens, and mimicking transition states of catalytic reactions, this process is similar to artificial catalytic Abs raised against stable analogs of chemical reactions transition states [53–60]. Moreover, the induction of anti-idiotypic Abs in AIDs with a primary antigen may result in some antibody catalytic activities. IgGs and/or IgAs, IgM abzymes hydrolyzing different oligopeptides, proteins, DNA, RNA, nucleotides, and polysaccharides were detected in the sera of patients with several AIDs. Healthy humans usually do not develop catalytic Abs or their activities are very low. The detection of abzymes has been recognized as the earliest indicator of autoimmune process [61, 62].

During pregnancy and immediately after delivery, women are often undergoing immune system changes similar to that of AIDs (autoimmune diseases) [61–63]. The published data suggest that pregnant women may be directly immunized through certain compounds of viral, bacterial, or food origin, which can efficiently stimulate production of different Abs. Immunization of animals by direct injection of antigens (mainly proteins) into the bloodstream or by oral administration 1–3 months before delivery leads to the production of Abs against corresponding proteins which later may be detected in the milk [64]. The generation of human milk Abs several months after the immunization indicates on the existence of a specific "immune memory" in lactating and in pregnant women. During the pregnancy, the autoimmunization of mothers may be similar to the processes occurring in blood of autoimmune patients. It has been reported that the level of DNA in the serum of normal women increases during the first 3 months of pregnancy and this resembles the situation in autoimmune diseases [65]. In addition, an increased level of apoptosis during the last 3 months of pregnancy has been demonstrated [66] together with the presence of low numbers of embryonic cells in the blood of pregnant women [67, 68]. This means that, in contrast to healthy humans, pregnant women may be efficiently immunized by compounds of various viruses and bacteria.

It was shown that IgGs from the sera of normal 2–7-month-old (CBAxC57BL)F1 and BALB/c mice and from the sera of 2–3-month-old autoimmune-prone MRL-lpr/lpr mice (conditionally healthy mice) are catalytically inactive [69–71]. During spontaneous development of deep system lupus erythematosus like pathology, the specific reorganization of immune system in these mice leads to the production of IgGs hydrolyzing DNA, ATP, and polysaccharides with low level of catalytic activity (conditionally prediseased mice). A significant increase in DNase, ATPase, and amylase activity of IgG is associated with the transition from predisease to deep disease condition and correlates with additional changes in differentiation and proliferation of bone marrow hematopoietic stem cells (HSC) and lymphocyte proliferation in different organs. Abzyme activities in the serum of pregnant females were comparable with those of predisease mice, but the profile of HSC differentiation and cell apoptosis levels in pregnant and predisease mice were quite different. Right after the beginning of lactation (4 days after delivery) and in a late time of lactation (14 days after delivery), the increase in cell apoptosis and significant changes in two different stages of the HSC differentiation profiles were observed; the first stage was accompanied with a significant increase and the second with a remarkable decrease in abzyme activity [69–71]. These data indicate that women blood and milk can contain various Abs against foreign and self-antigens (including abzymes) in high concentrations.

sIgAs and IgGs possessing DNase, RNase, amylase, and ATPase activities were found in blood serum and milk of pregnant and lactating females [63, 72–77]. It was shown that milk of clinically healthy human mothers contains quite unusual sIgA and IgGs possessing protein [78–80], lipid [81–83], and polysaccharide kinase activities [84–86]. The most impressive catalytic activities of human milk Abs are synthesizing ones (kinase activities), which are not detected for any serum antibody under any autoimmune disease. It was shown that IgGs and sIgAs phosphorylate not only casein but also approximately 15 other milk proteins [78–80]. It was shown that milk sIgA and IgG contain tightly bound lipids and oligosaccharides of unusual structure and these Abs effectively phosphorylate these ligands [81–85]. In contrast to canonical kinases, milk abzymes can transfer phosphate group to proteins, lipids, and oligosaccharides not only from ATP but also from all NTPs and dNTPs; the most surprising is that inorganic *ortho*phosphate is also very good substrate [78–85]. Canonical enzymes using *ortho*phosphate as a donor of phosphate groups are not yet found.

Many autoimmune pathologies can be "activated" or "triggered" in clinically healthy women during pregnancy and soon after childbirth [87, 88]. Independent of detectable autoimmune reactions during pregnancy woman may sometimes develop postnatal autoimmune pathologies such as system lupus erythematosus, Hashimoto's thyroiditis, phospholipid syndrome, polymyositis, and autoimmune myocarditis. One of the most frequently found postnatal autoimmune pathologies is Hashimoto's thyroiditis (1.9–16.7%) [88–90]. Various manifestations of these AIDs can be detected during the first 3–6 postnatal months. Interestingly, the relative activity of human blood Abs significantly increases after delivery and in the beginning of lactation. Moreover, catalytic activity of human milk Abs is 5–600-fold higher comparing the sera of the same women [63, 77]. Interestingly, the antibody DNase activity in blood of healthy pregnant women was four- to fivefold lower than that of pregnant women with pronounced autoimmune thyroiditis [63]. Thus, one cannot exclude that molecular mechanisms of immune system activation which leads to production of autoreactive or autoantibodies with and without catalytic activity are, to some extent, similar or overlapping in both autoimmune patients and human mothers.

Overall, pregnancy and especially the beginning of lactation may be considered as important periods associated with the production not only of different Abs against foreign antigens, auto-Abs, but also of Abs hydrolyzing foreign and self-antigens. Many abzymes from sera of autoimmune patients and sera and milk of human mothers possess the same activities including hydrolysis of DNA, RNA, nucleotides, and proteins. At the same time, they can have different biological functions. For example, DNase abzymes from sera of system lupus erythematosus [91], multiple sclerosis [92], and DNA-hydrolyzing Bence-Jones proteins from multiple myeloma patients [93] are cytotoxic, cause nuclear DNA fragmentation, and induce cell death by apoptosis, at the same time, DNase abzymes from human milk are not cytotoxic. Antiviral and antibacterial defenses of gut mucosal membranes are among the most important functions of human milk Abs. Immune exclusion may be achieved by entrapping viruses in immune complexes and its transcytosis and excretion into the lumen by inhibiting pathogens adherence to mucosal surface and by agglutination of microorganisms or interference with bacterial flagella [94]. According to our data, in contract to destructive abzymes of autoimmune patients, abzymes of human milk have a clear protective role against different harmful antigens and different infections [95, 96].

7. Fab arm exchange of human milk antibodies

Abs usually considered as the products of clonal B cell populations each producing Abs which recognize a single antigen. There is a common belief that mammalian biological fluids contain monovalent IgG molecules with two identical antigen-binding sites and stable structures. The Fab arm exchange of Abs was first described for isotype 4 of immunoglobulin G (IgG4) [97–100]. As the result of posttranslational modification, IgG4 molecule exchange with half molecules (HL fragments) becomes bispecific and provides anti-inflammatory activity. Mutagenesis studies revealed that for this activity CH3-domain is critical, and the in vitro experiments have shown that adding 0.1–10 mM of reduced glutathione (GSH) is sufficient to induce Fab arm exchange of IgG4 [98]. Later it was shown that natalizumab and other monoclonal Abs also exchange by Fab arms in vivo with endogenous human IgG4 in treated individuals that makes IgG4 unsuitable for human immunotherapy [101].

Polyspecificity may be defined as binding of Ab molecule to a large panel of diverse antigens. It was shown that the large number of monoclonal Abs can bind to the variety of totally unrelated foreign and self-antigens. Therefore, it was proposed that the best explanation of Abs polyreactivity is the flexibility of the antigen-binding "pocket" which can change conformation and accommodate different unrelated antigens [102]. There is no doubt that due to change of the conformation, some monoclonal Abs can bind not only specific antigen with high affinity but also some cognate and even foreign compounds with lower affinity. It is known that canonical enzymes can sometimes interact nonspecifically with foreign ligands demonstrating lower affinity comparing to specific substrates, but they usually cannot catalyze conversion of molecules of noncognate compounds. For example, it was shown that affinity of many sequence-specific enzymes for specific and nonspecific DNAs differs more often ~tenfold and rarely up to 100-fold. However, these enzymes catalyze the conversion of only specific DNAs [43].

Human milk IgGs and sIgAs were separated by affinity chromatography on DNA cellulose [103, 104]. The fractions of Abs with different affinity for DNA were analyzed in the hydrolysis of DNA, ATP, and oligosaccharides as well as in phosphorylation of proteins, lipids, and oligosaccharides. It was surprising that all IgG and sIgA fractions including those eluted under the conditions destroying strong complexes of antigen-binding sites with specific antigens (3 M NaCl, 2 M $MgCl_2$) have not only efficiently hydrolyzed DNA but also ATP and oligosaccharides as well as phosphorylated casein, oligosaccharides, and lipids. The same preparations of IgGs and sIgAs were separated on ATP-Sepharose, casein-Sepharose, and lipid-saturated silica gel; in all cases all antibody catalytic activities were distributed all over the profiles of all chromatograms (**Figure 5**).

These results cannot be explained with the current concept of one type of two (IgGs) or more (sIgAs) identical antigen-binding site of Abs. Using this concept, it was impossible even to explain the high affinity of several fractions of Abs for several different affinity sorbents and moreover the catalytic polyreactivity of different fractions. But the presence of two different antigen-binding sites in one IgG or up to four sites in one sIgA molecule can easily explain the nature of binding polyspecificity and catalytic polyreactivity. Our data indicated that human

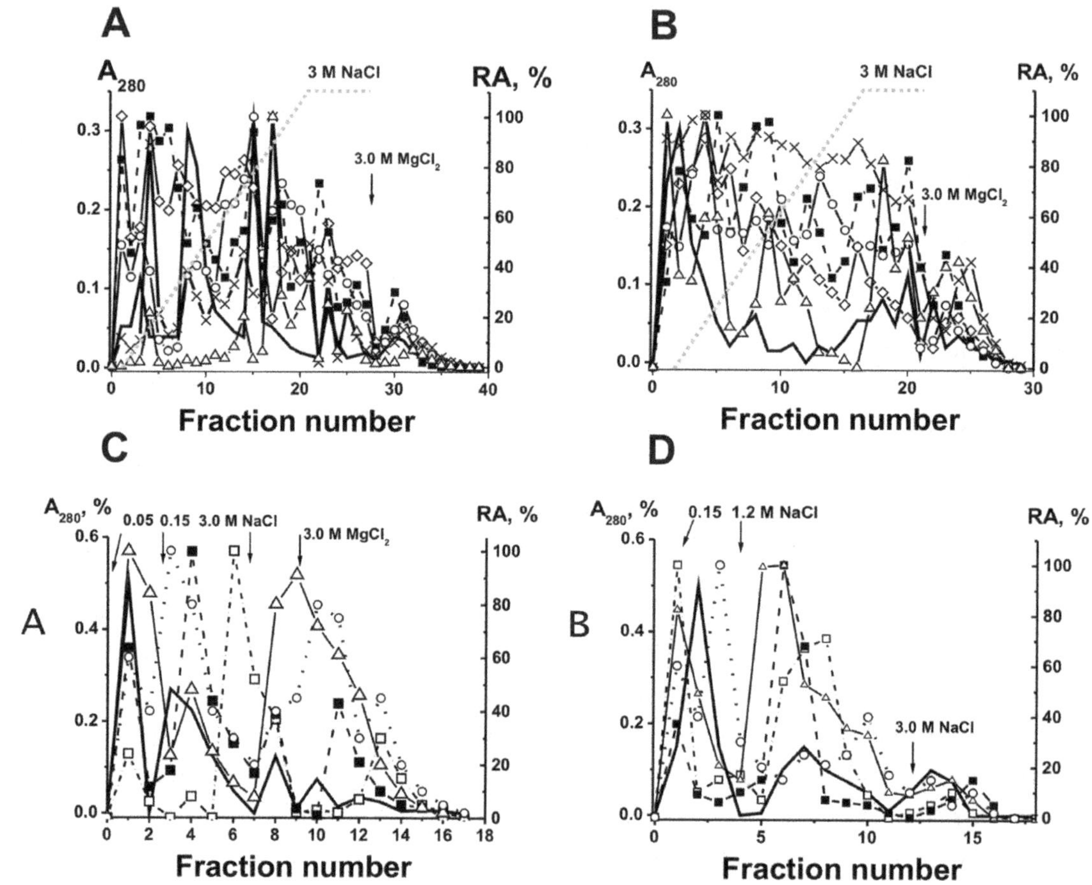

Figure 5. Affinity chromatography of human milk polyclonal IgGs on DNA cellulose (A), ATP-Sepharose (B), casein-Sepharose (C), and lipid-resin (D): (—), absorbance at 280 nm; symbols correspond to the relative catalytic activities (RA) in the hydrolysis of DNA (Δ), ATP (◉), and oligosaccharides (×); phosphorylation of lipids (■) and polysaccharides (◉) tightly bound to IgGs (A and B), hydrolysis of DNA (Δ), ATP (◉), and phosphorylation of casein (◉) and lipids tightly bound with Abs (■) (C and D). Depending on the RA and reaction analyzed, the reaction mixtures were incubated for 0.5–2 h and then the RAs were normalized to the standard conditions and the RA of the fraction with the highest activity was taken for 100%. The average error in the initial rate determination from two experiments in each case did not exceed 7–10%.

milk contains not only monofunctional catalytic Abs but also hybrid bifunctional IgGs and polyfunctional sIgAs with different combinations of HL fragments possessing distinct affinities and catalytic activities [103, 104].

To establish in vivo Abs Fab arm exchange in human milk, IgGs and sIgAs were separated by affinity chromatography on anti-κ-L-Sepharose and anti-λ-L-Sepharose bearing immobilized monoclonal Abs to human κ- and λ-light chains. The IgG fraction having affinity for anti-κ-L-Sepharose was re-chromatographed on anti-λ-L-Sepharose and vice versa. The relative percentage of Abs having affinity only for anti-κ-L-Sepharose, only to anti-λ-L-Sepharose, and to both of these sorbents was evaluated. In addition, using ELISA it was confirmed that IgGs having affinity only for anti-λ-L-Sepharose do not contain κ-IgGs, while IgGs with affinity for anti-κ-L-Sepharose are free of λ-IgGs. But IgGs interacting with both affinity sorbents demonstrated positive response to mouse Abs against human λ- and κ-Abs [103]. Similar results were obtained in the case of milk sIgAs [104]. It was shown that milk IgG preparations contain ~33%

of Abs only with κ-light chains, ~13% only with λ-chains, and ~54% of IgGs with both κ- and λ-light chains simultaneously proving the bispecific nature of milk IgG molecules. Nearly the same results were obtained in case of sIgA: ~48% of Abs contained only κ-light chains, ~35% contained only λ-chains, and ~17% were presented by chimeric molecules simultaneously containing both types of the light chains [104]. Significantly lower quantity of bispecific sIgAs comparing to IgGs can appear due to secretory component and/or J-chain which may hinder the exchange of sIgA by HL fragments. It is well known that B-lymphocytes or plasmatic cells producing Abs express only one gene coding light chain—of κ- or λ-type—as the other copies of light chain genes are removed during cell maturation. The exceptions to this rule are described in transformed hybridoma, leukemia, and myeloma cells and appear as the result of abnormalities; they are not observed in human milk donors [105, 106].

It was shown that chimeric bispecific human milk IgGs are presented mostly by IgG1 (74%) and lower amounts of IgG2–IgG4 (5–16%). Our data provided the evidence that in milk not only IgG4 but all IgG subclasses and also sIgA molecules can exchange with HL fragments, but not with individual light or heavy chains [103, 104].

The experiments on monoclonal IgG4 have shown that exchange with Fab arms in vitro occurs in the presence of reduced glutathione [99, 101]. The exchange of HL fragments between IgG and between sIgA molecules in vitro was studied using FITC and ^{32}P-labeled Abs. Nonlabeled and labeled IgG and sIgA fractions with different affinities for DNA were eluted from DNA cellulose with gradient of NaCl (0–3 M) and 8 M urea (initial chromatography). First, in control experiments the fractions of ^{32}P/FITC-labeled and ^{32}P/FITC-unlabeled Abs exhibiting lower and higher affinities for DNA were mixed in the presence of only reduced glutathione (GSH) or only milk plasma, incubated for 24 h, and subjected to re-chromatography on DNA cellulose (e.g., **Figure 6**). The radioactive (fluorescent) label was eluted with the same concentrations of salt (or urea) as in the case of the initial chromatography indicating the impossibility of the HL fragment exchange in these conditions (**Figure 6B** and **C**).

But in experiments when mixtures of 0.6 M [^{32}P]IgGs and 0.15 M IgGs were incubated in the presence of both GSH and milk plasma after re-chromatography on DNA cellulose, the radioactive label was mainly distributed between three peaks. The main part of the total ^{32}P-label moved to two IgG peaks with low affinity, and one corresponded to the initially nonradioactive IgGs 0.15 M IgGs (e.g., **Figure 6A**). Reverse situation was observed in the case of 0.15 M [^{32}P]IgGs and 0.6 M IgGs: as the result of Fab arm exchange, the most of ^{32}P label was detected in the IgG peaks and eluted with high NaCl concentration (≥0.6 M). Similar results were obtained in the case of IgGs labeled with FITC. Overall, after the exchange, 25–60% of labeled IgGs changed the affinity for DNA cellulose [103]. In the case of sIgA HL fragments, exchange led to the transition of only 11–20% of Ab from one fraction to another [104]. This is consistent with a lower content of chimeric κλ-sIgAs in human milk, and it may be due to hampering of the exchange with HL fragments by secretory and J-components of sIgA.

Thus, it has been shown that human milk contains unknown factor(s) stimulating the IgG and sIgA Fab arms exchange in the presence of reduced glutathione. Since the intact human milk contains GSH and other reducing compounds in significant concentrations, HL-fragment exchange can occur in vivo directly in human milk [103, 104]. In addition, it cannot be excluded

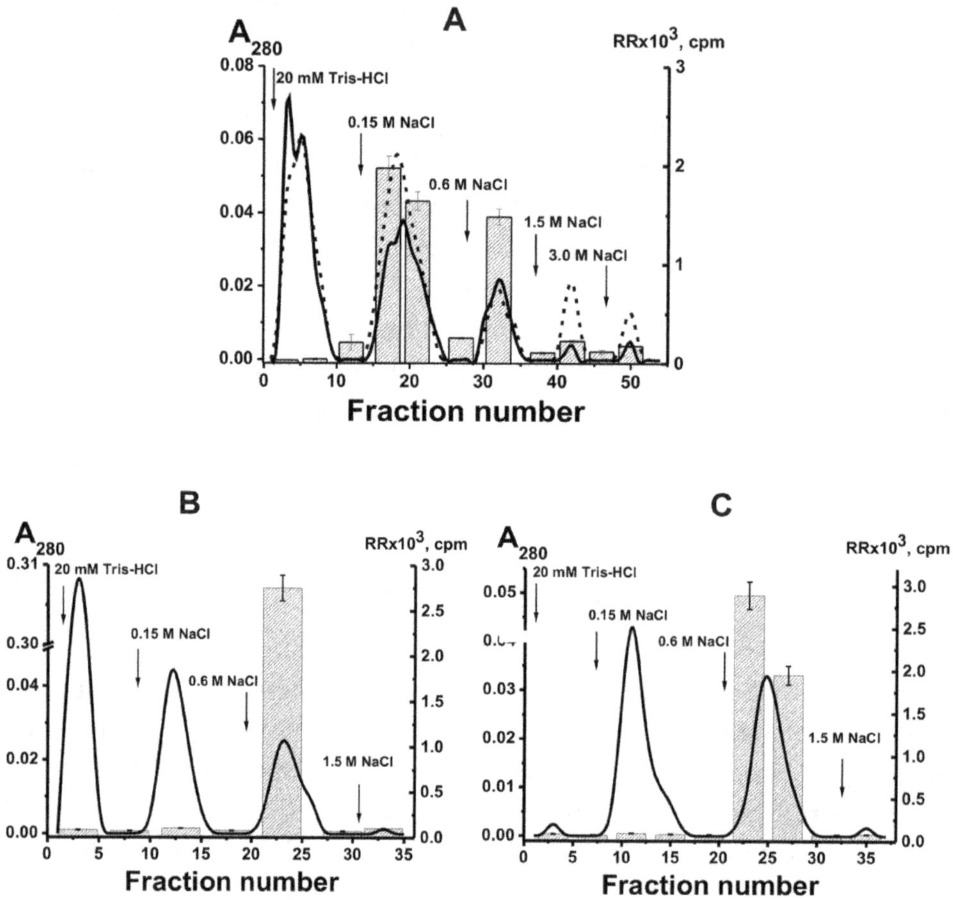

Figure 6. Affinity chromatography of nonmodified and ^{32}P-labeled polyclonal IgGs on DNA cellulose: (---) and (—), absorbance of IgGs at 280 nm before and after phosphorylation using γ-[^{32}P]ATP, respectively; the bars correspond to the relative radioactivity (RR) of [^{32}P]IgG fractions (A–C). Analysis of a relative efficiency of half-molecule exchange under different conditions between nonmodified IgGs and [^{32}P]IgGs having different affinity for DNA cellulose (A–C). Before chromatography the IgG preparations eluted from DNA cellulose by 0.15 M NaCl (0.15 M IgGs) were incubated with [^{32}P]IgGs eluted by 0.6 M NaCl (0.6 M [^{32}P]IgG) in the presence of GSH and milk plasma containing no Abs (A) or in the presence of only GSH (B) or only milk plasma containing no Abs (C).

that during the penetration of IgAs through the specific milk barrier, the secretory component (S) and the joint chain (J) can combine molecules of dimeric H_2L_2 λλ-IgAs, κκ-IgAs, and chimeric κλ-IgAs against different antigens forming many different variants of $(HL)_4SJ$ sIgA molecules. Therefore, some chimeric molecules of sIgAs can in principle contain from two up to four HL fragments to various antigens interacting with high affinity for different sorbents and catalyzing various chemical reactions.

It was shown recently that similarly to human milk Abs placenta, Abs undergo to extensive half-molecule exchange and IgG pools in average consist of ~43.5% κκ-IgGs and 41.6% of λλ-IgGs, while 15.0% of the IgGs contained both κ- and λ-light chains [107]. κκ-IgGs and λλ-IgGs contained, respectively, IgG1 (47.7% and 34.4%), IgG2 (36.3% and 44.5%), IgG3 (7.4% and 11.8%), and IgG4 (7.5% and 9.1%), while chimeric κλ-IgGs consisted of 43.5% IgG1, 41.0% IgG2, 5.6% IgG3, and 7.9% IgG4. The relative content of chimeric IgGs (~15.0%) in placenta is significantly lower than that of the milk (up to 54%). It can be supposed that the observed phenomenon may appear due to a lower content of factor(s) stimulating the exchange in placenta comparing with milk.

8. Complexes of human milk, containing antibodies and lactoferrin: structure, stability, and potential functions

Since the most biological processes are performed by protein complexes and many components of human milk are multifunctional and cooperate with other factors to produce specific effects modulating growth and development of neonates [108], identification and characterization of human milk protein complexes are important for understanding sophisticated functions of human milk. The complexes of LF with other proteins may be the most widely represented in human milk due to LF polyfunctionality and ability to form oligomers, as it was described above.

It was shown that LF interacts with calmodulin, a ubiquitous 17-kDa regulatory calcium-binding protein localized in the cytoplasm and nucleus of activated cells [109], with casein micelles [110], with β-lactoglobulin and albumin [111], and with ceruloplasmin [112]. All these complexes are relatively unstable.

Different methods to analyze whether in human milk LF can form any very stable protein complexes with other milk proteins were used [113]. Using gel filtration of milk proteins on Sepharose 4B column, there was a purified stable high molecular mass (~1000 kDa) multiprotein complex (SPC) from 15 preparations of human milk (**Figure 7A** and **7B**). By the LS and gel filtration, the stability of SPC was shown in the presence of high concentrations of NaCl and $MgCl_2$; however, the SPC efficiently dissociated in 2 M $MgCl_2$, 0.5 M NaCl, and 10 mM dithiothreitol—the conditions of immune complex dissociation (**Figure 7C**).

Obviously, it is unlikely that such stable complex is an occasional protein association. The amount of SPCs in total milk protein content varied from 6 to 25%. With electrophoretic analysis and MALDI TOF MS data, it was shown that all 15 SPCs contained LF and α-lactalbumin as major proteins, while human milk albumin and β-casein were present in moderate or minor amounts; a different content of IgGs and sIgAs depending on milk samples was observed. In addition, this complex contains several proteins/peptides with relatively low molecular masses (3–9 kDa) which have not yet been identified. All SPCs efficiently hydrolyzed DNA and maltoheptaose. Some fresh SPC preparations contained not only intact LF but also small amounts of its fragments which appeared in all SPCs during their prolonged storage. The fragments similarly to intact LF possessed DNase and amylase activities. It may be a consequence of a presence in SPCs of any canonical proteases in very low amounts or manifestations of the proteolytic activity directly by LF.

As it was mentioned above, oligomers of electrophoretic homogeneous LF completely dissociate during the gel filtration in the presence of NaCl [35]. However, human milk SPCs contain several other proteins and are very stable, requiring high concentrations of $MgCl_2$, NaCl, and DTT for effective dissociation [113]. These data suggest that large and small proteins

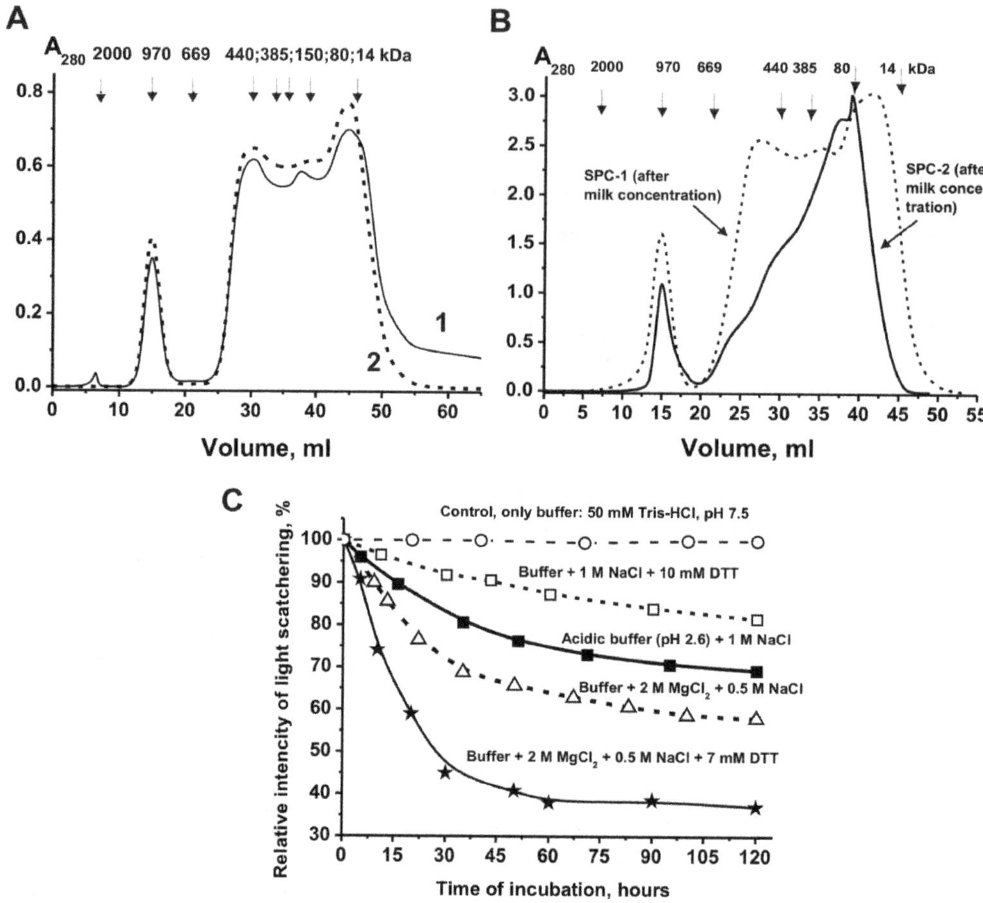

Figure 7. Isolation and analysis of milk protein complex stability. Gel filtration of untreated human milk (1 ml) of first donor (profile 1) and its plasma after removal of fat-lipid fraction (profile 2) on a Sepharose 4B column: (–) and (---), absorbance at 280 nm (A_{280}) corresponding to milk and plasma, respectively (A). Concentrated milk plasma (1 ml) of two donors was gel filtrated: (---) and (–); absorbance at 280 nm (A_{280}) corresponds to different donors (B). Changes of the relative LS intensity were measured in the equimolar mixture of SPCs from 15 milk samples (0.5 mg/ml) under different conditions (C); all conditions are shown in panel C. The highest relative LS at zero time of the incubation was taken for 100% (C).

in the SPCs are mainly interacting by electrostatic forces. At the same time, SPCs can be partially destroyed in the presence of 8 M urea which basically breaks the hydrogen bonds. Since DTTs increase the SPC dissociation, the binding of some proteins with disulfide bonds should be considered important for high stability of the SPCs. The presence of stable supramolecular complex in human milk exhibiting DNA and oligosaccharide hydrolyzing activity can be another reason for the extreme diversity of the human milk biological functions. Taking into consideration the LF functions listed above, complexes of LF with other milk proteins may be important for LF in vivo activity.

9. Conclusion

Obviously, human milk proteome is not a simple mixture of proteins and peptides. Milk proteins interact with each other, form supramolecular complexes, and catalyze various processes—hydrolytic, synthetic, and exchange reactions. The biological roles of milk proteins and protein complexes are very diverse: milk proteins may hydrolyze foreign (viral, bacterial) DNA and oligosaccharides of antigenic epitopes, and dietary milk proteins (casein) may increase the diversity of antigen-binding sites combinations in one molecule and regulate and enhance infant immune response. Some new binding sites for different ligands and catalytic centers with new enzymatic functions may be formed at interfaces of the LF molecules with other proteins, as well as in contact zones of different proteins incorporated to the possible supramolecular protein structures. We believe that the whole set of different biological functions of the individual milk proteins and their various complexes, extremely unusual abzymes with hydrolytic and synthetic function, as well as complexes of proteins, Abs, and abzymes with nucleic acids, lipids, and oligosaccharides, are still very poorly studied.

Acknowledgements

This research was made possible in part by grants: Russian Foundation for Basic Research (16-04-00603 a, 16-04-00604 a, 16-34-00163 mol_a, 16-34-60066 mol_a_dk) and funds from the SB RAS (59.1.2 “Specific oligomer protein complexes and Abs in norm and pathology”—0309-2014-0006; 1.7.15 “Molecular mechanisms of immunoglobulin and defence system proteins functions in prokaryotes and human”).

Author details

Sergey E. Sedykh, Valentina N. Buneva and Georgy A. Nevinsky*

*Address all correspondence to: nevinsky@niboch.nsc.ru

SB RAS Institute of Chemical Biology and Fundamental Medicine, Novosibirsk State University, Novosibirsk, Russia

References

[1] Anderson BF, Baker HM, Norris GE, Rice DW, Baker EN. Structure of human lactoferrin: crystallographic structure analysis and refinement at 2.8 Å resolution. J Mol Biol. 1989;209: 711–734.

[2] Van Berkel PH, van Veen HA, Geerts ME, de Boer HA, Nuijens JH. Heterogeneity in utilization of N-glycosylation sites Asn624 and Asn138 in human lactoferrin: a study with glycosylation-site mutants. Biochem J. 1996;319:117–122.

[3] Groves ML. The isolation of a red protein from milk. J Am Chem Soc. 1960;82:3345–3350. DOI: 10.1021/ja01498a029

[4] Montreuil J, Tonnelate J, Mullet S. Preparation and properties of lactotransferrin of human milk. Biochim Biophys Acta. 1960;45:413–421. DOI: 10.1016/0006-3002(60)91478-5

[5] Johansson B. Isolation of an iron-containing red protein from human milk. Acta Chem Scand. 1960;14:510–512. DOI: 10.3891/acta.chem.scand.14-0510

[6] Masson PL, Heremans JF, Schonne E. Lactoferrin, an iron-binding protein in neutrophilic leukocytes. J Exp Med. 1969;130:643–658.

[7] Brock JH. Transferrins. In: Harrison PM, editor. Metalloproteins. Palgrave Macmillan, UK; 1985. p. 183–263.

[8] De Sousa M, Brock JH. Iron in immunity, cancer and inflammation. 1st ed. John Wiley & Sons, New York; 1989. p. 434.

[9] Bennett RM, Merrit MM, Gabor G. Lactoferrin binds to neutrophilic membrane DNA. Br J Haematol. 1986;63:105–117.

[10] Masson PL, Heremans JF. Lactoferrin in milk from different species. Comp Biochem Physiol B. 1971;39:119–129. DOI: 10.1016/0305-0491(71)90258-6

[11] Kawakami H, Lönnerdal B. Isolation and function of a receptor for human lactoferrin in the human fetal intestinal brushborder membranes. Am J Physiol. 1991;261:G841–G846.

[12] Ashida K, Sasaki H, Suzuki YA, Lönnerdal B. Cellular internalization of lactoferrin in intestinal epithelial cell. Biometals. 2004;17:311–315. DOI: 10.1023/B:BIOM.0000027710.13543.3f

[13] Brock JH. Lactoferrin – 50 years on. Biochem Cell Biol. 2012;251:245–251. DOI: 10.1139/O2012-018

[14] Zimecki M, Mazurier J, Spik G, Kapp JA. Human lactoferrin induces phenotypic and functional changes in murine splenic B cells. Immunology. 1995;86:122–127.

[15] Ramaswamy H, Swamy CV, Das MR. Purification and characterization of a high molecular weight ribonuclease from human milk. J Biol Chem. 1993;268:4181–4187.

[16] Furmanski P, Li ZP, Fortuna MB, Swamy CV, Das MR. Multiple molecular forms of human lactoferrin. Identification of a class of lactoferrins that possess ribonuclease activity and lack iron-binding capacity. J Exp Med. 1989;170:415–429.

[17] He J, Furmanski P. Sequence specificity and transcriptional activation in the binding of lactoferrin to DNA. Nature. 1995;373:721–724.

[18] Fleet JC. A new role for lactoferrin: DNA binding and transcription activation. Nutr Rev. 1995;53:226–227.

[19] Mantel C, Miyazawa K, Broxmeyer HE. Physical characteristics and polymerization during iron saturation of lactoferrin, a myelopoietic regulatory molecule with suppressor activity. Adv Exp Med Biol. 1994;357:121–132.

[20] Bezault J, Bhimani R, Wiprovnick J, Furmanski P. Human lactoferrin inhibits growth of solid tumors and development of experimental metastases in mice. Cancer Res. 1994;54:2310–2312.

[21] Sawatzki G, Rich IN. Lactoferrin stimulates colony stimulating factor production in vitro and in vivo. Blood Cells. 1989;15:371–385.

[22] Gahr M, Speer CP, Damerau B, Sawatzki G. Influence of lactoferrin on the function of human polymorphonuclear leukocytes and monocytes. J Leukocyte Biol. 1991;49:427–433.

[23] De Sousa M, Breedvelt F, Dynesius-Trentham R, Trentham D, Lum J. Iron, iron-binding proteins and immune system cells. Ann N Y Acad Sci. 1988;526:310–322.

[24] Amouric M, Marvaldi J, Pichon J, Bellot F, Figarella C. Effect of lactoferrin on the growth of a human colon adenocarcinoma cell line-comparison with transferrin. In Vitro. 1984;20:477–482.

[25] Garre C, Bianchi-Scarra G, Sirito M, Musso M, Ravazzolo R. Lactoferrin binding sites and nuclear localization in K562(S) cells. J Cell Physiol. 1992;153:477–482.

[26] Arnold RR, Cole MF, McGhee FM. A bactericidal effect for human lactoferrin. Science. 1977;197:263–265.

[27] Yi M, Kaneko S, Yu DV, Marakami S. Hepatitis C virus envelope proteins bind lactoferrin. J Virol. 1977;71:5997–6002.

[28] Ellison RT, Giehl TJ, Laforse FM. Damage of the outer membrane of enteric gram-negative bacteria by lactoferrin and transferring. Infect Immun. 1988;56:2774–2780.

[29] Semenov DV, Kanyshkova TG, Buneva VN, Nevinsky GA. Interaction of human milk lactoferrin with ATP. Biochemistry (Moscow). 1998;63:67–75.

[30] Semenov DV, Kanyshkova TG, Buneva VN, Nevinsky GA. Human milk lactoferrin binds ATP and dissociates into monomers. Biochem Mol Biol Int. 1999;47:177–184.

[31] Kanyshkova TG, Semenov DV, Buneva VN, Nevinsky GA. Human milk lactoferrin binds two DNA molecules with different affinities. FEBS Lett. 1999;451:235–237. DOI: 10.1016/S0014-5793(99)00579-7

[32] Bagby G, Bennet R. Feedback regulation of granulopoiesis: polymerization of lactoferrin abrogates its ability to inhibit CSA production. Blood. 1982;60:108–112.

[33] Van Berkel P, Geerts M, van Veen H, Mericskay M, de Boier H, Nuijens J. N-terminal stretch Arg2, Arg3, Arg4 and Arg5 of human lactoferrin is essential for binding to

heparin, bacterial lipopolysaccharide, human lysozyme and DNA. Biochem J. 1997;328:145–151.

[34] Kanyshkova T, Babina S, Semenov D, Isaeva N, Vlassov A, Neustroev K, Kul'minskaya A, Buneva V, Nevinsky G. Multiple enzymic activities of human milk lactoferrin. Eur J Biochem. 2003;270:3353–3361. DOI: 10.1046/j.1432-1033.2003.03715.x

[35] Nevinskii AG, Soboleva SE, Tuzikov FV, Buneva VN, Nevinsky GA. DNA, oligosaccharides, and mononucleotides stimulate oligomerization of human lactoferrin. J Mol Recognit. 2009;22:330–342. DOI: 10.1002/jmr.952

[36] Babina S, Tuzikov F, Tuzikova N, Buneva V, Nevinsky G. The effect of nucleotides on oligomeric state of human lactoferrin. Mol Biol. 2006;40:137–150.

[37] Ward PP, Paz E, Conneely OM. Multifunctional roles of lactoferrin: a critical overview. Cell Mol Life Sci. 2005;62:2540–2548. DOI: 10.1007/s00018-005-5369-8

[38] Legrand D, Pierce A, Elass E, Carpentier M, Mariller C, Mazurier J. Lactoferrin structure and functions. Adv Exp Med Biol. 2008;606:163–194. DOI: 10.1007/978-0-387-74087-4_6

[39] Lönnerdal B, Iyer S. Lactoferrin: molecular structure and biological function. Annu Rev Nutr. 1995;15:93–110.

[40] Babina SE, Nevinsky GA. Lactoferrin interacts with nucleotides. Nucleosides Nucleotides Nucleic Acids 2004;23:1043–1046. DOI: 10.1081/NCN-200026064

[41] Babina SE, Semenov DV, Buneva VN, Nevinskii GA. Human milk lactoferrin hydrolyzes nucleoside-5′-triphosphates. Mol Biol. 2005;39:513–520.

[42] Nevinsky GA. Important role of weak interactions in long DNA and RNA molecule recognition by enzymes. Mol Biol. 1995;29:6–19.

[43] Nevinsky GA. Structural, thermodynamic, and kinetic basis of DNA and RNA-dependent enzymes functioning: important role of weak nonspecific additive interactions between enzymes and long nucleic acids for their recognition and transformation. In: Uversky VN, editor. Protein Structures: Kaleidoscope of Structural Properties and Functions. Kerala: Research Signpost; 2003. p. 133–222.

[44] Nevinsky GA. The role of weak specific and nonspecific interactions in enzymatic recognition and conversion of long DNAs. Mol Biol. 2004;38:636–662.

[45] Guschina TA, Soboleva SE, Nevinsky GA. Recognition of specific and nonspecific DNA by human lactoferrin. J Mol Recognit. 2013;26:136–148. DOI: 10.1002/jmr.2257

[46] Kirpota OO, Endutkin AV, Ponomarenko MP, Ponomarenko PM, Zharkov DO, Nevinsky GA. Thermodynamic and kinetic basis for recognition and repair of 8-oxoguanine in DNA by human 8-oxoguanine-DNA glycosylase. Nucleic Acids Res. 2011;39:4836–4850. DOI: 10.1093/nar/gkq1333

[47] Hendrixson DR, Qiu J, Shewry SC, Fink DL, Petty S, Baker EN, Plaut AG, St Geme JW.

Human milk lactoferrin is a serine protease that cleaves Haemophilus surface proteins at arginine-rich sites. Mol Microbiol. 2003;47:607–617.

[48] Legrand D, Mazurier J, Metz-Boutigue M, Jolles J, Jolles P, Montreuil J, Spik J. Characterization and localization of an iron-binding 18-kDa glycopeptide isolated from the N-terminal half of human lactotransferrin. Biochim Biophys Acta. 1984;787:90–96.

[49] Legrand D, Salmon V, Coddeville B, Benaissa M, Plancke Y, Spik G. Structural determination of two N-linked glycans isolated from recombinant human lactoferrin expressed in BHK cells. FEBS Lett. 1995;365:57–60.

[50] Suck D. DNA recognition by DNase I. J Mol Recognit. 1994;7:56–70.

[51] Bernardi G. Spleen acid deoxyribonuclease. In: Boyer PD, editor. The Enzymes. 3rd ed. New York, NY: Academic Press; 1971. p. 271–287.

[52] Babina SE, Kanyshkova TG, Buneva VN, Nevinsky GA. Lactoferrin is the major deoxyribonuclease of human milk. Biochemistry (Moscow). 2004;69:1006–1015.

[53] Pollack SJ, Jacobs JW, Schultz PG. Selective chemical catalysis by an antibody. Science 1986;234:1570–1573.

[54] Tramontano A, Janda KD, Lerner RA. Catalytic antibodies. Science 1986;234:1566–1570.

[55] Tramontano A, Janda KD, Lerner RA. Chemical reactivity at an antibody binding site elicited by mechanistic design of a synthetic antigen. Proc Natl Acad Sci U S A 1986;83:6736–6740.

[56] Lerner RA, Tramontano A. Antibodies as enzymes. Trends Biochem Sci. 1987;12:427–438.

[57] Stewart JD, Benkovic SJ. Recent developments in catalytic antibodies. Int Rev Immunol. 1993;10:229–240.

[58] Martin AB, Schultz PG. Opportunities at the interface of chemistry and biology. Trends Cell Biol. 1999;9:24–28.

[59] Nevinsky GA, Semenov DV, Buneva VN. Catalytic antibodies (abzymes) induced by stable transition-state analogs. Biochemistry (Moscow). 2000;65:1233–1244.

[60] Keinan EE, editor. Catalytic antibodies. Germany: Wiley–VCH Verlag GmbH and Co. KgaA; 2005. p. 586. DOI: 10.1002/3527603662

[61] Nevinsky GA. Natural catalytic antibodies in norm and in HIV-infected patients. In: Fyson Hanania Kasenga, editor. Understanding HIV/AIDS Management and Care – Pandemic Approaches the 21st Century. Rijeka, Croatia: InTech; 2011. p. 151–192.

[62] Nevinsky GA, Buneva VN. Autoantibodies and natural catalytic antibodies in health, multiple sclerosis, and some other diseases. Adv Neuroimmune Biol. 2012;3:157–182. DOI: 10.3233/NIB-2012-012042

[63] Buneva VN, Kudryavtseva AN, Gal'vita AV, Dubrovskaya VV, Khokhlova OV, Kalinina IA, Galenok VA, Nevinsky GA. Dynamics of antibody nuclease activity in blood of women during pregnancy and lactation. Biochem (Moscow). 2003;68:890–900.

[64] Fey HR, Burtler R, Marti F. The production in the pregnant cow of anti-human immunoglobulin to be used for the antiglobulin test. Vox Sang. 1973;25:245–253.

[65] Mestecky J, McGhee JR. Immunoglobulin A (IgA): molecular and cellular interactions involved in IgA biosynthesis and immune response. Adv Immunol. 1987;40:153–245.

[66] Kazakov VI, Bozhkov VM, Linde VA, Repina MA, Mikhailov VM. Extracellular DNA of pregnant women blood. Tsitologiia (Moscow). 1995;37:232–235.

[67] Newman J. How breast milk protects newborns. Sci Am. 1995;273:76–79.

[68] Mikhailov VM, Linde VA, Rozanov IM, Kottsova NA, Susloparov LA, Konycheva EA. DNA synthesis and content in human decidual cells at different stages of differentiation based on flow cytometry data. Tsitologiia (Moscow). 1992;34:67–71.

[69] Andryushkova AA, Kuznetsova IA, Orlovskaya IA, Buneva VN, Nevinsky GA. Antibodies with amylase activity from the sera of autoimmune-prone MRL/MpJ-lpr mice. FEBS Lett. 2006;580:5089–5095. DOI: 10.1016/j.febslet.2006.08.036

[70] Andryushkova AS, Kuznetsova IA, Orlovskaya IA, Buneva VN, Nevinsky GA. Nucleotide-hydrolyzing antibodies from the sera of autoimmune-prone MRL-lpr/lpr mice. Int Immunol. 2009;21:935–945. DOI: 10.1093/intimm/dxp060

[71] Andryushkova AS, Kuznetsova IA, Buneva VN, Toporkova LB, Sakhno LV, Tichonova MA, Chernykh ER, Orlovskaya IA, Nevinsky GA. Formation of different abzymes in autoimmune-prone MRL-lpr/lpr mice is associated with changes in colony formation of haematopoetic progenitors. J Cell Mol Med. 2007;11:531–551. DOI: 10.3233/NIB-2012-012042

[72] Savel'ev AN, Kanyshkova TG, Kulminskaya AA, Buneva VN, Eneyskaya EV, Filatov MV, Nevinsky GA, Neustroev KN. Amylolytic activity of IgG and sIgA immunoglobulins from human milk. Clin Chim Acta. 2001;314:141–152.

[73] Kanyshkova TG, Semenov DV, Khlimankov D, Buneva VN, Nevinsky GA. DNA-hydrolyzing activity of the light chain of IgG antibodies from milk of healthy human mothers. FEBS Lett. 1997;416:23–26.

[74] Nevinsky GA, Kanyshkova TG, Semenov DV, Vlassov AV, Gal'vita AV, Buneva VN. Secretory immunoglobulin A from healthy human mothers' milk catalyzes nucleic acid hydrolysis. Appl Biochem Biotechnol. 2000;83:115–129.

[75] Buneva VN, Kanyshkova TG, Vlassov AV, Semenov DV, Khlimankov D, Breusova LR, Nevinsky GA. Catalytic DNA- and RNA-hydrolyzing antibodies from milk of healthy human mothers. Appl Biochem Biotechnol. 1998;75:63–76.

[76] Semenov DV, Kanyshkova TG, Kit YY, Khlimankov DY, Akimzhanov AM, Gorbunov DA, Buneva VN, Nevinsky GA. Human breast milk immunoglobulins G hydrolyze nucleotides. Biochemistry (Moscow). 1998;63:944–951.

[77] Semenov DV, Kanyshkova TG, Karataeva NA, Krasnorutskii MA, Kuznetsova IA, Buneva VN, Nevinsky GA. Catalytic nucleotide-hydrolyzing antibodies in milk and serum of clinically healthy human mothers. Med Sci Monit. 2004;10:BR23–BR33.

[78] Kit Yu, Semenov D, Nevinskii GA. Do catalytically active antibodies exist in healthy people?. Mol Biol. 1995;29:893–906.

[79] Kit YuYa, Semenov DV, Nevinsky GA. Phosphorylation of different human milk proteins by human catalytic secretory immunoglobulin A. Biochem Mol Biol Int. 1995;39:521–527.

[80] Nevinsky GA, Kit Y, Semenov DV, Khlimankov D, Buneva VN. Secretory immunoglobulin A from human milk catalyses milk protein phosphorylation. Appl Biochem Biotechnol. 1998;75:77–91.

[81] Gorbunov DA, Semenov DV, Shipitsin MV, Nevinsky GA. Unusual phospholipids of human breast milk. Dokl Biochem Biophys. 2001;377:62–64.

[82] Gorbunov DV, Semenov DV, Shipitsin MV, Kit YY, Kanyshkova TG, Buneva VN, Nevinsky GA. Phosphorylation of minor lipids of human milk tightly bound to secretory immunoglobulin A. Rus J Immunol. 2000;5:267–278.

[83] Gorbunov DV, Karataeva NA, Buneva VN, Nevinsky GA. Lipid kinase activity of antibodies from milk of clinically healthy human mothers. Biochim Biophys Acta 2005;1735:153–166.

[84] Karataeva NA, Gorbunov D, Prokudin IV, Buneva VN, Kulminskaya AA, Neustroev KN, Nevinsky GA. Human milk antibodies with polysaccharide kinase activity. Immunol Lett. 2006;103:58–67.

[85] Karataeva NA, Buneva VN, Nevinsky, GA. Polysaccharide kinase activity of human milk IgG antibodies. Biochemistry (Moscow). 2006;71:1488–1504.

[86] Karataeva NA, Nevinsky A. Enzymes phosphorylating lipids and polysaccharides. Biochemistry (Moscow). 2007;72:357–379.

[87] Amino N, Mori H, Iwatani Y, Tanizawa O, Kawashima M, Tsuge I, Ibaragi K, Kumahara Y, Miyai K. High prevalence of transient post-partum thyrotoxicosis and hypothyroidism. N Engl J Med. 1982;306:849–852.

[88] Tanaka A, Lindor K, Ansari A, Gershwin ME. Fetal microchimerisms in the mother: immunologic implications. Liver Transpl. 2000;6:138–143.

[89] Dayan CM, Daniels GH. Chronic autoimmune thyroiditis. N Engl J Med. 1996;335:99–107.

[90] Freeman R, Rosen H, Thysen B. Incidence of thyroid dysfunction in an unselected post-partum population. Arch Intern Med. 1986;146:1361–1364.

[91] Kozyr AV, Sashchenko LP, Kolesnikov AV, Zelenova NA, Khaidukov SV, Ignatova AN, Bobik TV, Gabibov AG, Alekberova ZS, Suchkov SV, Gnuchev NV. Anti-DNA autoantibodies reveal toxicity to tumor cell line. Immunol Lett. 2002;80:41–47.

[92] Nevinsky GA, Buneva VN. Catalytic antibodies in healthy humans and patients with autoimmune and viral pathologies. J Cell Mol Med. 2003;7:265–276.

[93] Sinohara H, Matsuura K. Does catalytic activity of Bence-Jones proteins contribute to the pathogenesis of multiple myeloma? Appl Biochem Biotechnol. 2000;83:85–94.

[94] van de Perre P. Transfer of antibody via mother's milk. Vaccine. 2003;21:3374–3376.

[95] Nevinsky GA, Buneva VN. Natural catalytic antibodies – abzymes. In: Keinan E, editor. Catalytic Antibodies. Weinheim: Wiley–VCH Verlag GmbH and Co. KgaA; 2005. p. 503–567.

[96] Nevinsky GA. Natural catalytic antibodies in norm and in autoimmune diseases. In: Brenner KJ, editor. Autoimmune Diseases: Symptoms, Diagnosis and Treatment. Nova Science Publishers, New York; 2010. p. 1–107.

[97] Rob C, Aalberse RC, Schuurman J. IgG4 breaking the rules. Immunology. 2002;105:9–19.

[98] van der Neut Kolfschoten M, Schuurman J, Losen M, Bleeker WK, Martínez-Martínez P, Vermeulen E, den Bleker TH, Wiegman L, Vink T, Aarden LA, De Baets MH, van de Winkel JG, Aalberse RC, Parren PW. Anti-inflammatory activity of human IgG4 antibodies by dynamic Fab arm exchange. Science. 2007;317:1554–1557. DOI: 10.1126/science.1144603

[99] Schuurman J, Van Ree R, Perdok GJ, Van Doorn HR, Tan KY, Aalberse RC. Normal human immunoglobulin G4 is bispecific: it has two different antigen-combining sites. Immunology. 1999;97:693–698.

[100] Rispens T, den Bleker TH, Aalberse RC. Hybrid IgG4/IgG4 Fc antibodies form upon 'Fab-arm' exchange as demonstrated by SDS-PAGE or size-exclusion chromatography. Mol Immunol. 2010;47:1592–1594.

[101] Labrijn AF, Buijsse AO, van den Bremer ET, Verwilligen AY, Bleeker WK, Thorpe SJ, Killestein J, Polman CH, Aalberse RC, Schuurman J, van de Winkel JG, Parren PW. Therapeutic IgG4 antibodies engage in Fab-arm exchange with endogenous human IgG4 in vivo. Nat Biotechnol. 2009;27:767–771. DOI: 10.1038/nbt.1553

[102] Notkins AL. Polyreactivity of antibody molecules. Trends Immunol. 2004;25:174–179.

[103] Sedykh SE, Buneva VN, Nevinsky GA. Human milk IgGs contain various combinations of different antigen-binding sites resulting in multiple variants of their bispecificity. PLoS One. 2012;7:e42942. DOI: 10.1371/journal.pone.0042942

[104] Sedykh SE, Buneva VN, Nevinsky GA. Human milk sIgA molecules contain various combinations of different antigen-binding sites resulting in a multiple binding specificity of antibodies and enzymatic activities of abzymes. PLoS One. 2012;7:e48756. DOI: 10.1371/journal.pone.0048756

[105] Diaw L, Siwarski D, DuBois W, Jones G, Huppi K. Double producers of kappa and lambda define a subset of B cells in mouse plasmacytomas. Mol Immunol. 2000;37:775–781.

[106] Kwan SP, Max EE, Seidman JG, Leder P, Scharff MD. Two kappa immunoglobulin genes are expressed in the myeloma S107. Cell. 1981;26:57–66.

[107] Lekchnov EA, Sedykh SE, Dmitrenok PS, Buneva VN, Nevinsky GA. Human placenta: relative content of antibodies of different classes and subclasses (IgG1–IgG4) containing lambda- and kappa-light chains and chimeric lambda–kappa-immunoglobulins. Int Immunol. 2015;27:297–306. DOI: 10.1093/intimm/dxv003

[108] Lonnerdal B. Nutritional and physiologic significance of human milk proteins. Am J Clin Nutr. 2003;77:1537S–1443S.

[109] Gifford JL, Ishida I, Vogel HJ. Structural characterization of the interaction of human lactoferrin with calmodulin. PLoS One. 2012;7:e51026.

[110] Anema SG, de Kruif CG. Interaction of lactoferrin and lysozyme with casein micelles. Biomacromolecules. 2011;12:3970–3976.

[111] Lampreave F, Pineiro A, Brock JH, Castillo H, Sanchez L, Calvo M. Interaction of bovine lactoferrin with other proteins of milk whey. Int J Biol Macromol. 1990;12:2–5.

[112] Sabatucci A, Vachette P, Vasilyev VB, Beltramini M, Sokolov A, Pulina M, Salvato B, Angelucci CB, Maccarrone M, Cozzani I, Dainese E. Structural characterization of the ceruloplasmin: lactoferrin complex in solution. J Mol Biol. 2007;371:1038–1046.

[113] Soboleva SE, Dmitrenok PS, Verkhovod TD, Buneva VN, Sedykh SE, Nevinsky GA. Very stable high molecular mass multiprotein complex with DNase and amylase activities in human milk. J Mol Recognit. 2015;28:20–34. DOI: 10.1002/jmr.2409

10

An Important Milk Enzyme: Lactoperoxidase

Zeynep Koksal, Ilhami Gulcin and Hasan Ozdemir

Abstract

It has been acknowledged since ancient times that milk and dairy products have a vital role in nutrition and contribute considerably to human health. Because of its content, Because of its content, it has many important effects on the life that include immunoglobulins, enzymes, hormones, growth factors, antibacterial agents, fat acids, vitamins, and minerals. Approximately 70 indigenous enzymes have been reported in normal bovine milk, including lactoperoxidase. Lactoperoxidase LPO is a natural constituent of milk, saliva, and tears. It also exists in all mammary milk. LPO is an iron heme group basic glycoprotein, with a molecular weight of approximately 78 kDa. The LPO enzyme catalyzes the $H_2O_2 + SCN^- \rightarrow OSCN^- + H_2O$ reaction. Hydrogen peroxide and hypothiocyanate are indispensable for antimicrobial activity. The biological significance of LPO is involved in the natural host defense system against pathogenic microorganisms.

Keywords: milk, protein, lactoperoxidase, enzyme, LPO system, LPO inhibition

1. Introduction

1.1. Peroxidases

Peroxidases (POD: H_2O_2-Oxydoreductase E.C.1.11.1.7) are oxidoreductase enzymes, which catalyze the reactive oxygen species generated during metabolism, and are converted into harmless molecules [1]. These exhibit antioxidant characteristics and catalyze the oxidation of organic and inorganic substrates with hydrogen peroxide being the electron acceptor [2]. Those enzymes are present in eukaryotes, prokaryotes, and photosynthetic cells [3, 4].

LPO is generally found in mammalians such as human [5, 6], bovine [7], buffalo [8], goat [9], sheep [10], llama, cow, camel, and mice milk [6, 11], saliva [12], tears [13], and mammary, salivary, and lachrymal glands [6, 14].

Peroxidases are frequently used in the studies of metabolic reactions, enzymatic functions, protein structures [15] and in clinical diagnoses, microanalytic applications, and the food and drug industry [16, 17]. Mammalian POD enzymes are localized in milk, saliva, and tears as lactoperoxidases (LPO) [18] and in leukocytes and platelets as myeloperoxidases [6].

Prosthetic groups of peroxidase are protoheme and are connected to the apoprotein loosely in contrast to many hemoprote [14]. Firstly, protein portion is synthesized in the organisms bearing peroxidases. However, enzymes are not functionally active. The enzyme gets activated by both apoprotein and hem groups [3, 19]. The general formula of the reaction catalyzed by peroxidases is shown below [3].

$$ROOH + AH_2 \rightarrow ROH + H_2O + A$$

The H_2O_2 formed during the metabolism having oxidizing property must be quickly removed. The catalase and peroxidase enzymes exhibiting antioxidant properties play this role in cells [20]. The amount of hydrogen peroxide in the cells is removed by catalase in peroxisomes. In other parts of the cells, the peroxidase enzymes utilize various aromatic components as the substrate to [21] neutralize H_2O_2 [22].

The characteristics of peroxidase enzyme that is isolated from milk are similar to animal and human peroxidases [23]. It displays 55, 54, and 45% similarity with myeloperoxidase (MPO), eosinophilperoxidase (EPO), and thyroidperoxidase (TPO), respectively. Following xanthine oxidase, lactoperoxidase is the most common enzyme in milk and it is also commonly present in whey which is the liquid remaining after milk has been curdled and strained. Each lactoperoxidase enzyme contains an iron molecule. The conformation of the protein is stabilized by a chelated calcium ion [24–26].

The main function of the LPO enzyme is to catalyze the oxidation of thiocyanate with H_2O_2 to hypothiocyanite having antimicrobial activity [27, 28].

$$SCN^- + H_2O_2 \xrightarrow{LPO} OSCN^- + H_2O$$

The combined action of these three components, which constitute the 'lactoperoxidase system', was defined by Reiter and coworkers [29–31]. The biological significance of the system is protection of the lactating mammary gland and the intestinal tract of newborn infants, thereby providing a natural host defence system against invading microorganisms [23]. The LPO catalyzes the oxidation of some halides (I_2 and Br_2 but not Cl_2) to yield the most oxidizable form of I^-, that is, I_2 [26].

$$H_2O_2 + 2ABTS^{+} \xrightarrow{LPO} 2ABTS^{+} + 2H_2O$$

The LPO-H_2O_2-SCN^- system shows bacteriostatic effect and thus prevents bacterial growth and development, whereas the LPO-H_2O_2-I_2 system shows bactericidal potency thus killing bacteria. Both SCN^- and I_2 have shown strong bactericidal effect when present within the system [32, 33].

1.2. Lactoperoxidase enzyme

Lactoperoxidase (LPO, E.C. 1.11.1.7) is one of the crucial enzymes in milk with oxidoreductase activity. The peroxidase isolated from milk was given the name lactoperoxidase [23] and was the first enzyme reported to be found in milk [34]. The main function of the enzyme is to catalyze the oxidation of molecules in the presence of hydrogen peroxide and to help production of products with a wide antimicrobial activity. Pseudohalogens, thiocyanates, or halogens should function as second substrates for the enzyme to exhibit such antimicrobial effects [23, 29].

The LPO system shows a significant protective effect in bovine milk. The activation of the system depends on the concentration of the two reactants, thiocyanate and hydrogen peroxide. In the presence of hydrogen peroxide, the system catalyzes the transformation of thiocyanate into hypothiocyanate, which has an antibacterial nature [35–37]. The end products of these compounds are oxidized and hence are safe for human health.

Many studies show that this system destroys several bacterial and fungal strains [38–42]. Lactoperoxidase has a broad antifungal activity [43, 44]. Mastitis is a bacterial inflammation in mammals. The effects of different concentrations of thiocyanate-H_2O_2 medium on several antibacterial and antifungal strains were studied to solve this dairy industry issue [45, 46]. They are capable of reducing bacterial growth by damaging the cell membranes and inhibiting activities of several cytoplasmic enzymes.

The LPO enzyme, a glycoprotein consisting of 8–10% carbohydrate, comprises a chain containing 612 amino acids. It consists of a single polypeptide chain of molecular weight approximately 78 kDa [47]. It is a basic protein containing heme as its prosthetic group with an isoelectric pH value of 9.2 [24–26]. Furthermore, it is very active in acidic pH [48]. It is fairly voluminous as a LPO molecule [49]. The Ca^{2+} ion stabilizes the enzyme. The Ca^{2+} ion disappears under pH 5.0 and thus reduces the stability of the enzyme [26].

The biocidal activity of the LPO results from the products of the chemical reactions that it catalyzes. Hypothiocyanate, which is the main product of the reaction, interacts with the thiol groups of various proteins, which is critical for the survival of pathogens. The impact of LPO on bacteria results from the oxidation of sulfhydryl. The oxidation of the -SH groups makes the bacterial cytoplasmic membrane lose its ability to transport glucose, potassium ions, amino acids, and peptides [45].

The biological significance of this enzyme results from the fact that it has a natural protection system against the invasion of microorganisms. Besides this antiviral effect, it is reported that it protects animal cells against various damages and peroxidative effects [23, 29–31]. Lactoperoxidase is a significant agent of the defense system against pathogen microorganisms from the digestive system of neonatal babiess. The LPO enzyme functions as a natural compound of the non-immune biological defense system of mammals and it catalyzes the oxidation of the thiocyanate ion into the antibacterial hypothiocyanate [52].

Although peroxidases have similar catalytic mechanisms, they are distinguished for their ability to oxidize halides and pseudohalides. For example, only myeloperoxidase is able to oxidize bromine, iodine, and chlorine at neutral pH. Lactoperoxidase, on the contrary, can only oxidize iodine and thiocyanate but it either cannot or can hardly oxidize brome under the same conditions (**Figure 1**) [53].

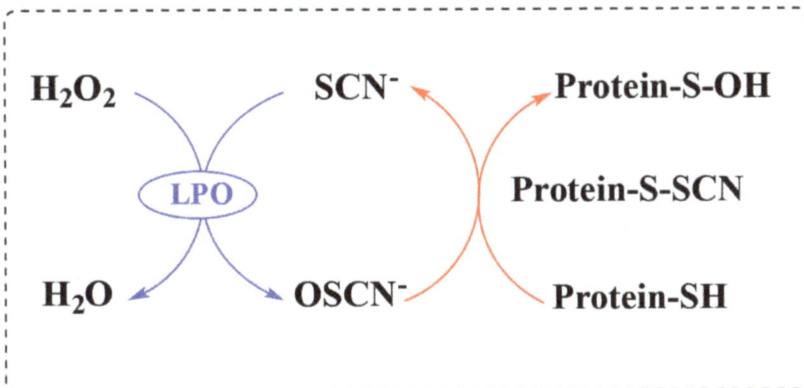

Figure 1. The activation mechanism of the peroxidase enzyme.

In the first step of this mechanism, the peroxidase enzyme reacts with one equivalent of the peroxide to form an Fe (IV) containing compound I that is the porphyrin cation radical. In the second step, the cation radical takes a proton of the substrate and reduced (substrate)-Fe (IV) form, the substrate becomes radical to lost a proton. The compound II takes a proton from the substrate and return the first reducing form. Also radicalic substrate which are formed with interacting each other are polymerization [54].

The LPO enzyme is covalently bound to the heme group proteins, with the bond occurring between the hydroxyl group of the heme group and the carboxyl group of the protein [14]. Approximately 10% of the molecule is composed of carbohydrates and the molecule contains five main potential glycosylation regions and 15 semi-cysteine residues [11, 23, 45]. The heme group at the catalytic center is protoporphyrin IX, which is covalently bound to the polypeptide chain along with the disulfide bridge. The iron compound, which is a part of the heme group, constitutes 0.07% of LPO. The calcium ion is tightly bound to the enzyme, which ensures the molecular conformation and structural integrity of the enzyme [55].

1.3. Lactoperoxidase system in milk

The activation of the natural antibacterial system has been adopted for protecting raw milk. This system is defined as the lactoperoxidase/thiocyanate/hydrogen peroxide (LPO) system.

The antibacterial effect of the system on milk is based on the oxidation of SCN^- ions catalyzed by the lactoperoxidase enzyme in the presence of H_2O_2. The short-lived components, $OSCN^-$ ions, generated in this oxidation reaction have bacteriostatic effect [56, 57]. The system is recommended in developing countries where there are no sufficient cooling facilities for collection of raw milk and their transportation to processing centers [57].

1.4. Antibacterial effect of LPO

LPO oxidized the -SH groups of the enzymes in bacteria such as hexokinase and glyceraldehyde-3-phosphate dehydrogenase and tends to lose biological functions of these enzymes. As a result, the bacterial cytoplasmic membranes are damaged structurally, and glucose, purine, pyrimidine, and amino acid uptake as well as protein, DNA, and RNA synthesis are blocked. Thus, bacteria growth and proliferation is prevented (**Figure 2**) [58].

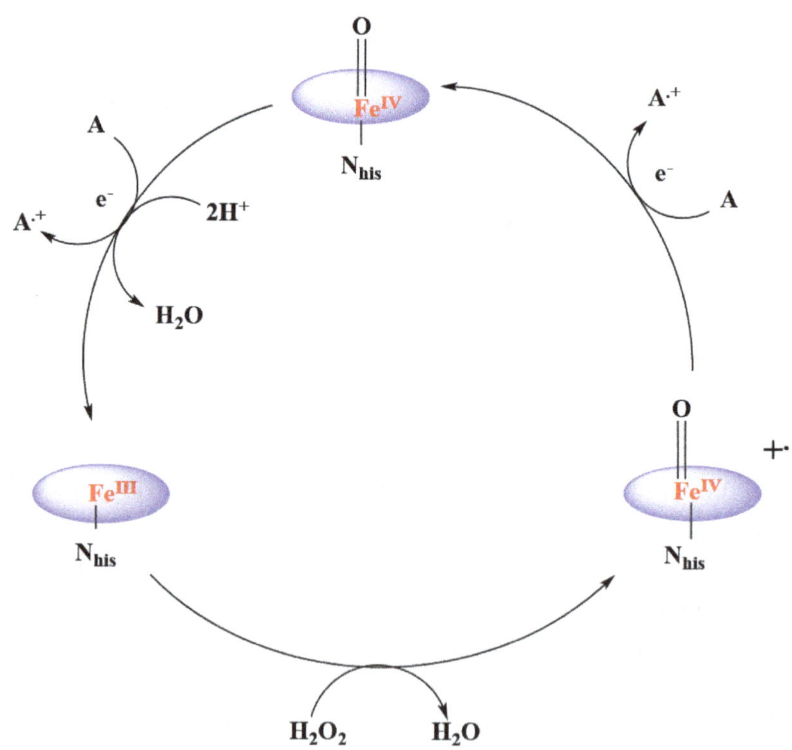

Figure 2. Antibacterial effect of the lactoperoxidase enzyme.

1.5. Lactoperoxidase system and health

Milk and milk products area rich sources of mineral, protein, vitamin, nutritional elements [56, 57]. Lactoperoxidase showed antifungal activity in apple juice and salt solution. In vitro findings also show that the LPO/H_2O_2/halide system has a strong virucidal activity against HIV-1 [59].

In conditions of iodine deficiency, the level of SCN ions in milk plays a vital role in thyroid function. The studies showed that when milk containing 19 ml/L thiocyanate ion and iodine at 0.1 mg/l concentration is consumed, the thyroid functions of patients with iodine deficiency did not have any negative symptoms [60].

1.6. Determination of the lactoperoxidase activity with spectrophotometer

The measurement of the activity is based on the oxidation of 2, 2'-azino-bis(3-ethylbenzthiazoline-6-sulfonic acid) (ABTS) chromogenic substrate by H_2O_2 and observation of the increase in the absorbance caused by the resultant colored compound at 412 nm and pH: 6.0 [61]. For determination of LPO activity with spectrophotometric assay, 2.8 mL of ABTS (1 mM) and 0.1 mL of H_2O_2 (3.2 mM) were pipetted in the spectrophotometer tube of 3 mL. Enzyme solution of 0.1 mL was added and the tube was turned upside down before it was placed in the spectrophotometer. The increase in the absorbance was observed against blank at 412 nm for 3 min and recorded every 60 sec. Phosphate tampon (0.1 M) at pH: 6.0 was used as blank, instead of the enzyme, while all other solutions were used at the same rates. The following formula was used to determine the activity:

$$A = \varepsilon.b.c \qquad c = A/\varepsilon.b \qquad V = c.D_f$$

A	: Absorbance (absorbance read at the end of 1 minute)
b	: Path length (1 cm)
c	: Concentration (μmol/mL)
ε	: Extinction coefficient (32400 $M^{-1} \times cm^{-1}$)
D_f	: Dilution coefficient
V	: Velocity of the reaction (μmol/mL.min.)

1.7. Unit of enzyme activity

One LPO unit (EU) is defined as the amount of LPO that catalyzes the oxidation of 1 μmol of substrate (ABTS) per min at 20°C [52].

1.8. Substrates

2,2′-azino-bis(3-etilbenztiazolin-6-sülfonik asit (ABTS) [62], p-phenylenediamine [63], pyrogallol, guaiacol, catechol, phenols, aromatic amines, ascorbates, epinephrine, and tetramethylbenzidine [6, 14, 47, 52, 61].

1.9. Purification procedures

To date, several chromatographic methods have been reported about purification and characterization of the LPO enzyme from bovine milk [14]. For example, CM-Cellulose [64], CM-Sephadex ion-exchange chromatography [46, 64], Sephadex G-100 gel filtration chromatography [6, 64], hydrophobic affinity chromatography on Phenyl-Sepharose CL-4B [5], and Toyopearl-SP cation exchange chromatography [64] are among the methods used in the purification of LPO enzyme from bovine milk. LPO was purified in one stage using the affinity technique and sulfanilamide was used as the ligand [65].

1.10. Kinetic studies

The K_m and V_{max} values are suitable parameters for kinetic studies. K_m is a substrate concentration at which half of the enzyme active sites are filled. V_{max} is an expression of the catalytic activity of the enzyme.

To find K_m and V_{max} values for LPO, activity was measured at 20°C, 412 nm, at pH 6.0, for five different substrate concentrations. For this purpose, generally 0.2–1.5 mL volume from the stock solution of the substrates was used. The total volume was made up to 2.8 mL with a buffer solution and then 0.1 mL enzyme and 0.1 mL H_2O_2were added. The K_m and V_{max} values were calculated from the Lineweaver-Burk graph [65, 66].

1.11. Procedure of enzyme inhibition

The K_i and IC_{50} values show the inhibitory effect on enzyme. The K_i and IC_{50} values depend on the inhibitory mechanism. IC_{50} is the inhibitor concentration required for 50% inhibition and K_i value is the constant.

To investigate the inhibitory effects of some inhibitors on LPO and to determine the IC_{50} values, the LPO activity was measured in the presence of five different concentrations of inhibitor. For example, the experimental procedure for inhibitor: A control sample without inhibitor was taken as 100% and an activity-[Inhibitor] plot was drawn. To determine the K_i, three different concentrations were used for x inhibitor. ABTS was also used as a substrate at five different concentrations. Lineweaver-Burk plots (1/V-1/[S]) were obtained for inhibitor; the K_i and the inhibition type were calculated from these plots [65].

1.12. Selection of the LPO inhibitors and the ligand

The inhibitors of the LPO enzyme were identified in studies on the enzyme [52, 67]. Non-selective monoamine reuptake inhibitors consist of opipramol, lofepramime, dibenzepin, protriptyline, melitracen, butriptyline, dimetacrine, dosulepin, and quinipramine; selective serotonin reuptake inhibitors include alaproclate and etoperidone; non-selective monoamine oxidase inhibitors comprise moclobemide, toloxatone, and isocarboxazid; and other antidepressants are viloxazine, minaprine, bifemelane, oxaflozane, and medifoxamine (**Table 1**).

Many inhibitors such as sulfanilamides [65], propofol and derivatives [68, 69], some anesthetic drugs [70], some bacteria [46], some phenolic acid compounds and phenolics [71, 77], avermectins [73], adrenaline, melatonin, serotonin and norepinephrine [45, 73, 74], fungi and bacteria [58], antibiotics [75], hydrazines [52], and some thiocarbamide compounds [67] are assayed and reported as a LPO inhibitor in the literature.

Inhibitor	IC_{50}	K_i	Inhibition type	Publication
L-Adrenaline	34.5 mM	2.26 mM	Noncompetitive	Sisecioğlu et al. [74]
Ceftazidime pentahydrate	0.048 mM	0.018 ±0.0035 mM	Competitive	Sisecioğlu et al. [76]
Prednisolone	0.053 mM	0.019 ±0.0005 mM	Competitive	
Amikacin sulfate	0.26 mM	0.04 ±0.015 mM	Competitive	
Ceftriaxone sodium	0.29 mM	0.10 ±0.055 mM	Competitive	
Teicoplanin	1.016 mM	0.13 ±0.022 mM	Competitive	
Melatonin	1.46 mM	0.82 ±0.28	Competitive	Sisecioğlu et al. [75]
Serotonin	1.29 mM	0.26 ±0.04	Competitive	
Norepinephrine	67.2 mM	62 mM	Noncompetitive	Sisecioğlu et al. 2010b
L-Ascorbic acid (Vitamin Q)	2.03 mM	0.508 ±0.257 mM	Competitive	Sisecioğlu et al. [43]
Menadione sodium Bisulfate (Vitamin K3),	0.025 mM,	0.0107 ±0.0044 mM,	Competitive	
Folic acid	0.0925 mM	0.0218 ±0.0019 mM	Competitive	
2,6-Dimethylphenol	836.67 nM	4442 nM		Koksal et al. 2014
2,6-Di-Tbutylphenol	10 nM	9 nM	Competitive	
Di(2,6-Dimethylphenol)	6.86 nM	0.53 nM.	Competitive	
Di(2,6-Di-Tbutylphenol)	185 nM	48.33 nM	Competitive	
Di(2,6-Diisopropylphenol)	154 nM	19.33 nM	Competitive	
Sulphanilamide	0.84 nM	3.57 nM	Competitive	Atasever et al. [65]
Emamectin benzoate	4.33 μM	6.82±2.60 μM	Competitive	Koksal et al. [73]
Eprinomectin	16.90 μM	4.80±1.95 μM	Competitive	
Moxidectin-Vetranal	99.00 μM	61.31±9.89 μM	Competitive	
Abamectin	138.60 μM	103.73±34.03 μM	Competitive	
Doramectin	173.20 μM	80.14±29.38 μM	Competitive	
Ivermectin	231.00 μM	519.97±47.62 μM	Noncompetitive	
Caffeic acid	393.61 nM	430.033±79.04 nM	Competitive	Gulcin et al. [72]
Ketamine	0.29 mM	0.019 ± 0.031	Noncompetitive	Ozdemir et al. [70]
Bupivacaine	0.155 mM	0.015 ±0.021 mM	Noncompetitive	

Table 1. Inhibitors of lactoperoxidase enzyme.

1.13. Concentration

LPO is the second most abundant whey enzyme in bovine milk, [31, 77] and its concentration is approximately 30 mg/L [23]. The peroxidase activity in cow milk is 20 times richer than in human milk and contains 1.2–19.4 units/mL LPO [78]. The mean LPO enzyme activity varies in different species, for example, 1.4 units/mL in cow, 0.34–2.38 units/mL in lamp, 1.5–4.45 units/mL in goat, 0.794 units/mL in buffalo, 22.0 units/mL in pig, and 0.06–0.97 units/mL in human [78].

1.14. Application fields

There is a growing interest in the purification of LPO with increasing applications. The LPO system's natural biological functions are preferred against antimicrobial chemicals. Lactoperoxidase has many fields of application. It is widely used especially in milk-processing facilities in the milk industry [79]. LPO is used in milk and cheese to reduce the microflora [23]. The LPO enzyme derived from various animal sources has a significant role in the suppression of bacterial growth and helps bacterial inhibition. Inhibition of bacterial growth by the bovine LPO is attributed to the peroxidase system, which contains H_2O_2 and thiocyanate [9]. The antimicrobial effect of the LPO system occurs naturally in milk. LPO has a bacteriostatic effect on gram-positive and gram-negative bacteria. The antibacterial studies on the LPO enzyme purified from camel milk show that the LPO-thiocyanate and peroxide system leads to significant inhibition of pathogenic bacteria.

Most popular applications of the system include food production for preservation of raw milk, pasteurized milk, and cheese milk during storage and/or transportation to the processing plants. The system can be used to avoid the suppression of acidity in yoghurt. In the absence of refrigeration, LPO system is preferable [80] and the system can be used to extend the shelf-life of pasteurized, raw, and cheese milk [79, 81]. This is used for the preservation of emulsions and cosmetics.

The LPO system can ensure an extensive spectrum of antimicrobial properties against bacteria and yeasts when it is composed of LPO, H_2O_2, SCN^-, and I_2. Hence, it is preferred in oral care products and cosmetic preservation [82, 83]. The system is used against fish pathogenic bacteria in aquaculture with strong bacterial effects.

Author details

Zeynep Koksal[1], Ilhami Gulcin[2*] and Hasan Ozdemir[2]

*Address all correspondence to: igulcin@atauni.edu.tr and igulcin@yahoo.com

1 Department of Chemistry, Faculty of Sciences, Istanbul Medeniyet University, Istanbul, Turkey

2 Department of Chemistry, Faculty of Sciences, Atatürk University, Erzurum, Turkey

References

[1] Davies KJ. Oxidative stress: the paradox of aerobic life. Biochemical Society Symposium. 1995;61:1–31.

[2] Hussain S, Slikker W, Ali SF. Age related changes in antioxidant enzymes, superoxide dismutase, catalase, glutathione peroxidase and glutathione in different region of mouse brain. International Journal of Developmental Neuroscience 1995;13:811–817. DOI: 10.1016/0736-5748(95)00071-2

[3] Van Huystee RB. Some molecular aspects of plant peroxidase biosynthetic studies. Annual Review of Plant Physiology. 1987;38:205. DOI: 10.1146/annurev.pp.38.060187.001225

[4] Tanakaa T, Satoa S, Kumuraa H, Shımazakıa K. Expression and characterization of bovine lactoperoxidase by recombinant baculovirus. Bioscience, Biotechnology and Biochemistry. 2003;67:2254–2261. DOI:10.1271/bbb.67.2254

[5] Langbakk B, Flatmark T. Lactoperoxidase from human colostrum. Biochemical Journal. 1989;592:627–631.

[6] Shin K, Hirotoshi H, Bo L. Purification and quantification of lactoperoxidase in human milk with use of immunoadsorbents with antibodies against recombinant human lactoperoxidase. The American Journal of Clinical Nutrition. 2001;73:984–989.

[7] Xiao L, Li C, Xin S. Isolation and purification of lactoperoxidase in bovine milk. Zhongguo Rupin Gongye. 2007;35:23–25.

[8] Ozdemir H, Hacibeyoglu HI, Uslu H. Purification of lactoperoxidase from creek-water buffalo milk and investigation of kinetic and antibacterial properties. Preparative Biochemistry and Biotechnology. 2002;32:143–155. DOI:10.1081/PB-120004127

[9] Jacob BM, Antony KE, Sreekumar B, Haridas M. Thiocyanate-mediated antifungal and antibacterial property of goat milk lactoperoxidase. Life Sciences. 2000;66:2433–2439. DOI:10.1016/S0024-3205(00)80003-X

[10] Joseph PV, Jacob C, Maridas M. Purification of sheep lactoperoxidase from colostrum: characterization and antifungal studies. International Journal of Chemical Sciences. 2004;2:120–125.

[11] Elagamy E, Ruppanner R, Ismail A, Champagne CP, Assal R. Antibacterial and antiviral activity of camel milk protective proteins. Journal of Dairy Research. 1992;59:169–175. DOI: 10.1017/S0022029900030417

[12] Welk A, Rudolph P, Kreth J, Schwahn C, Kramer A, Below H. Microbicidal efficacy of thiocyanate hydrogen peroxide after adding lactoperoxidase under saliva loading in the quantitative suspension test. Archives of Oral Biology. 2011;56:1576–1582. DOI: 10.1016/j.archoralbio.2011.04.016

[13] Galijasevic S; Hodzic E. Melatonın mediates classic lactoperoxidases catalytic mechanism. International Journal of Pharmaceutics. 2014;4:1–6.

[14] Doumonted C, and Rousset B. Identificatıon, purification and characterization of a nonhome lactoperoxidase in bovine milk. Journal of Biological Chemistry. 1983;258:14166.

[15] Hiraga S, Sasaki K, Ito H, Ohashi Y, Matsui H. A large family of class III plant peroxidases. Plant and Cell Physiology. 2001;42:462–468. DOI: 10.1093/pcp/pce061.

[16] Kalin R, Atasever A, Ozdemir H. Single-step purification of peroxidase by 4-aminobenzohydrazide from Turkish blackradish and Turnip roots. Food Chemistry. 2014;150:335–340. DOI:10.1016/j.foodchem.2013.10.125

[17] Kwak SS, Kim SK, Park IH, Liu JR. Enhancement of peroxidase activity by stress-related chemicals in sweet potato. Phytochemistry. 1996;43:565. DOI: 10.1016/0031-9422(96)00315-9

[18] Singh AK, Singh N, Sharma S, Shin K, Takase M, Kaur P, Srinivasan A, Singh TP. Inhibition of lactoperoxidase by its own catalytic product: crystal structure of the hypothiocyanate-inhibited bovine lactoperoxidase at 2.3-A° resolution. Biophysical Journal. 2009;96:646–654. DOI:10.1016/j.bpj.2008.09.019

[19] Fric F. Oxidative enzymes. Heitefuss R. et al. editors. Encyclophedia of Plant Physiology, 4, 6-17, New York. Physiological Plant Pathology Volume 4 of the series Encyclopedia of Plant Physiology. New York, Springer; 1976. 617–631 pp.

[20] Halliwell B, Gutteridge JMC, editors. Free radicals in biology and medicine. 2nd edn. Clarendon;1989. 86 p.

[21] Robert KM, Petter AM, Darly KG, Victor WR. Harper's biochemistry. Appleton and Lange. Stamford, CT; 2000. 869–870 pp.

[22] Scandalios JG. Oxygen stress and superoxide dismutases. Plant Physiology.1993;101:1–7.

[23] Reiter B, HaÈrnulv G. Lactoperoxidase antibacterial system: natural occurrence, biological functions and practical applications. Journal of Food Protection. 1984;47:724–732.

[24] Pourtois M, Binet C, Van Tieghem N, Courtois PR, Vandenabbeele A, Thirty L. Saliva can contribute in quick inhibition of HIV infectivity. AIDS. 1991;5:598–600.

[25] Atamer M, Kocak C, Cimer A, Odabasi S, Tamucay B, Yamaner N. Some quality characteristics of Kasar cheese manufactured from milk preserved by activation of lactoperoxidase/thiocyanate/hydrogen peroxide (LP) system. Milchwissenschaft. 1999;54:553–556.

[26] Kussendrager KD, van Hooijdonk ACM. Lactoperoxidase: physico-chemical properties, occurrence, mechanism of action and applications. British Journal of Nutrition. 2000;84:19–25. 10.1017/S0007114500002208

[27] Hoogendoorn H. Activation of the salivary peroxidase antimicrobial system: clinical studies. Pruitt KM, Tenovuo JO. Editors. In: The Lactoperoxidase System, Chemistry and Biological Significance. New York: Marcel; 1985. 217±227 pp.

[28] Pruitt KM, Jorma OT. The lactoperoxidase system. Chemistry and Biological Significance. Immunology Series (27), Proteins and Enzymes. New York: Dekker; 1985. 257 p.

[29] Reiter B, Perraudin JP. Lactoperoxidase: biological functions. In: Peroxydases in Chemistry and Biology. Boca Raton: CRC Press; 1991. 143–180 pp.

[30] Wolfson LM, Sumner SS. Antimicrobial activity of the lactoperoxidase system: a review. Journal of Food Protection. 1993;56:887–892.

[31] de Wit JN, van Hooydonk ACM. Structure, functions and applications of lactoperoxidase in natural antimicrobial systems. Netherlands Milk & Dairy Journal. 1996;50: 227±244.

[32] Thomas EL. Products of lactoperoxidase-catalyzed oxidation of thiocyanate and halides. KM Pruitt, JO Tenovuo, editors. In: The Lactoperoxidase System, Chemistry and Biological Significance. New York: Marcel Dekker; 1985. 31±53 p.

[33] Guthrie WG. A novel adaptation of a naturally occurring antimicrobial system for cosmetic protection. SOFW-Journal. 1992;118:556–562.

[34] Arnold C. Einige neue Reactionen der Milch. Archiv der Pharmazie. 1881;219:41–42. DOI: 10.1002/ardp.18812190108

[35] Gulcin I. Antioxidant activity of caffeic acid (3,4-dihydroxycinnamic acid). Toxicology. 2006a;217:213–220. DOI:10.1016/j.tox.2005.09.011

[36] Gulcin I, Mshvildadze V, Gepdiremen A, Elias R. Screening of antioxidant and antiradical activity of monodesmosides and crude extract from Leontice smirnowii Tuber. Phytomedicine. 2006b;13:343–351. DOI 10.1016/j.phymed.2005.03.009[26]

[37] Haddain MS, Ibrahim SA, Robinson RK. Preservation of raw milk by activation of the natural lactoperoxidase systems. Food Control. 1996;7:149–152. DOI: 10.1016/0956-7135(96)00023-0

[38] Gulcin I. Comparison of in vitro antioxidant and antiradical activities of L-tyrosine and L-Dopa. Amino Acids. 2007;32:431–438. DOI: 10.1007/s00726-006-0379-x

[39] Gulcin I. Antioxidant activity of L-Adrenaline: an activity-structure insight. Chemico-Biological Interactions. 2009;179:71–80. DOI:10.1016/j.cbi.2008.09.023

[40] Gulcin I. Antioxidant properties of resveratrol: A structure-activityinsight. Innov. Food Sci. Emerg. 2010;11:210–218. DOI:10.1016/j.ifset.2009.07.002

[41] Gulcin I Antioxidant activity of caffeic acid (3,4-dihydroxycinnamic acid). Toxicology. 2006a; 217: 213–220.

[42] Gulcin I, Elias R, Gepdiremen A, Taoubi K, Koksal E. Antioxidant secoiridoids from fringe tree (Chionanthus virginicus L.). Wood Science and Technology. 2009;43:195–212. DOI 10.1007/s00226-008-0234-1

[43] Sisecioğlu M, Cankaya M, Ozdemir H. Effects of some vitamins on lactoperoxidase enzyme activity. International Journal for Vitamin and Nutrition Research. 2009;79:188–194. DOI: http://dx.doi.org/10.1024/0300-9831.79.3.188

[44] Jacob BM, Monoj NK, Haridas M. Antibacterial property of goat milk lactoperoxidase. Indian Journal of Experimental Biology.1998;31:808.

[45] Sisecioglu M, Gulcin I, Cankaya M, Atasever A, Ozdemir H. The effects of norepinephrine on lactoperoxidase enzyme. Scientific Research and Essays. 2010b;5:1351–1356.

[46] Uguz MT, Ozdemir H. Purification of bovine milk lactoperoxidase and investigation of antibacterial properties at different thiocyanate mediated. Applied Biochemistry and Microbiology. 2005;41:397–401.

[47] Ozdemir H, Aygul I, Kufrevioglu OI. Purification of lactoperoxidase from bovine milk and investigation of the kinetic properties. Preparative Biochemistry and Biotechnology. 2001;31:125–134. DOI: 10.1081/PB-100103378

[48] Wever R, Kast WM, Kasinoedin JH, Boelens R. The peroxidation of thiocyanate catalysed by myeloperoxidase and lactoperoxidase. Biochimica et Biophysica Acta. 1982;709:212–219. DOI:10.1016/0167-4838(82)90463-0

[49] Sievers G. Structure of milk lactoperoxidase. A study using circular dischroism and difference absorption sperctroscopy. Biochimica et Biophysica Acta. 1980;624: 249.

[50] Pakkanen R, Aalto J. Growth factors and antimicrobial factors of bovine colostrum. International Dairy Journal. 1997;7:285–297. DOI:10.1016/S0958-6946(97)00022-8

[51] Scammel AW. Production and uses of colostrum. Journal of Dairy Technology. 2001;56: 74–82.

[52] Kumar R, Bhatla KL. Purification, crystallization and preliminary x-ray crystallographic analysis of lactoperoxidase from buffalo milk. Acta Crystallographica. 1995;51:1094. DOI: 10.1107/S0907444995004422

[53] Boots JW, Floris R. Review lactoperoxidase: from catalytic mechanism to practical applications. International Dairy Journal. 2006;16:1272–1276.

[54] Veitch NC. Horseradish peroxidase: a modern view of a classic enzyme. Phytochemistry. 2004;65:249–259.

[55] Booth KS, Kimura S, Lee HC, Ikeda-Saito M, Caughey WS. Bovine myeloperoxidase and lactoperoxidase each contain a high affinity binding site for calcium. Biochemical and Biophysical Research Communications. 1989;160:879±902. DOI: 10.1016/0006-291X(89)92519-9

[56] Smithers GW. Whey and whey proteins—from 'gutter-to-gold'. International Dairy Journal. 2008;18:695–704. DOI:10.1016/J.IDAIRYJ.2008.03.008

[57] Marshall K. Therapeutic applications of whey protein. Alternative Medicine Review. 2004;9:136–156.

[58] Sisecioglu M, Kirecci E, Cankaya M, Ozdemir H, Gulcin I, Atasever A. The prohibitive effect of lactoperoxidase system (LPS) on some pathogen fungi and bacteria. African Journal of Pharmacy and Pharmacology. 2010;4:671–677.

[59] Seifu E, Buys EM, Donkin EF. Significance of the lactoperoxidase system in the dairy industry and its potential applications: a review. Trends in Food Science & Technology. 2005;16:137–154. DOI:10.1016/j.tifs.2004.11.002

[60] Özen AE, Kılıç M. Peynir altı suyundan elde edilen serum proteinlerinin fonksiyonel özellikleri. Gıda Teknolojileri Elektronik Dergisi. 2007;3:45–49.

[61] Shindler JS, Bardsley WG. Steady-state kinetics of lactoperoxidase with ABTS as chromogen. Biochemical and Biophysical Research Communications. 1975;67:1307.

[62] Pruitt KM, Kamau DN. The lactoperoxidase system of bovine and human milk. In: DS Robinson, NAM Eskin, Editors. Oxidative Enzymes in Foods. London: Elsevier Applied Science; 1991. 133–174p.

[63] Aurand LW, Roberts WM, Cardwell JT. A method for the estimation of peroxidase activity in milk. Journal of Dairy Science. 1956;39:568–573.

[64] Shimazaki K, Nishio M, Kawano N. Separation of biologically active proteins from whey. Rakuno Kagaku, Shokuhin no Kenkyu. 1988;37: 45–51.

[65] Atasever A, Ozdemir H, Gülcin I, Küfrevioglu OI. One-step purification of lactoperoxidase from bovine milk by affinity chromatography. Food Chemistry. 2013;136: 864–870. DOI: 10.1016/j.foodchem.2012.08.072

[66] Gulcin I, Yildirim A. Purification and characterization of peroxidase from Brassica oleracea var. Acephala. Asian Journal of Chemistry. 2005;17:2175–2183.

[67] Doerge DR. Mechanism-based inhibition of lactoperoxidase by thiocarbamide goitro genens. Biochemistry. 1986;16:472. DOI: 10.1021/bi00410a026

[68] Sisecioglu M, Çankaya M, Gulcin I, Ozdemir H. The inhibitory effect of propofol on bovine lactoperoxidase. Protein and Peptide Letters. 2009;16:46–49. DOI: 10.2174/092986609787049394

[69] Köksal Z, Usanmaz H, Özdemir H, Gulcin I, Güney M. Inhıbıtıon effects of some phenolic and dimeric phenolic compounds on bovine lactoperoxidase LPO enzyme. International Journal of Academic Research. 2014;6:2.

[70] Ozdemir H, Uguz MT. In vitro effects of some anaesthetic drugs on lactoperoxidase enzyme activity. Journal of Enzyme Inhibition and Medicinal Chemistry. 2005;20:491–495. DOI:10.1080/14756360500225045

[71] Sarikaya SBO, Sisecioglu M, Cankaya M, Gulcin I, Ozdemir H. Inhibition profile of a series of phenolic acids on bovine lactoperoxidase enzyme. Journal of Enzyme Inhibition and Medicinal Chemistry. 2014:1–5.

[72] Gulcin I, Scozzafava A, Supuran CT, Akıncıoğlu H, Koksal Z, Turkan F, Elwasel S. The effect of caffeic acid phenethyl ester (CAPE) metabolic enzymes including acetylcholinesterase, butyrylcholinesterase, glutathione S-transferase, lactoperoxidase and carbonic anhydrase isoenzymes I, II, IX and XII. Journal of Enzyme Inhibition and Medicinal Chemistry. 2015:1–7. DOI: 10.3109/14756366.2015.1094470

[73] Koksal Z, Kalin R, Gülcin İ, Ozdemir H, Atasever A. The effect of some avermectins on lactoperoxidase from bovine milk. Food Proporties. 2015; DOI: 10.1080/10942912.2015.1076457.

[74] Sisecioğlu M, Gulcin I, Cankaya M, Ozdemir H. The inhibitory effects of L-adrenaline on lactoperoxidase enzyme purified from bovine milk. International Food Properties. 2012;15:1190–1199.

[75] Sisecioğlu M, Cankaya M, Gulcin I, Ozdemir H. Interactions of melatonin and serotonin with lactoperoxidase enzyme. Journal of Enzyme Inhibition and Medicinal Chemistry. 2010a;25:779–783.

[76] Sisecioğlu M, Uguz MT, Cankaya M., Ozdemir H, Gulcin I. Effects of ceftazidime pentahydrate, prednisolone, amikacin sulfate, ceftriaxone sodium and teicoplanin on bovine milk lactoperoxidase activity. Internatıonal Journal of Pharmacology. 2011;7:79–83. DOI: 10.3923/ijp.2011.79.83

[77] Pruitt KM, Kamau DN. The lactoperoxidase system of bovine and human milk. In: DS Robinson, NAM Eskin, Editors. Oxidative Enzymes in Foods. London: Elsevier Applied Science; 1991. 133–174p.

[78] Seifu E, Buys EM, Donkin EF. Significance of the lactoperoxidase system in the dairy industry and its potential applications: a review. Trends in Food Science & Technology. 2005;16:137–154. DOI:10.1016/j.tifs.2004.11.002

[79] Barrett NE, Grandison AS, Lewis MJ. Contribution of the lactoperoxidase system to the keeping quality of pasteurised milk. Journal of Dairy Research. 1999;66:73–80. DOI: 10.1017/S0022029998003252

[80] IDF. Code of practices for the preservation of raw milk by the lactoperoxidase system. Bulletin of the International Dairy Federation. 1988;234:1–15.

[81] Martinez CE, Mendoza PG, Alacron FJ, Garcia HS. Reactivation of the lactoperoxidase system during raw milk storage and its effect on the characteristics of pasteurised milk. Journal of Food Protection. 1988;51:558–561.

[82] Boots MicroCheck. Natural biocide comes fresh from the dairy. Cosmetics & Toiletries Manufacturers & Suppliers (CTMS). 1992;6(2):35.

11

Measurement of Casein Micelle Size in Raw Dairy Cattle Milk by Dynamic Light Scattering

Peter Hristov, Ivan Mitkov, Daniela Sirakova,
Ivan Mehandgiiski and Georgi Radoslavov

Abstract

The particle size of milk influences its microstructure and defines the qualities of dairy products, such as colloidal stability and texture. Moreover, differences in casein micelle size may affect milk processing, especially cheese making. Hence, the size of casein micelle is an important characteristic of raw milk and determines the yield of dairy products. The aim of the present research is to estimate the casein micelle size in the raw milk of dairy cattle by dynamic light scattering. The obtained results may be used for genetic elaboration of the breed, as well as to increase the competitiveness of the milk industry by selection of animals with higher casein micelle size.

Keywords: dairy cattle, casein micelle size, dynamic light scattering, DLS, cow milk

1. Introduction

It is well known that the casein fraction of bovine milk exists as polydisperse, large, roughly spherical colloidal particles, 50–600 nm in diameter (average ~150 nm), called "casein micelles" [1]. The size, form, and structure of the casein micelle are of great importance for the milk industry especially for cheese making, yellow cheese, etc. [2]. Casein micelle contains an average of 3.4 g H_2O per gram dry matter, which consists approximately of 93% protein and about 7% of inorganic component (Ca^{2+}), formed phosphoprotein complexes. The casein micelle in milk consists of four caseins: αs1- (CSN1S1), αs2- (CSNS2), β- (CSN2), and k-casein (CSN3). The latter is important for the stability and properties of the casein micelle. Although k-casein is in a relatively low amount of the casein system (12–15% of whole casein), it is soluble in the presence of Ca^{2+}, whereas the remaining 85% of casein are precipitated by Ca^{2+}. CSN3 stabilizes up to 10 times its weight of the Ca^{2+}-sensitive caseins via the formation of micelles [3] (**Figure 1**).

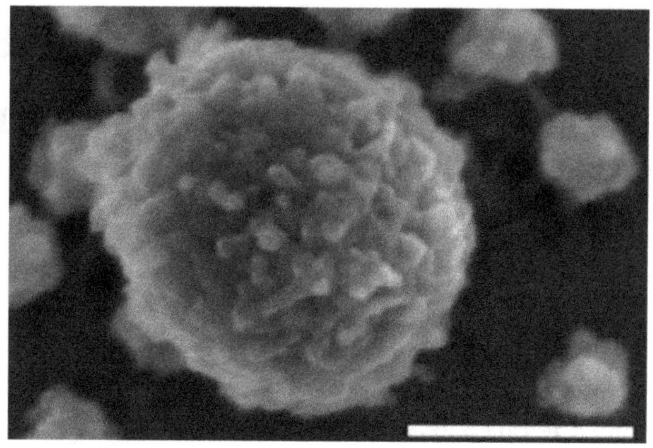

Figure 1. Electron micrograph of an individual casein micell [4].

One of the most important issues concerning the casein micelle is how the casein particles (micelles) in milk are stabilized and destabilized by the action of various agents and conditions. Thus, various hypotheses of casein micelle structure are proposed.

2. Casein micelle structure

2.1. Coat-core models

The first model of casein micelle structure was proposed by Waugh and Nobel in 1965 [5]. This model is based on the casein solubility in Ca^{2+} solutions. The model describes the formation of low-weight-ratio complexes of αs1- and k-caseins in the absence of calcium. The monomers of αs1- or β-caseins interact with a charged phosphate loop (**Figure 2**), then begin to aggregate

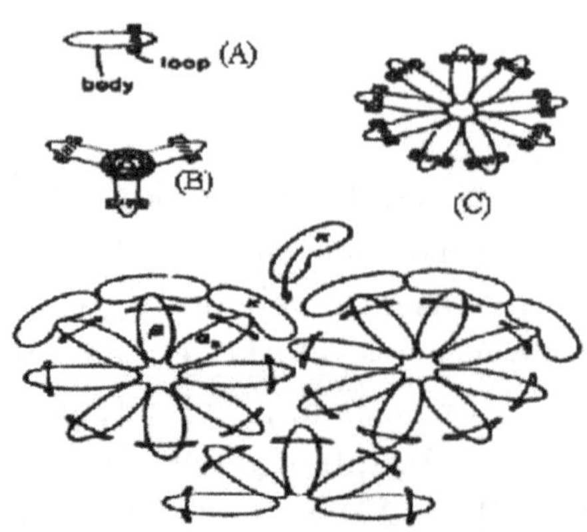

Figure 2. Structure of casein micelle proposed by Waugh [6]. (A) Monomer model for αs1- or β-caseins with charged loop. (B) A tetramer of αs1-casein monomers. (C) Planar model for a core polymer of αs1- and β-caseins [7].

to a limiting size (the caseinate core) while the calcium ions are added. Precipitation of the caseinate is stopped after a monolayer of the low-weight αs1-k-casein complexes is formed. This coat complex has the k-casein monomers spread out completely on the surface [8].

2.2. Sub-micelle (subunit) models

The first sub-micelle model was proposed by Morr in 1967 [9]. Morr considered that αs1-, β-, and k-casein monomers form small uniform submicelles. The submicelles, estimated by sedimentation velocity studies, are stabilized by hydrophobic bonding and calcium caseinate bridges, and the submicelles are aggregated and held together by colloidal calcium phosphate linkages with a micelle structure covered by αs1- and k-casein [7, 10].

The model, described by Slattery and Evard [11] and Slattery [12], falls in the last category. In this model, the caseins first aggregate via hydrophobic interaction into subunits of 15–20 molecules each. The pattern of interaction is such that it brings about a variation in the k-casein content of these submicelles. The k-casein congregate on the micelle surface, those submicelles poor (like αs and β-casein) or totally deficient in k-casein are located in the interior of the micelle (**Figure 2**). This model does not explain what provokes the segregation of the k-casein or why k-casein molecules, having preferred to associate with their own kind to form aggregate patches in the k-rich submicelles, should then associate with the other caseins to complete the building of the submicelle. Another criticism of this model is the late entry of calcium phosphate into the assembly process by Slattery and Evard [11]. Separating the caseins from calcium and phosphate until this point in the assembly process is not really possible, since both calcium and phosphate are involved in the phosphorylation of the protein chain which occurs post-translation [13] immediately and presumably before the association of the chains into submicelles.

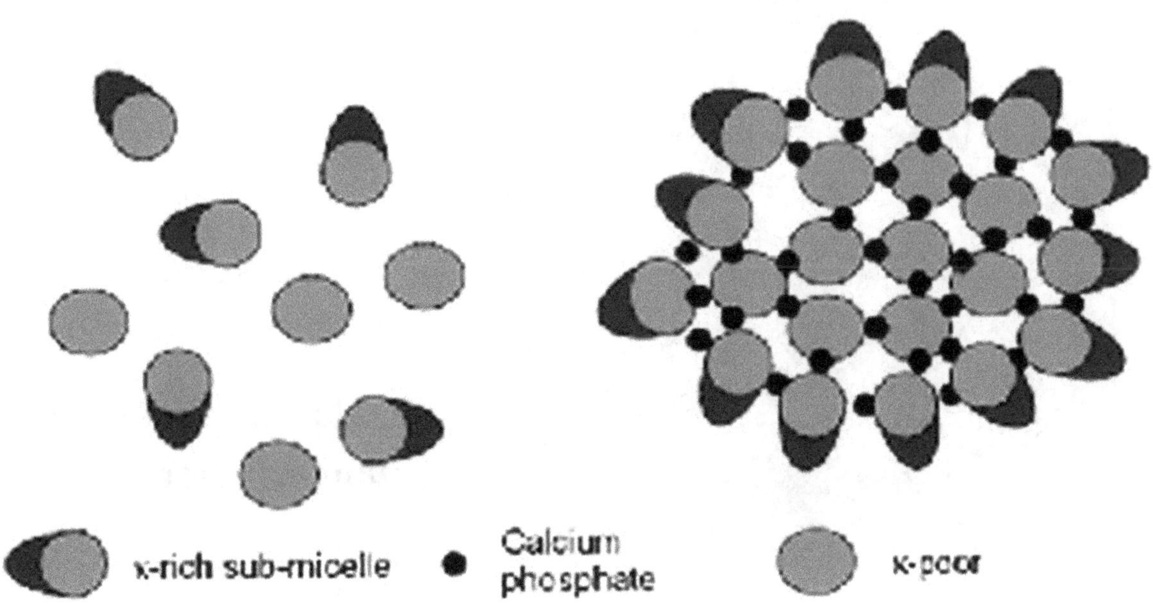

Figure 3. Representation of the casein micelle dual binding model proposed by Horne [16].

2.3. Internal structure models

The last category of models is based on the properties of the isolated casein constituents, causing or directing the formation of the internal structure of the casein micelle [7].

The first internal structure model was suggested by Rose in 1969 [14]. He assumed that β-casein monomers begin to self-associate into chain-like polymers. Subsequently, αs1-caseins molecules are attached to the β-casein polymers, while k-caseins interact with αs1-caseins, forming aggregates of limited size. Upon forming the micelle structure, colloidal calcium phosphate acts as a stabilizing agent and cross-links the network. In addition, these micelle networks are oriented in such a way that β-casein is directed internal, while the k-casein is directed external [8].

An alternative internal structural model is proposed by Holt [15] with the help of transmission electron micrographs. This model emphasized the role of hydrophobic interactions in giving rise to submicelles, the Holt model relies solely on the interactions between the caseins and calcium phosphate to hold the micelle together. In this model, the calcium phosphate is in the form of nanoclusters and the interaction sites on the caseins are the phosphoseryl clusters of the calcium-sensitive caseins (**Figure 3**). Because aS1- and aS2-caseins have more than two such clusters, arguably in the case of aS1-casein, they are able to cross-link the nanoclusters into extended 3-dimensional network structures. The dual-binding in Holt models is in the size of the nanocluster and in the number of phosphate clusters (or casein molecules) that the surface of the nanocluster can accommodate (**Figure 4**).

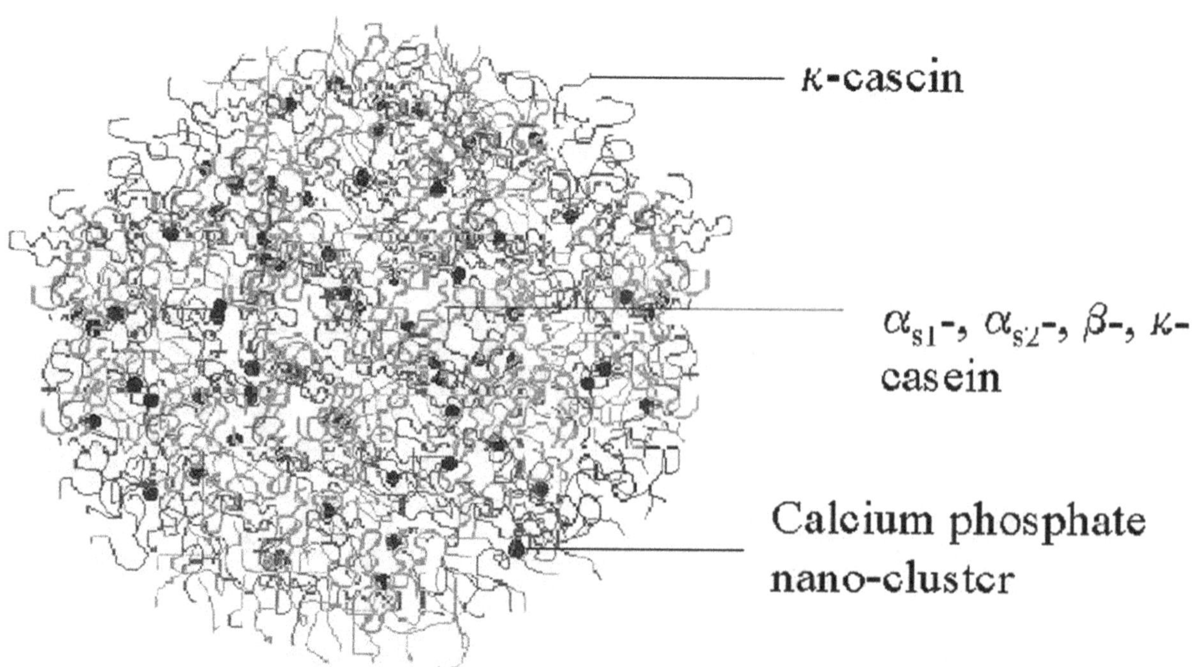

Figure 4. Schematic representation of Holt's model. The calcium phosphate formed nanocluster as interact with phosphoseryl residue of caseins [16].

3. Dynamic light scattering

Dynamic light scattering (DLS), sometimes referred to as quasi-elastic light scattering (QELS) or as photon correlation spectroscopy (PCS), is a non-invasive, well-established technique for measuring the size and size distribution of molecules and particles typically in the submicron region, and with the latest technology lower than 1 nm.

A laser or any other monochromaric light source is shot through a polarizer and into a sample. The scattered light is then sent through a second polarizer where it is collected by a photomultiplier. The resulting image is then projected onto a screen and this phenomenon is known as speckle pattern (**Figure 5**) [17].

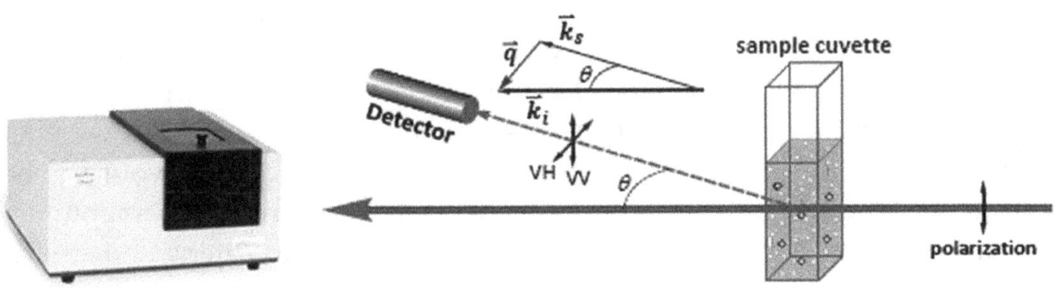

Figure 5. Principle of dynamic light scattering. www.brookhaveninstruments.com.

All of the molecules in the solution are being hit with the light and all of the molecules diffract the light in all directions [18]. The diffracted light from all of the molecules can either interfere constructively (shown as light regions) or destructively (shown as dark regions). The process is repeated at short time intervals. The resulting set of speckle patterns is then analyzed by an autocorrelator in order to compare the intensity of light at each spot over time. The polarizers that are being used can be set up in two geometrical configurations. One of the polarizers is set in a vertical/vertical (VV) geometry, while the second polarizer allows light through which is in the same direction as the primary polarizer. On the contrary, in vertical/horizontal (VH) geometry, the second polarizer allows light which is not in same direction as the incident light.

3.1. Practical realization

DLS is used for characterization of the various particle sizes (PSs), including proteins, carbohydrates, micelles, polymers, and nanoparticles. If the system is not dispersing in size, the average effective diameter of the particles can be determined [19]. This measurement depends on many factors as size of the particle core, size of surface structures, particle concentration, and the type of ions present in the medium. The diffusion coefficient of the particles can be determined since DLS typically measures fluctuations in scattered light intensity due to diffusing particles. DLS software of commercial instruments displays the distribution of particle population at different diameters [20]. In a monodisperse system, there should be only one population, while a polydisperse system would exhibit multiple particle populations. If a sample contains more than one size population, then either the CONTIN analysis should be applied for PCS instruments or the power spectrum method should be applied for Doppler shift instruments [21]. DLS can be used for convenient stability studies. Conducting periodical DLS measurements of a sample can demonstrate whether the particles aggregate over time. This can be achieved by checking for an increase in the particle hydrodynamic radius. If particles actually aggregate, a larger population of particles with a larger radius should be observed. In addition, some DLS machines can analyze stability depending on temperature by controlling the temperature in situ.

4. Measurement of casein micelle size by dynamic light scattering (DLS)

The average casein micelle size varies widely between milk samples of individual cows. Casein micelle size is also variable and can range between 154 and 230 nm in milks from individual cows [22]. Moreover, micelles as small as 125 nm and as large as 487 nm have been reported in fractionated bulk milk [23]. Casein micelle size may be influenced by cow genetics, e.g., casein protein variants [24–26], protein post-translational modification involving phosphorylation and glycosylation of casein molecules, and the levels of mineral compounds, such as calcium, in the milk [27]. Micelle size may also be influenced by farming and environmental factors such as feed [28] and season [29]. On the other hand, casein micelle size is of an important significance concerning influence the renneting behavior and the texture of manufactured dairy products [30]. The main used techniques for the analysis of PS and particle size distribution (PSD) are DLS, nanoparticle tracking analysis (NTA), scanning electron microscopy (SEM), size exclusion chromatography (SEC), cell electrophoresis, analytical ultracentrifugation (AUC), etc. Various analytical techniques may give different results [31–34]. DLS is the most user-friendly and it gives relatively accurate and consistent results of protein samples which can be obtained in a short period of time [35]. The size of casein micelle is estimated in raw milk samples, pasteurized milk, bulk milk, milk powder, etc. Here we will highlight determining of casein micelle size in raw milk by the DLS method.

Research of Bijl et al. [25] have revealed the casein micelle size in raw milk in 50 Holstein-Friesian cows and 54 Montbéliarde cows. Initially, all animals were genotyped for the k-casein gene. On the basis of average casein micelle size, the authors defined two types of micelles: small (170.6 ± 9.1 nm) and large (206.7 ± 5.0 nm). The results showed no significant differences between the two types of micelles and caseins concentration in milk. On the other hand, there was a positive influence between k-casein and casein micelle size. The mean casein micelle size in CSN3 AA cows was 203.5 ± 14.6 nm), while in cow with AB genotype, it was 173.1 ± 5.4 nm. This study was the first to report a correlation between casein micelle size and glycosylation of k-CN. The authors concluded that changes in the structure of k-casein clusters caused by glycosylation can influence micellar stabilization during or after casein micelle formation in the mammary gland, and thereby influence casein micelle size.

Another study by de Kruif and Huppertz [22] has revealed the casein micelle size in individual milk samples in 18 Holstein-Friesian cows by DLS measurement. Data from **Figure 4** showed clear differences between hydrodynamic radius (Rh) between cows but do not vary as a function of stage of lactation. To study the effect of different lactations on casein micelle size, the authors have collected the individual milk samples from 68 cows.

The results revealed that Rh did not change during milking, lactation, or even over a period of 3 years. This suggests that casein micelle size is strongly genetically determined and is extremely constant in the milks of individual cows. In this study, casein composition of each individual milk samples was determined by RP-HPLC. The results showed that first there were no significant differences between casein composition, as a function of stage of lactation, and second, casein composition did not influence on the casein micelle size (**Figure 6**).

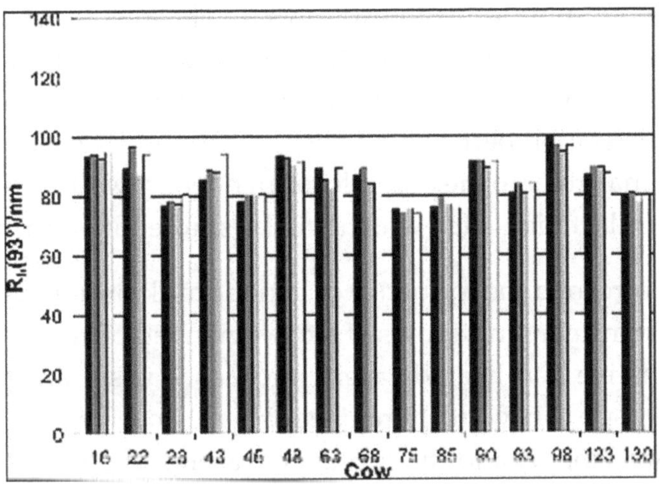

Figure 6. Hydrodynamic radius at a scattering angle of 93° in the milk of 15 individual cows taken at 4 different time points during a single lactation [22].

The study of Hristov et al. [26] aimed to compare the size of casein micelle in individual milk samples in dependence of kappa casein (CSN3) genetic polymorphism in 16 cows of Bulgarian Rhodopean cattle breed. The three defined kappa casein genotypes AA, AB, and BB were used for determining the casein micelle size by DLS. On the other hand, individual milk samples of each cow were assessed for protein and fat content. The results showed differences in the size and polydispersity of the casein micelles in milk from cows with different genotypes. The hydrodynamic radii of micelles at a scattering angle of 90°C varied from 80 to 120 nm and polydispersity varied from 0.15 to 0.37. The authors concluded that the casein micelle size of CSN3 AA cows (~120 nm) exceeds in about 60% of cows with AB (~80 nm) and BB genotype (~70 nm). In contrast, protein and fat content in milk cannot be correlated to casein micelle size. The obtained results could be useful for improving technological properties of milk and the yield of milk products (**Figure 7**).

Beliciu and Moraru [36] have analyzed the effect of the solvent on the accuracy of casein micelle PS determination by DLS at different temperatures and how to establish a clear protocol for these measurements. DLS analyses were performed at 6, 20, and 50°C in raw and pasteurized skimmed milk as sources of casein micelles. The pH, ionic concentration, refractive index, and viscosity of all solvents were determined. The solvents were evaluated by DLS to ensure that they did not have a significant influence on the results of the PS measurements. The authors concluded that when an ultrafiltered permeate was used as a solvent, the PS and polydispersity of casein micelles decreased as temperature increased. The effective diameter of casein micelles from raw skimmed milk diluted with ultrafiltered permeate was 176.4 ± 5.3 nm at 6°C, 177.4 ± 1.9 nm at 20°C, and 137.3 ± 2.7 nm at 50°C. Overall, the results of this study suggest that the most suitable solvent for the DLS analyses of casein micelles was casein-depleted ultrafiltered permeate. Dilution with water led to micelle dissociation, which significantly affected the DLS measurements, especially at 6 and 20°C.

A similar investigation was carried out by Mootse et al. [37]. The authors aimed to study casein micelle size in individual Estonian Holstein dairy cows during 1-year period by DLS. The main results of this study can be summarized as follows: 1) The average mean intensity (mode) of casein micelle particle size distribution (CM PSD) in raw milk of Estonian Holstein dairy cows was 171.13 nm and its variation (range 135–210 nm) resembled statistically normal distribution. 2) There was a weak correlation between average mode and its variation in milk samples of individual cows which may refer to the possible influence of cows' physiological status, disease incidences, and stages of lactation, etc., which will be studied in a further research.

Devold et al. [28] have screened the influence of genetic milk protein variants on mean size of native and heated casein micelles in 58 cows of Norwegian red cattle breed. The results showed that the mean size of native and heated casein micelles was significantly influenced by the following parameters: group of cows (different feeding regime), genotype of αS1-casein (native mean size only) and k-casein, pH and the content of casein, whey protein, and casein number (**Table 1**).

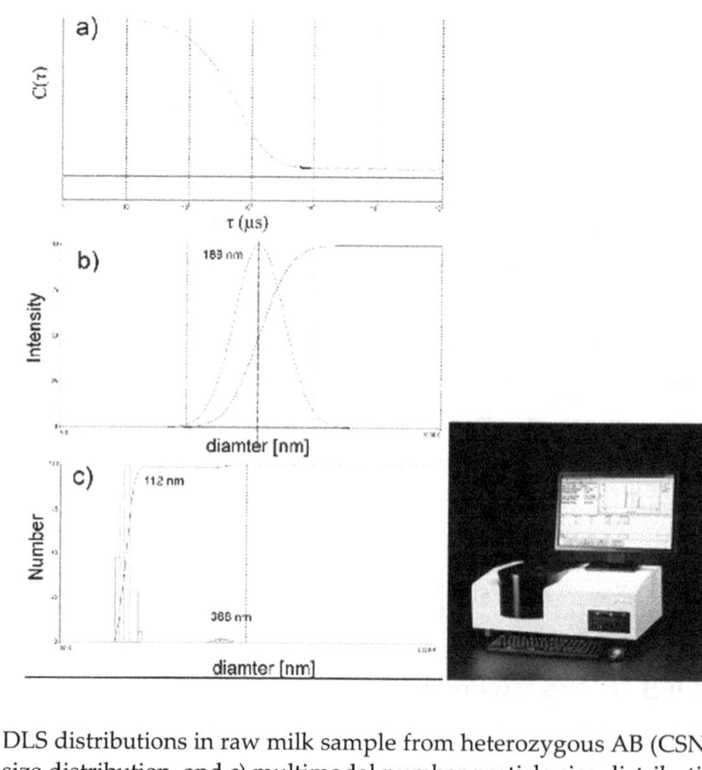

Figure 7. Representative DLS distributions in raw milk sample from heterozygous AB (CSN3) cow: a) correlation function, b) intensity particle size distribution, and c) multimodal number particle size distribution [26].

Variable	Mean size of casein micelles		
	Native	Heated	Change
αs1-CN	(BB > BC)[a]	c	c
β-CN	c	c	c
α-CN	(AA, AE > AB)[b]	(AA, AE > AB)[a]	c
β-lg	c	c	c
Fed. reg.	(Gr.3E > Gr.1)[a]	(Gr.3E > Gr.1)[a]	c
W.O.Lact.	c	c	c
Casein (%)	b	b	c
WP (%)	b	b	c
Casein no.	b	a	c
pH	a	a	c
Ca^{2+}	c	c	a
Ca	c	c	c
Mg	c	c	c
Citrate	c	c	c

[a] $p < 0.05$.
[b] $p < 0.01$.
[c] N.S. = non-signifcant [28].

Table 1. Variation for mean size of native and heated casein micelles and for heat-induced change in size.

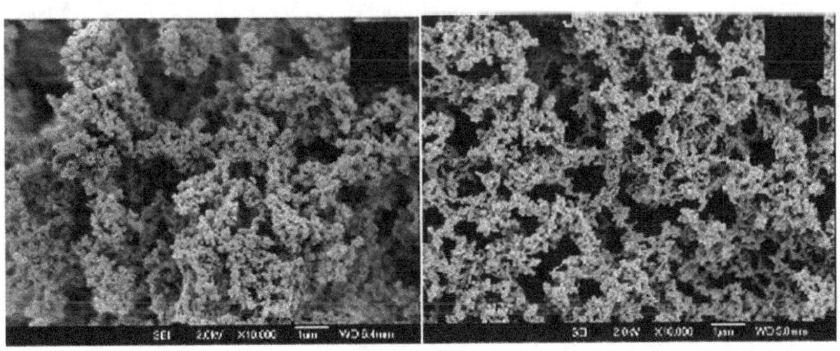

Figure 8. Scanning electron microscopy (SEM) imaging of the casein gel network produced from left, small casein micelle, and right, b large casein micelle milks. Images are at 10,000× magnification, scale bar = 1 μm [30].

Logan et al. [30] investigated the combined effects of milk fat globule size (MFG) and casein micelle size in bulk milk on the onset of gelation and the maximum rate of gelation. The results showed that casein micelle size is the major determinant of rennet gel strength. Milk fat globules could conceivably enhance the rheological properties of the rennet gel depending on their size in relation to the pore size of the casein micelle network. When the gels were formed with milks of large casein micelle, the size of MFG did not affect the gel properties, probably because both large and small MFG were smaller than the pore size of the large casein micelle network. On the other hand, when the gels were formed with milks of small casein, the large MFG enhanced the gel firmness more than the small MFG. It is possible that when the size of large MFG fits well with the pore size of the casein micelle network, a synergistic effect on rennet elastic gel network is provided (**Figure 8**).

Finally, the effect of gross composition, protein composition, total and ionic calcium content, phosphorous content, and casein micelle size on chymosin-induced gelation was determined in milk from 98 Swedish Red cows by Gustavsson et al. [36]. The obtained results showed that protein content, ionic calcium concentration, total calcium content, and casein micelle size were the most important factors explaining the variation of gelation properties in this sample set of Swedish red cows. Furthermore, it was shown that composite effect of beta and kappa casein genetic variants have an effect on casein micelle size and it was suggested that this could be the reason for previously published differences in gelation properties between the composite genotypes in the present data set. Non-coagulating milk is a problem in Swedish red cows and the present study showed that non-coagulating milk is more common in cows in their first parity than in cows in their second parity.

5. Conclusion

Many products in the food industry, including milk, ice cream, mayonnaise, jam, etc., are colloids or have been produced via colloids. Dispersed particles determine, through their concentration, distribution, size, and structure, the physicochemical and organoleptic quality properties of a product. Additionally, the complex interactions between colloidal particles and with other ingredients will contribute to the rheology, texture, stability, appearance, and many other food characteristics. Techniques to measure the particle properties of the colloid systems allow for better understanding of these complex relations and further improvement food processing techniques and recipes. Furthermore, the implementation of such sensing-technologies in a food plant could enable online monitoring of the colloids or the derived products during the production process and promote early detection of an altering product quality.

Author details

Peter Hristov[1*], Ivan Mitkov[1], Daniela Sirakova[1], Ivan Mehandgiiski[2] and Georgi Radoslavov[1]

*Address all correspondence to: peter_hristoff@abv.bg

1 Department of Animal Diversity and Resources, Institute of Biodiversity and Ecosystem Research, Bulgarian Academy of Sciences, Sofia, Bulgaria

2 Experimental and stockbreeding station, Agricultural Academy, Smolyan, Bulgaria

References

[1] Fox PF, Brodkorb A. The casein micelle: Historical aspects, current concepts and significance. International Dairy Journal. 2008;18:677–684. DOI: 10.1016/j.idairyj.2008.03.002.

[2] Di Stasio L, Mariani P. The role of protein polymorphism in the genetic improvement of milk production. Zootecnica e Nutrizione Animale. 2000;26:69–90.

[3] De Kruif CG, Holt C. Casein micelle structure, functions and interactions. In: Fox PF, McSweeney PLH, editors. Advanced Dairy Chemistry. Proteins, 3rd ed. New York: Kluwer Academic/Plenum Publishers; 2003. p. 233–276. DOI: 10.1007/978-1-4419-8602-3_5.

[4] Dalgleish DG, Spagnuolo P, Goff HD. A possible structure of the casein micelle based on high-resolution field-emission scanning electron microscopy. International Dairy Journal. 2004;14:1025–1031. DOI: 10.1016/j.idairyj.2004.04.008.

[5] Waugh DF, Noble RW. Casein micelles. Formation and structure II. Journal of the American Chemical Society. 1965;87:2246–2257. DOI: 10.1021/ja01088a026.

[6] Waugh DF. Formation and structure of casein micelles. In: McKenzie HA, editor. Milk Proteins: Chemistry and Molecular Biology. New York, NY: Academic Press; 1971, Vol. II, pp. 3–85. ISBN: 978-0-12-485202-0.

[7] Wong NP. Fundamental of Dairy Chemistry, 3rd ed. New York: Van Nostrand Reinhold; 1988, pp. 481–492.

[8] Phadungath C. Casein micelle structure: A concise review. Journal of Science and Technology. 2005;27(1): 201–212.

[9] Morr CV. Effect of oxalate and urea upon ultracentrifugation properties of raw and heated skim milk casein micelles. Journal of Dairy Science. 1967;50:1744–1751. DOI: 10.3168/jds.S0022-0302(67)87710-5.

[10] Rollema HS. Casein association and micelle formation. In: Fox PF, editor. Advanced Dairy Chemistry, Vol. 1: Proteins. Essex: Elsevier Science Publisher, Ltd.; 1992, pp. 111–140. ISBN: 9780306472718.

[11] Slattery CW, Evard R. A model for the formation and structure of casein micelles from subunits of variable composition. Biochimica and Biophysica Acta. 1973;317(2):529–538.

[12] Slattery CW. Casein micelle structure: An examination of models. Journal of Dairy Science. 1976;59:1547–1556.

[13] Summers WC. Felix d'Herelle and the origins of molecular biology. Medical History. 2001;45(2):294–295.

[14] Rose D. A proposed model of micelle structure in bovine milk. Dairy Science Abstracts. 1969;31:171–175.

[15] Holt C. Structure and stability of the bovine casein micelle. In: Anfinsen CB, Edsall JD, Richards FR, Eisenberg DS, editors. Advances in Protein Chemistry, Vol. 43. San Diego: Academic Press; 1992, pp. 63–151. ISBN: 978-0-12-034288-4.

[16] Horne DS. Casein micelles as hard spheres: Limitations of the model in acidified gel formation. Colloids and Surfaces A: Physicochemical and Engineering. 2003;213:255–263. DOI: 10.1016/S0927-7757(02)00518-6.

[17] Goodman J. Some fundamental properties of speckle. Journal of the Optical Society of America. 1976;66:1145–1150. DOI: 10.1364/josa.66.001145.

[18] Urban C, Schurtenberger P. Characterization of turbid colloidal suspensions using light scattering techniques combined with cross-correlation methods. Journal of Colloid and Interface Science. 1998;207(1):150–158. DOI: 10.1006/jcis.1998.5769.

[19] Pusey PN. Suppression of multiple scattering by photon cross-correlation techniques. Current Opinion in Colloid and Interface Science. 1999;4:177. DOI: 10.1016/S1359-0294(99)00036).

[20] Rodríguez-Fernández J, Pérez-Juste J, Liz-Marzán LM, Lang PR. Dynamic light scattering of short au rods with low aspect ratios. The Journal of Physical Chemistry. 2007;111:5020–5025. DOI: 10.1021/jp067049x.

[21] Provencher S. Contin: A general purpose constrained regularization program for inverting noisy linear algebraic and integral equations. Computer Physics Communications. 1982;27(3):229. DOI: 10.1016/0010-4655(82)90174-6.

[22] Kruif de CG, Huppertz T. Casein micelles: Size distribution in milks from individual cows. Journal of Agricultural and Food Chemistry. 2012;60:4649–4655. DOI: 10.1021/jf301397w.

[23] Niki R, Kim GY, Kimura T, Takahashi K, Kohyama K, Nishinari K. Physical properties and microstructure of rennet gels from casein micelles of different sizes. Milchwissenschaft. 1994;49(6):325–329.

[24] Lodes A, Krause I, Buchberger J, Aumann J, Kloster-Meyer H. The influence of genetic variants of milk proteins on the compositional and technological properties of milk: 1. Casein micelle size and the content of non-glycosylated k-casein. Milchwissenschaft. 1996;51:268–373. DOI: 10.3168/jds.2014-9285.

[25] Bijl E, de Vries R, van Valenberg H, Huppertz T, van Hooijdonk T. Factors influencing casein micelle size in milk of individual cows: Genetic variants and glycosylation of k-casein. International Dairy Journal. 2014;34:135–141. DOI: 10.1016/j.idairyj.2013.08.001.

[26] Hristov P, Neov B, Sbirkova H, Teofanova D, Radoslavov G, Shivachev B. Genetic polymorphism of kappa casein and casein micelle size in the Bulgarian Rhodopean

cattle breed. Biotechnology in Animal Husbandry. 2014;30(4):561–570. DOI: 10.2298/BAH1404561H

[27] Glantz M, Devold TG, Vegarud GE, Lindmark Månsson H, Stålhammar H, Paulsson M. Importance of casein micelle size and milk composition for milk gelation. Journal of Dairy Science. 2010;93(4):1444–1451. DOI: 10.3168/jds.2009-2856.

[28] Devold TG, Brovold MJ, Langsrud T, Vegarud GE. Size of native and heated casein micelles, content of protein and minerals in milk from Norwegian Red Cattle—effect of milk protein polymorphism and different feeding regimes. International Dairy Journal. 2000;10(5):313–323. DOI: 10.1016/S0958-6946(00)00073-X.

[29] Holt C, Baird L. Natural variations in the average size of bovine casein micelles: I. Milks from individual Ayrshire cows. Journal of Dairy Research. 1978;45(03):339–345. DOI: 10.1017/S0022029900016563.

[30] Logan A, Day L, Pin A, Auldist M, Leis A, Puvanenthiran A, Augustin MA. Interactive effects of milk fat globule and casein micelle size on the renneting properties of milk. Food and Bioprocess Technology. 2014;7(11):3175–3185. DOI: 10.1007/s11947-014-1362-2.

[31] Anema SG, Lowe EK, Stockmann R. Particle size changes and casein solubilisation in high-pressure-treated skim milk. Food Hydrocolloids. 2005;19:257–267. DOI: 10.1016/j.foodhyd.2004.04.025.

[32] Tran Le T, Saveyn P, Hoa HD, Van der Meeren P. Determination of heat-induced effects on the particle size distribution of casein micelles by dynamic light scattering and nanoparticle tracking analysis. International Dairy Journal. 2008;18:1090–1096. DOI: 10.1016/j.idairyj.2008.06.006.

[33] Dejan A. Dynamic Light Scattering and Application to Proteins in Solutions. Ljubljana: University of Ljubljana Faculty of Mathematics and Physics, Department of Physics; 2010, p. 19.

[34] Raza H, Claire G, Joël S. Revealing casein micelle dispersion under various ranges of NaCl: Evolution of articles size and structure. World Academy of Science, Engineering Technology. 2011;51:972–982.

[35] Vasco F, Andrea H, Wim J. Critical evaluation of nanoparticle tracking analysis (NTA) by NanoSight for the measurement of nanoparticles and protein aggregates. Pharmacology Research. 2010;27(5):796–810. DOI: 10.1007/s11095-010-0073-2.

[36] Gustavsson F, Glantz M, Buitenhuis AJ, Lindmark-Månsson H, Stålhammar H, Andrén A, Paulsson M. Factors influencing chymosin-induced gelation of milk from individual dairy cows: Major effects of casein micelle size and calcium. International Dairy Journal. 2014;39:201–208. DOI: 10.1016/j.idairyj.2014.06.011.

[37] Mootse H, Pisponen A, Pajumägi S, Polikarpus A, Tatar V, Sats A, Poikalainen V.

Investigation of casein micelle particle size distribution in raw milk of estonian holstein dairy cows. Agronomy Research. 2014;12(3):153–158.

[38] Beliciu CM, Moraru CI. Effect of solvent and temperature on the size distribution of casein micelles measured by dynamic light scattering. Journal of Dairy Science. 2009;92(5):1829–1839. DOI: 10.3168/jds.2008-1467.

Permissions

All chapters in this book were first published in MPFSBPHA, by InTech Open; hereby published with permission under the Creative Commons Attribution License or equivalent. Every chapter published in this book has been scrutinized by our experts. Their significance has been extensively debated. The topics covered herein carry significant findings which will fuel the growth of the discipline. They may even be implemented as practical applications or may be referred to as a beginning point for another development.

The contributors of this book come from diverse backgrounds, making this book a truly international effort. This book will bring forth new frontiers with its revolutionizing research information and detailed analysis of the nascent developments around the world.

We would like to thank all the contributing authors for lending their expertise to make the book truly unique. They have played a crucial role in the development of this book. Without their invaluable contributions this book wouldn't have been possible. They have made vital efforts to compile up to date information on the varied aspects of this subject to make this book a valuable addition to the collection of many professionals and students.

This book was conceptualized with the vision of imparting up-to-date information and advanced data in this field. To ensure the same, a matchless editorial board was set up. Every individual on the board went through rigorous rounds of assessment to prove their worth. After which they invested a large part of their time researching and compiling the most relevant data for our readers.

The editorial board has been involved in producing this book since its inception. They have spent rigorous hours researching and exploring the diverse topics which have resulted in the successful publishing of this book. They have passed on their knowledge of decades through this book. To expedite this challenging task, the publisher supported the team at every step. A small team of assistant editors was also appointed to further simplify the editing procedure and attain best results for the readers.

Apart from the editorial board, the designing team has also invested a significant amount of their time in understanding the subject and creating the most relevant covers. They scrutinized every image to scout for the most suitable representation of the subject and create an appropriate cover for the book.

The publishing team has been an ardent support to the editorial, designing and production team. Their endless efforts to recruit the best for this project, has resulted in the accomplishment of this book. They are a veteran in the field of academics and their pool of knowledge is as vast as their experience in printing. Their expertise and guidance has proved useful at every step. Their uncompromising quality standards have made this book an exceptional effort. Their encouragement from time to time has been an inspiration for everyone.

The publisher and the editorial board hope that this book will prove to be a valuable piece of knowledge for researchers, students, practitioners and scholars across the globe.

List of Contributors

Anindya Mukhopadhya and Torres Sweeney
School of Veterinary Medicine, University College Dublin, Dublin, Ireland

Donata Marletta and Salvatore Bordonaro
Department of Agriculture, Food and Environment (Di3A), University of Catania, Italy

Flavio Tidona
Council for Research in Agriculture and Agricultural Economy Analysis (CREA), Research Centre for Fodder Production and Dairy Casearie (CREA-FLC), Lodi, Italy

Adham M. Abdou and Hend A. Elbarbary
Food Control Department, Faculty Veterinary Medicine, Benha University, Moshtohor, Kaliobiya, Egypt

Guillermo Docena, Paola Smaldini, Renata Curciarello and Angela María Candreva
Institute of Immunologic and Physiopathologic Studies - IIFP, School of Sciences, The National University of La Plata, La Plata, Argentina

R. Anusha and O.S. Bindhu
Department of Biochemistry, Centre for Post Graduate Studies, Jain University, Bangalore, India

Maria Elisabetta Baldassarre, Raffaella Panza and Nicola Laforgia
Department of Biomedical Science and Human Oncology, Neonatology and NICU Section, University "ALDO MORO", Bari, Italy

Mohd Younus Bhat and Tanveer Ali Dar
Department of Clinical Biochemistry, University of Kashmir, Srinagar, J&K, India

Laishram Rajendrakumar Singh
Dr. B.R. Ambedkar Centre for Biomedical Research, University of Delhi, North Campus, New Delhi, India

P.K. Theil
Aarhus University, Tjele, Denmark

W.L. Hurley
University of Illinois, Urbana-Champaign, USA

Sergey E. Sedykh, Valentina N. Buneva and Georgy A. Nevinsky
SB RAS Institute of Chemical Biology and Fundamental Medicine, Novosibirsk State University, Novosibirsk, Russia

Zeynep Koksal
Department of Chemistry, Faculty of Sciences, Istanbul Medeniyet University, Istanbul, Turkey

Ilhami Gulcin and Hasan Ozdemir
Department of Chemistry, Faculty of Sciences, Atatürk University, Erzurum, Turkey

Peter Hristov, Ivan Mitkov, Daniela Sirakova and Georgi Radoslavov
Department of Animal Diversity and Resources, Institute of Biodiversity and Ecosystem Research, Bulgarian Academy of Sciences, Sofia, Bulgaria

Ivan Mehandgiiski
Experimental and stockbreeding station, Agricultural Academy, Smolyan, Bulgaria

Index

A

Acute Phase Reaction, 112, 122
Adipophilin, 159, 162
Allergenic Proteins, 52, 55, 59, 63
Anaphylaxis, 54, 56, 63, 65, 71, 113
Angiotensin-converting Enzyme, 84, 110
Anti-inflammatory Activity, 5, 9-10
Antibacterial Activity, 31-34, 36, 39, 44-48, 116, 125
Antibacterial Agent, 21

B

Bacillus Subtilis, 32, 87
Bifidobacterium, 7-8, 13-14, 39, 78-79, 100, 118, 136
Bioactive Peptides, 2, 11-17, 21-22, 73-74, 82, 97-101, 103-106, 110
Butyrophilin, 159, 162

C

Candida Albicans, 7-8, 34, 50, 87
Casein Micelle, 22, 143-148, 150-151, 153-154, 156, 219-222, 224-232
Casein Micelle Size, 22, 146, 219, 224-228, 230-231
Catalytic Antibodies, 174, 185, 199, 202
Cationic Iron-binding Protein, 29-30
Chymotrypsin, 2-3, 7, 9, 30, 46, 77-78, 87, 89
Clostridium, 39, 49, 56
Colostrum, 18, 21, 45-46, 48, 80, 85, 87, 103, 158-163, 165-173, 213, 216

D

Dietary Protein, 92, 109, 113, 158, 163-167
Disulfide Bonds, 60, 120, 148, 151, 194
Donkey Caseins, 18, 22
Dynamic Light Scattering, 219, 223-224, 230-232

E

Electron Micrograph, 220
Enzymatic Hydrolysis, 1, 3, 14, 31, 42, 75, 77-78, 85, 107
Enzyme Inhibition, 210, 217-218
Eosinophil Cationic Protein, 112, 116, 124, 137-141

F

Fab Arm Exchange, 174, 189-191, 202
Faecal Calprotectin, 112-113, 117-121, 128, 132-133
Faecal Markers, 112-113, 115-116, 119, 131
Food Intolerances, 56, 67
Food Preservatives, 29-30, 88

G

Gastrointestinal Tract, 1-2, 4-8, 13, 39, 49, 53, 56, 77, 95, 123
Gel Filtration, 32, 176-178, 183, 193-194, 209
Glycoprotein, 9, 21, 29-30, 43, 143, 162, 174, 204, 206
Gram-negative Bacteria, 7-8, 21, 30, 35-37, 121, 123, 212
Gut Health, 1, 6
Gut Microbiota, 1, 6, 9, 13-14, 118

H

Hydrogen Peroxide, 204-207, 213-214
Hypothiocyanate, 204, 206-207, 214

I

Immunoglobulins, 9, 20-22, 53, 59-61, 158, 160-162, 167, 170-171, 201
Immunomodulatory, 1, 9, 12, 14, 21, 30, 41, 64, 67, 73, 75, 88-89, 107, 115, 126, 162
Immunomodulatory Activity, 1, 41
Immunotherapy, 52, 58, 62-63, 66, 68, 71, 88, 189
Infant Nutrition, 18, 22, 95
Inflammatory Diseases, 40, 52, 116

L

Lactadherin, 159
Lactobacillus, 7-9, 12, 14-15, 17, 28, 78, 86, 97, 99-100, 110, 120, 133, 136
Lactoferrin, 2, 6-8, 14-15, 20-22, 27-29, 42-51, 60-61, 84, 103, 108, 160, 162, 170, 174-176, 178, 193, 195-199, 203
Lactoperoxidase, 6, 21-22, 29, 32, 35, 44, 46, 204-209, 211, 213-218

M

Mass Spectrometry, 25, 32, 62, 80, 110
Microbial Fermentation, 3
Milk Albumin, 174-175, 193
Milk Allergens, 26, 52, 59-60, 65
Milk Enzyme, 204
Milk Hydrolysates, 1-2, 4-10, 100
Myeloperoxidase, 127, 205, 207, 216

N

Neutrophil Granulocytes, 116, 124
Norepinephrine, 210-211
Novel Food, 18

Nutraceuticals, 13, 73-75, 93, 95-97, 107

O

Oral Tolerance, 52, 71

P

Paracellular Pathway, 81

Pathogenic Microorganisms, 7-9, 33, 204

Peroxidase, 124, 205-207, 211-217

Phosphoproteins, 142, 149

Polydisperse, 219, 223

Porcine Pepsin, 31-32, 39

Protein Concentrations, 162-163

Protein Hydrolysates, 2, 4-6, 9, 12-14, 43, 73, 83, 101, 103

Proteins Of Colostrum, 158-159

S

Spectrophotometer, 209

Staphylococcus Aureus, 7, 23, 34, 86, 134

Supramolecular Complexes, 194

T

Thermostability, 33, 45

Transcytosis, 80, 188

Transferrin, 29-30, 42, 45-46, 197

U

Ultracentrifugation, 153, 224, 229

W

Whey Protein, 7-10, 16, 20, 60, 78, 83-84, 87, 89-90, 92, 94-95, 99-100, 102-103, 108-110, 143, 160, 217, 226